Michael A. Fullen and John A. Catt

SOIL MANAGEMENT
PROBLEMS AND SOLUTIONS

Routledge
Taylor & Francis Group
LONDON AND NEW YORK

Reprint 2019

First published 2004 by Hodder Arnold, a member of the Hodder Headline Group
Distributed in the USA by Oxford University Press

Published 2014 by Routledge
2 Park Square, Milton Park, Abingdon, Oxon OX14 4RN
711 Third Avenue, New York, NY, 10017, USA

Routledge is an imprint of the Taylor & Francis Group, an informa business

British Library Cataloguing in Publication Data
A catalogue record for this book is available from the British Library

Library of Congress Cataloging-in-Publication Data
A catalog record for this book is available from the Library of Congress

ISBN: 978-0-367-24247-3 (hbk)

Typeset in 10/13 Acaslon Regular by Charon Tec Pvt. Ltd, Chennai, India

Digitally Printed at Replika Press Pvt. Ltd.
For sale in India, Pakistan, Nepal, Bhutan, Bangladesh and Sri Lanka only.

In loving memory of William Kenny, who in his magnificent garden in County Durham taught his young grandson the magic and mystery of soil.

MAF

In memory of the late Frank Kenworthy, a talented and inspiring teacher of geology, who first attracted me to the study of soil.

JAC

Contents

Foreword

It is difficult to overstate the serious consequences for humanity of damage to soil. If agricultural soils are affected by erosion, salt, chemical pollutants, nutrient depletion or organic matter decline, how will food, fibre and fuel be produced for the 9 billion people expected to inhabit our planet by the mid-twenty-first century? If chemical pollutants or pathogens from waste disposal practices pass through soils to water, how will sufficient clean water be supplied? And how can the effects of mismanagement be corrected?

The importance of soils to human development and the ecological functioning of the planet are gradually being appreciated and many governments and bodies such as the EU are writing soil protection strategies. But it will not be easy to implement the good intentions in these plans, as so many necessary human activities have the potential to damage soils. In many places the main threats are from the impacts of industrial activity, urban expansion or overintensive agriculture. In other areas, poverty leads to inappropriate farming and land use. Climatic change is becoming an increasingly serious pressure throughout the world.

Implementing good soil management will require a large number of people, with a knowledge of soils and their functions, working in numerous roles including land managers, farmers, policy makers, regulators, planners, consultants or advisers in the public, private and NGO sectors. Fortunately, many students now receive at least some teaching about soils – some in specialist soil science degrees but, increasingly, as modules within broadly based courses including geography, natural resource management, environmental studies and many more. This book will be a great asset to any student wanting to learn about the use and management of soil, whether at undergraduate or postgraduate level. It complements other soil science textbooks because it is based on practical issues of soil management. Each chapter focuses on an issue where there are significant threats from human or natural causes, either to the soil itself or to related parts of the environment. For each threat the authors provide information on the causes of the problem, practices to avoid it, ways of correcting the damage and sufficient (but not too much) theoretical background. They often give a fascinating historical perspective on the issue but, importantly, give concise summaries of the most up-to-date research findings, with key references for those who wish to know more.

I commend this book to any student wishing to learn about the management of soil and related aspects of our environment. It will be equally valuable to students aiming to become soils specialists and to those for whom soils will be just one part of their portfolio of skills. And, for anyone further advanced in their career, this book will provide a rapid update to their knowledge; they will wish it had been written when they were students!

Professor David Powlson
Agriculture and Environment Division
Rothamsted Research, UK

Preface

The book introduces soil management issues and is suitable mainly for second- and third-year students on degree courses in Environmental Sciences, Geography and Agriculture. While teaching soil management courses, both authors noted that no single book covered all the main aspects of soil management. In the early twenty-first century, society faces many problems and challenges associated with soil management that need practical and sustainable solutions. This book develops an overview of problems of soil management and strategies for their solution. Such a broad remit does not permit detailed reviews. However, each chapter includes suggestions for further reading and so provides pathways to more detailed and advanced studies.

Chapter 1 reviews the scale and nature of soil management problems. In **Chapter 2** patterns and processes of water and wind erosion are examined, along with techniques for soil conservation. **Chapter 3** addresses the inter-related problems of desertification and salinization of arid soils. This includes a review of the extent and severity of the problems and the relative importance of human and natural causes. The issue of the amelioration of arid soils is considered, including the feasibility of desert reclamation and soil desalinization. Optimal use of soil requires careful consideration of soil water and in **Chapter 4** soil water management issues are investigated, specifically irrigation, drainage and their environmental consequences. **Chapter 5** explores the related problems of chemical pollution of soil, water and the atmosphere. These include agricultural and industrial sources of pollution. The problems posed by nitrate, phosphate, pesticides and pathogenic micro-organisms are also reviewed. **Chapter 6** considers the various natural and industrial causes of soil acidification, and the effects of acidification on plant, animal and human health. The chapter concludes with a review of procedures to combat acidification. **Chapter 7** considers the nature of soil structure and how it can be modified. This includes discussion of the benefits and problems of zero and conventional tillage practices. The problems of soil compaction and the reclamation and restoration of quarries, landfill sites and mine-spoil are also considered. **Chapter 8** considers the importance of soil organic matter conservation, including the dynamics of organic matter in soils, its loss through agricultural activities and methods of increasing its abundance in soil. The value of crop residues and implications of peat wastage are also considered. Climatic change is currently receiving intense investigation and soils play an important role, as discussed in **Chapter 9**. Soils affect the global carbon dioxide, methane and nitrous oxide cycles and budgets, and possibilities for managing soils to minimize emissions of these 'greenhouse' gases are discussed. However, the history and likely causes of recent and earlier climatic change in the past suggest that our future climate may well be influenced in unexpected ways by natural factors as well as by increasing greenhouse gases. The concluding **Chapter 10** summarizes the prospects in the twenty-first century, in particular the changing problems and solutions in the face of growing global population and global warming. The effects of urbanization on the extent and quality of soils and suitable techniques for soil reclamation, rehabilitation, restoration and recreation are examined. The chapter concludes with a consideration of relationships between soils and environmental health and the potential for improved soil management for habitat creation.

Dr Michael A. Fullen is Reader in Soil Science at The University of Wolverhampton. His interests include soil erosion, soil conservation and desertification. Most of his fieldwork is conducted in Europe and Southeast Asia.

Professor John A. Catt is Honorary Professor of Geography at University College London. Previously he was Principal Scientist at Rothamsted Experimental Station. His interests include Quaternary Geology and soil–crop interactions.

List of Tables

List of Illustrations

List of Figures

List of Plates

Acknowledgements

Sincere thanks are expressed to colleagues for their helpful comments and suggestions in the preparation of this book. These include:

Dr Lyn Besenyei, Dr Eleanor Cohn, Dr Jackie Hooley, Dr Alison McCrea, Professor David Mitchell, Professor Jean Poesen, Dr Margaret Oliver and Professor Ian Trueman (MAF) and Angela Arnold, Dudley Christian, Mark Maslin and Hugh Prince (JAC). Special thanks are due from both authors to Dr Colin Booth for his patient help in collating the manuscript and preparing figures.

Permissions

The authors and publishers thank the following for permission to use copyright material in this book:

J. Wiley and Sons Ltd., Chichester, for Figure 7.9 (p. 122) and Table 7.1 (p. 110) from Thomas, D.S.G. and Middleton, N.J., 1994, *Desertification: Exploding the Myth,* Chichester: J. Wiley; and for Figure 4.1 (p. 111) from Rhoades, J.D., 1990, 'Soil salinity – causes and controls', in Goudie, A.S. (ed.), *Techniques for Desert Reclamation,* Chichester: J. Wiley, 109–34.

The Institute of Desert Research of the Chinese Academy of Sciences (IDRAS), Lanzhou, P.R. China, for permission to reproduce the maps in Figures 3.2 and 3.3 by Zhu Zhenda (ed.), 1958 and 1981, *The Map of Developmental Degree of Desertification in Daqinggou, Keerqin (Horqin) Steppe, Inner Mongolia, China,* Lanzhou: IDRAS.

The Geographical Magazine for Figure 3.5, taken from p. 27 of Fullen, M.A. and Mitchell, D.J., 1991, 'Taming the shamo dragon', *The Geographical Magazine* 63, 26–9.

The International Soil Reference and Information Centre, Wageningen, for Tables 1.1 and 1.2, taken from pp. 29–32 of Oldeman, L.R., Halleling, R.T.A. and Sombroek, W.G., 1990, *World Map of the Status of Human-Induced Soil Degradation,* Wageningen: ISRIC/UNEP in cooperation with Winand Staring Centre-International Soil Science Society (ISSS)-FAO-ITC (The International Institute for Geo-Information Science and Earth Observation).

The Macaulay Institute for Table 1.4, taken from page 11 of the Soil Survey of Scotland, 1984, *Organization and Methods of the 1:250,000 Soil Survey of Scotland,* The Macaulay Institute for Soil Research Aberdeen: University Press Aberdeen.

The Institute of Desert Research of the Chinese Academy of Sciences for permission to reproduce Table 3.3, taken from Zhu Zhenda, Liu, S. and Di, X., 1988, *Desertification and Rehabilitation in China,* Lanzhou: The International Centre for Education and Research on Desertification Control, and Dr A. McCrea for Table 10.1.

The National Soil Resources Institute, Cranfield University, for Figures 6, 11, 13 and 21 from Hall, D.G.M., Reeve, M.J., Thomasson, A.J. and Wright, V.F., 1977, *Water Retention, Porosity and Density of Field Soils*, Soil Survey Technical Monograph 9, Harpenden: Rothamsted Experimental Station; for Tables 11, 12 and data on pp. 65–66 from Hall *et al.*, 1977; and for Table 4 from Thomasson, A.J., (ed.), 1975, *Soils and Field Drainage*, Soil Survey Technical Monograph 7, Harpenden: Rothamsted Experimental Station.

The British Soil Science Society and Dr D.B. Davies, Editor of *Soil Use and Management,* for Figure 2 from Powlson, D.S., 1993, 'Understanding the soil nitrogen cycle', *Soil Use and Management* 9, 87; Figure 1 from T.R. Worthington and P.W. Danks, 1994, 'Nitrate leaching and intensive outdoor pig production', *Soil Use and Management* 10, ii; Figure 1 from Goulding, K.W.T., 2000, 'Nitrate leaching from arable and horticultural land', *Soil Use and Management* 16, 146; Figure 4 from Davies, D.B., 2000, 'The nitrate issue in England and Wales', *Soil Use and Management* 16, 143; Figure 2 from Bhogal, A., Young, S.D. and Sylvester-Bradley, R., 1997, 'Straw incorporation and immobilization of spring-applied nitrogen', *Soil Use and Management* 13, 114; Figure 1 from Blake, L., Johnston, A.E. and Goulding, K.W.T., 1994, 'Mobilization of aluminium in soil by acid deposition and its uptake by grass cut for hay – a chemical time bomb', *Soil Use and Management* 10, 52; Figure 1 from Smith, P., Milne, R., Powlson, D.S., Smith, J.U., Falloon, P. and Coleman, K., 2000, 'Revised estimates of the carbon mitigation potential of UK agricultural land', *Soil Use and Management* 16, 294; Table 5 from Williams, J.R., Chambers, B.J., Hartley, A.R., Ellis, S. and Guise, H.J., 2000, 'Nitrogen losses from outdoor pig farming units', *Soil Use and Management* 16, 241; Table 3 from Goulding, K.W.T., Poulton, P.R., Webster, C.P. and Howe, M.T., 2000, 'Nitrate leaching from the Broadbalk Wheat Experiment, Rothamsted, UK, as influenced by fertilizer and manure inputs and the weather' *Soil Use and Management* 16, 247; Table 3 from Vinten, A.J.A., Lewis, D.R., Ferlon, D.R., Leach, K.A., Howard, R., Svoboda, I. and Ogden, I., 2002, 'Fate of *Escherichia coli* and *Escherichia coli* O157 in soils and drainage water following cattle slurry application at 3 sites in southern Scotland', *Soil Use and Management* 18, 229; Table 1 from Gildon, A. and Rimmer, D.L., 1993, 'The use of soil in colliery spoil reclamation', *Soil Use and Management* 9, 150; and Tables 1 and 2 from Lawson, D.M., 1989, 'The principles of fertilizer use for sports turf', *Soil Use and Management* 5, 126.

CABI Publishing for Tables 2.2 (p. 12) and 9.1 (p. 156) from Smith, S.R., 1996, *Agricultural Recycling of Sewage Sludge and the Environment,* Wallingford: CABI.

Her Majesty's Stationery Office for Table on p. 5 of Ministry of Agriculture, Fisheries and Food, 2000, *Fertiliser Recommendations for Agricultural and Horticultural Crops,* MAFF Reference Book 209; and Appendix 2 (pp. 40–41) of MAFF, 1973, *Lime and Liming,* MAFF Reference Book 35, London: HMSO.

Elsevier Global Rights Department for Figure 10 from Bronswijk, J.J.B., Groenenberg, J.E., Ritsema, C.J., Vanwijk, A.L.M. and Nugroho, K., 1995, 'Evaluation of water management strategies for acid sulphate soils using a simulation model: a case study in Indonesia', *Agricultural Water Management* 27, 139; Figure 22 from Broecker, W.S. and Denton, G.H., 1990, 'The role of ocean–atmosphere reorganisations in glacial cycles', *Quaternary Science Reviews* 9, 330; and Figure 3 from Oleson, J.E. and Bindi, M., 2002, 'Consequences of climate change for European agricultural productivity, land use and policy', *European Journal of Agronomy* 16, 245.

The Royal Society of London and Professor G.R. Coope for Figure 2 from Coope, G.R., 1977, 'Fossil Coleopteran assemblages as sensitive indicators of climatic changes during the Deversion (Last) cold stage', *Philosophical Transactions of the Royal Society of London* B280, 321.

The American Society of Agronomy for Figure 3 from Heckrath, G., Brookes, P.C., Poulton, P.R. and Goulding, K.W.T., 1995, 'Phosphorus leaching from soils containing different P concentrations in the Broadbalk Experiment' *Journal of Environmental Quality* 24, 908.

Rothamsted Research for Plate 6.1.

1 Introduction

The soil is the common mother of all things, because she has always brought forth all things and is destined to bring them forth continuously.

Lucius Collumella, AD 60

Introduction

Columella was one of the earliest known writers on agriculture and clearly understood the importance of soil as a resource within his native Iberia (Spain). Although 71 per cent of the surface of our planet is covered by water, we still call it 'Earth'. Along with air and water, soil is essential for life on Earth. We need soil for the growth of our crops, for grazing our animals and for the growth of our timber.

Most of the Earth's surface is too hot, cold, wet, dry or steep for agricultural use. A surprisingly small proportion of the land surface can be used for crops. In 1987, the United Nations (UN) estimated that only 11.3 per cent of the land surface (13,077 million ha) could be used for arable and permanent crops. This resource base must support a growing world population, currently increasing by about 80 million people per year. The International Development Research Centre (IDRC), based in Ottawa, Canada, estimated the world population to be 6,159,463,956 and the area of productive land at 8,585,272,604 ha (values on 05 November 2001). The US Census placed the world population in mid-2003 at 6,302,486,693 and growing by an annual rate of 1.16 per cent. That is an extra 73,395,376 people in a year or, on average, an extra 8378 people per hour. These values are best estimates, but indicate the scale of the problem. Further information can be accessed from both the IDRC and US Census web sites, respectively at: http://www.idrc.ca/ and http://www.census.gov/ ipc/www/worldpop.html (accessed 16 February 2004).

1.1 SOIL DEGRADATION

It is imperative that the soil resource base is conserved for current and future generations. An old Chinese proverb asks the pertinent question 'once the skin is gone, where can the hair grow?'. The UN has expressed considerable concern over the status of the world's soils. A Commission, chaired by the Norwegian Prime Minister, Gro Harlem Brundtland, investigated soil degradation and produced the Report *Our Common Future* in 1987 (Brundtland, 1987). The United Nations Environment Programme (UNEP) was then charged with producing a world map of the current status of world

Table 1.1 Global assessment of soil degradation (millions of hectares)

Continent	Total land surface	Human-induced degradation	Water erosion	Wind erosion	Chemical deterioration	Physical deterioration
Asia	4256	748	440.6	222.2	73.2	12.1
Africa	2966	494	227.4	186.5	61.5	18.7
S. America	1768	243	123.2	41.9	70.3	7.9
Europe	950	219	114.5	42.2	25.8	36.4
Australasia	882	103	82.8	16.4	1.3	2.3
N. America	1885	95	59.8	34.6	0.1	0.9
C. America	306	63	46.3	4.6	6.9	5.0
Total	13,013	1965	1094.6	548.4	239.1	83.3

Source: Oldeman *et al.* (1990).

Table 1.2 Causative factors of soil degradation (millions of hectares)

Continent	Deforestation	Overgrazing	Agricultural mismanagement	Overexploitation	Bioindustrial activities
Asia	298	197	204	46	1
S. America	100	68	64	12	–
Australasia	12	83	8	–	–
Africa	67	243	121	63	–
N. & C. America	18	38	91	11	–
Europe	84	50	64	1	21
Total	579	679	552	133	22

Source: Oldeman *et al.* (1990).

soils in a project called the 'Global Assessment of Soil Degradation' (GLASOD). This project was mainly coordinated by Wageningen University in The Netherlands and resulted in the publication in 1990 of the *World Map of the Status of Human Induced Degradation* (Oldeman *et al.*, 1990). The GLASOD Report took a rather pessimistic view of the future, concluding 'the earth's soils are being washed away, rendered sterile or contaminated with toxic chemicals at a rate that cannot be sustained'.

GLASOD estimated that the loss of agricultural land through soil erosion was 6–7 million ha yr^{-1}, with an additional 1.5 million ha being lost through waterlogging, salinization and alkalinization. In this context, loss does not necessarily mean the land disappears, although locally it does because of marine transgressions. Usually, it means a deterioration in soil properties to the extent that the soil is no longer productive. Table 1.1 shows the main types of soil degradation reported in the GLASOD survey and their extent on different continents. Table 1.2 breaks these down into causative factors. These data must be viewed with caution as they are 'best estimates'. They could be challenged on several counts, including scientific rigour and accuracy, and attempts are under way to update and validate the data. However, they do clearly suggest that we have an important global problem.

Plate 1.1 The abandoned ancient city of Jiaohe, Xinjiang Province, China (photo M.A. Fullen)

History has some important lessons for us. The collapse of past civilizations resulted partly from land degradation. The fertile crescent of the Tigris–Euphrates basin (presently mainly Iraq) formed the basis of the ancient Sumerian civilization in Mesopotamia (Thomas and Middleton, 1994; Johnson and Lewis, 1995). Approximately 5000 years ago techniques for growing cereals were developed, providing the foundation of the region's cereal-based agricultural economy. Hence, the area has been described as the 'cradle of civilization'. Similar past civilizations included the Egyptian Empire, which developed in a corridor along the River Nile and the Han Chinese Empire, which had its focus on the Yellow River. Land degradation played a crucial role in the deterioration and eventual collapse of these civilizations.

Land degradation covers a complex series of processes, including soil erosion (by both wind and water), the expansion of desert-like conditions (desertification) and the contamination of soils with salts (salinization). It has been enmeshed in a complex series of social changes, including social unrest within and between political units and ethnic tensions. These often occurred at a time of notable climatic change. Illustrative of these changes is the ruined ancient city of Jiaohe, in Xinjiang Province, China (Plate 1.1). The city was the thriving capital community of western China. However, a combination of land degradation, mainly a result of soil erosion, desertification and salinization, increased climatic aridity, social tensions and invasions by Mongolian tribes led to the eventual abandonment of the city in the thirteenth century: http://www.travelchinaguide.com/attraction/xinjiang/turpan/ jiaohe.htm (accessed 16 February 2004).

As a global community we must learn from the lessons of the past. We live in a time of climatic changes, rapid increases in the global population and rapid decreases in the extent and quality of the soil resource base. Regional military conflicts continue, especially in the world's arid zones, and many are related to conflicts over water resources. The next nine chapters review the problems facing soil resources. They are meant not just to present a pessimistic view, but also to explore constructively ways in which we can tackle and resolve some of these issues.

1.2 SOIL SURVEY, SOIL CLASSIFICATION AND LAND EVALUATION

In order to understand and predict soil behaviour we must be able to assess soil properties, categorize and classify soils and then map their spatial distribution. In the field, soil surveyors examine a vertical 'slice' of soil in a pit, referred to as a soil profile. In the profile, soils are studied in layers or horizons. The nature and properties of each horizon and the relationships between horizons are considered. Some soil properties may be assessed in the field, with soil samples taken for subsequent laboratory analysis. Soil scientists who study the origin and development of soils are described as pedologists. Indeed, the term 'ped' frequently recurs in soil science, being taken from the Greek *pedos*, meaning 'soil'.

For rapid reconnaissance and to support profile analyses, pedologists may auger soil samples or take soil cores. These narrow sections and cores are then placed on plastic sheets, for study and sampling. Sometimes topsoils are sampled for specific analyses (e.g., for soil fertility). Samples are removed from specific depths (e.g., 0–5 cm in grassland soils or 0–20 cm in cultivated soils) and samples taken for analysis.

In the field, a number of properties are assessed. Each national soil survey produces soil survey field handbooks, which in essence are similar, though with subtle differences. These provide very precise procedures for the characterization of soil properties and sampling. For instance, Hodgson (1976) described the procedures used by the Soil Survey of England and Wales (now known as the 'National Soil Resources Institute', NSRI). Essential properties include:

- colour (described by reference to standardized colour chips in a Munsell Colour Chart);
- texture (the relative proportions of stones, sand, silt and clay in a sample);
- structure (the arrangement of particles into aggregated units);
- consistence (the degree and kind of cohesion of soil material).

Soil samples are then transported back to the laboratory for analysis. A particularly useful introduction to field and laboratory work in soil science is provided by Marsden and Allison (1992) and an informative guide to laboratory analytical techniques is provided by Rowell (1994).

In terms of physical soil analysis, particle size analysis is the most frequently used technique. Particles are divided into selected size ranges (FitzPatrick, 1986; Brady and Weil, 1999; Ashman and Puri, 2002). The fundamental distinction is between particles coarser than 2.0 mm, known as the 'coarse fraction' and consisting of stones, pebbles, cobbles and boulders, and the fine earth fraction,

Table 1.3 Soil particle size groups and their boundaries

Textural group	NSRI[a]		USDA[b]	
	(mm)	(μm)	(mm)	(μm)
Coarse fraction	>2.0	>2000	>2.0	>2000
Fine earth fraction	<2.00	<2000	<2.00	<2000
Sand	2.0-0.06	2000-60	2.0-0.05	2000-50
Silt	0.06-0.002	60-2	0.05-0.002	50-2
Clay	<0.002	<2	<0.002	<2

[a]NSRI, soil particle size classification system used in England and Wales.
[b]USDA, soil textural classification system used by the US Department of Agriculture.

with particles less than 2.0 mm diameter. For convenience, smaller particles are measured in microns (μm, 10^{-6} m or 10^{-3} mm). There are slight differences between different national systems (Table 1.3). Particle size analysis is usually achieved by a combination of sieving (for the coarse fraction and sands) and sedimentation in water for silts and clays. However, laser diffraction techniques are increasingly used because of their rapidity and replicability. In these techniques, a laser beam is passed through a dilute suspension of fine earths. The laser beam is diffracted and the greater the angle of diffraction the finer the particles. The diffraction spectra are computed against standards of known size, to give the particle size distribution (Loizeau *et al.*, 1994).

Using particle size analyses, we can identify soil particle size classes (i.e., stony, sandy, silty, clay-rich) (see Figure 4.2). Loams describe soils with a fairly balanced mix of sand, silt and clay particles. In terms of their properties, loams are often particularly useful for agriculture. They combine the desirable properties of sandy soils (well drained and quick to warm) with those of silty or clay soils (strong ability to retain water and exchange nutrients with plants).

Most analyses of soil chemical and fertility properties are carried out on fine earth fractions (particles <2 mm diameter). These methods most frequently measure nutrient content. Sometimes this is the total amount of a specified element. However, most elements are insoluble or only very slowly soluble and so are not readily available to plants. Therefore, many studies try to extract 'exchangeable', 'available' or 'labile' nutrients, which can be extracted using weak (usually mildly acidic) chemical solutions, such as ammonium acetate. However, it is difficult to assess the amounts of a particular nutrient available to plant roots, but extractants at least allow comparisons between different soil types and soil treatments, even if the values obtained are only approximate.

Nutrients in soils can be classified into two groups. Macronutrients are needed in large quantities to improve plant growth and include nitrogen (N), phosphorus (P), potassium (K), calcium (Ca), magnesium (Mg) and sulphur (S). Micronutrients are needed in small or 'trace' amounts for successful plant growth and include manganese (Mn), copper (Cu), nickel (Ni), zinc (Zn), molybdenum (Mo), boron (B), chlorine (Cl), iron (Fe), iodine (I), cobalt (Co), chromium (Cr), vanadium (V) and thallium (Tl) (FitzPatrick, 1986).

On the basis of field and laboratory data, pedologists can then classify soils. This is an important step, as soil management strategies need to be based on an understanding of the nature and properties of soils. This process is analogous to botanical work requiring a taxonomic classification of plants. However, whereas the Swedish botanist Carl Linnaeus provided a system of plant taxonomy that is still universally applied, no comparable system of classification exists for soils.

The first scientific (i.e., genetic) soil classification was proposed by the Russian soil scientist Vasilli Dokuchaev in the nineteenth century. Based at St Petersburg and Moscow Universities, Dokuchaev investigated relationships between soils and environmental factors on the Russian Steppes. In 1883 he identified five soil-forming factors: parent material, biota (plants and animals), relief, climate and time. Although the model has been subject to modification, Dokuchaev's concept remains at the core of pedology. Thus, the basis of many national soil survey systems is at least partly Russian in origin. This is true for the systems used in most European countries, Australia and Canada, in which many of the terms used are Russian (e.g., the Russian *chernozem* means 'black earth'). However, each national system has evolved somewhat differently, adapting to national conditions; consequently many problems arise when comparing them. For instance, the soil classification system of England and Wales is different from that of Scotland (Tables 1.4 and 1.5).

After the Russian Revolution, much of the impetus in soil survey work developed in the USA, under the leadership of Curtis F. Marbut. Marbut identified the fundamental distinction between pedalfers (leached soils, characteristic of humid climates) and pedocals (non-leached soils, characteristic of arid climates). The US Soil Survey began to develop a new classification system in

Table 1.4 Soil classification system of England and Wales (after Avery, 1980)

Major group[a]	Brief definition
1. Terrestial raw soils (5)	These occur in very recently formed material not significantly altered by soil forming processes. They have no pedogenic horizons other than a superficial organic or organo-mineral layer $<$5 cm thick, or a buried horizon below 30 cm depth.
2. Raw gley soils (2)	These occur in mineral material that has remained waterlogged since deposition. They may be unvegetated and are chiefly confined to intertidal flats and saltings that represent stages in the development of mature salt marshes.
3. Lithomorphic soils (7)	Shallow soils in which the only significant pedogenic process has resulted in the formation of an organic or organic-enriched mineral surface horizon. They are formed over bedrock or little altered, soft unconsolidated material at or within 30 cm depth.
4. Pelosols (3)	Slowly permeable clayey soils with no prominent mottled (gleyed) surface horizon at or above 40 cm depth. They crack deeply in dry seasons and have a coarse blocky or prismatic structure.
5. Brown soils (8)	Soils in which pedogenic processes have produced dominantly brownish or reddish subsurface horizons with no prominent mottling or greyish colours (gleying) above 40 cm depth. They are widespread, mainly on permeable materials, at elevations below approximately 300 m and are mostly in agricultural use.
6. Podzolic soils (5)	These are soils with a black, dark brown or ochreous subsurface horizon resulting from pedogenic accumulation of iron and aluminium or organic matter or some combination of these. They normally form as a result of acid weathering conditions and, under natural or semi-natural vegetation, have an unincorporated acid organic layer at the surface.
7. Surface-water gley soils (2)	These are seasonally waterlogged, slowly permeable soils, prominently mottled above 40 cm depth.
8. Groundwater gley soils (7)	These are soils, normally developed within or over permeable materials that have prominently mottled or uniformly grey subsoils resulting from periodic waterlogging by a fluctuating groundwater table.
9. Man-made soils (2)	These are soils formed in material modified or created by human activity. They result from abnormal management practices such as the addition of earth containing manures or refuse, unusually deep cultivation or the restoration of soil material following mining or quarrying.
10. Peat soils (2)	These are predominantly organic soils derived from partially decomposed plant remains that accumulated under waterlogged conditions.

[a]Values in parentheses denote the number of subgroups.

1960. It was periodically revised and is now known as the 'US Soil Taxonomy'. Cruickshank (1972) presented a useful review of the development of pedological thought.

The US system is hierarchical. All soils are divided initially into 12 Soil Orders (Entisols, Inceptisols, Alfisols, Spodosols, Ultisols, Oxisols, Mollisols, Aridisols, Vertisols, Histosols, Andosols and Gelisols). Each of the Orders is then divided into Suborders, Great groups, Subgroups, Families and Series (Brady and Weil, 1999). Therefore, the 12 Soil Orders are the largest taxa and all world soils can be placed into one of them. The Series is the local-scale mapping unit and in the USA there are approximately 16,000 of them. Various prefixes and suffixes are used with the names of Orders, etc., to help classify soils. This produces a terminology that is very precise and descriptive, but can be very cumbersome and unfriendly. The most appropriate introductory approach when using the system is to divide the full taxonomic term into its various components and translate their meanings using descriptive tables. Then the full term meanings can be reassembled to give a very precise description of each soil type.

The UN attempted to harmonize world soil classification by combining the best components of the Russian and US systems with those from other national systems. In 1974, the Food and Agriculture

Table 1.5 Soil classification system of Scotland (after Soil Survey of Scotland, 1984)

Division	Major soil group[a]
1. Immature soils	1.1 Lithosols (1)
	1.1 Regosols (2)
	1.2 Alluvial soils (3)
	1.3 Rankers (4)
2. Non-leached soils	2.1 Rendzinas (1)
	2.2 Calcareous soils (1)
3. Leached soils	3.1 Magnesian soils (1)
	3.2 Brown earths (2)
	3.3 Podzols (6)
4. Gleys	4.1 Surface-water gleys (6)
	4.2 Groundwater gleys (6)
5. Organic soils	5.1 Peats (4)

[a]Values in parentheses denote the number of subgroups.

Organization/United Nations Educational, Scientific and Cultural Organization (FAO/UNESCO) developed a world soil classification system, dividing all the world's soils into 26 Soil Units (FAO/UNESCO, 1974). This system continues to be developed and updated (Spaargaren, 1994). However, as with the US Taxonomy, the soil terms can be very confusing, especially when applied at a local scale. A particularly difficult one encountered during a visit to arid desert soils in Xinjiang Province, China, was a Yermipetrosaligypsicsolonchak!

Soil scientists debate whether the FAO system is better or worse than the US Soil Taxonomy. Largely it is a case of personal preference, but in soil surveys both systems are often used.

The soil classification system used in England and Wales is also hierarchical. At a national scale, soils are divided into ten major groups (Avery, 1980). Major groups are subdivided into groups (of which there are 43), subgroups ($n = 83$) and then series. There are about 700 recognized series in England and Wales (Hodgson and Thompson, 1985).

In Scotland, soils are again divided according to a hierarchical system. There are five divisions, each divided into major soil groups ($n = 12$), major soil subgroups ($n = 37$) and series (Soil Survey of Scotland, 1984). There are 530 recognized series in Scotland.

Based on soil survey field and laboratory data, soil maps are then constructed. There are several difficulties in constructing maps. The first is scale, which can vary from 1:5 million for global maps to 1:10,000 for local surveys. Maps for large areas must be at a small scale, and only a few of the more generalized soil types can be shown. In smaller areas, more detail can be presented at a larger scale. For instance, the National Soil Map of England and Wales covers six sheets at a scale of 1:250,000, and local maps are produced for selected 10 km × 10 km areas at the larger scale of 1:25,000. A second problem is where to draw soil boundaries, which usually indicate zones of rapid change in what is often a continuum of change. One frustrated soil surveyor described the problem as 'drawing lines that don't exist around areas that are all the same'!

The soil series is the most common local-scale mapping unit and is used for detailed mapping in many countries. The definition of series used by the NSRI is 'a group of soils similar in the character and arrangement of the horizons of the profile and developed under similar conditions from one type of parent material'. Thus, to belong to the same series, soils should have a similar arrangement of soil horizons and degree of soil development (pedogenesis) and should occur on the same underlying material (e.g., a certain rock or sediment type, such as river sediment or glacial deposits). Series are usually named after a location where they are widespread or were first identified and described.

A common purpose of soil survey is to produce land use capability classifications. This grades the land, becoming poorer with increasing number. Class 1 land is 'capable of producing a very wide range of crops' and Class 7 is 'land with very limited agricultural value'. The system in the UK is subject to modification, but Class 1–4 land is suitable for arable crops, Class 5 is suitable mainly as improved grassland, Class 6 is land suitable only as rough grazing and Class 7 land is of little agricultural value (Bibby, 1991). Land use capability subclasses can also be added to denote the nature of limitations (e.g., wetness (w), soil limitations (s), gradient and soil pattern limitations (g), susceptibility to erosion (e) or climatic limitations (c)).

Soil maps have various uses and applications, from general agricultural use to specialist geochemical maps. Traditionally, they are plotted as two-dimensional maps. However, electronic maps stored on CD-ROM and accessed by computer are increasingly used. Sometimes, these are three-dimensional, with survey information 'draped' over a three-dimensional digital terrain model of the landscape. For instance, Figure 1.1 shows a three-dimensional map of the agricultural soils of the Isle of Man. Further three-dimensional views and some movies can be accessed on the web site: http://iom.wlv.ac.uk/start.htm (accessed 16 February 2004).

Figure 1.1 A three-dimensional map of the agricultural soils of the Isle of Man

Summary

Soil is a vital resource for the global community. However, productive soils are both limited in extent and subject to a complex array of processes that can degrade their usefulness. It is essential that we recognize the extent and severity of current soil management problems and explore suitable strategies for soil management. As an initial evaluation stage, we must employ techniques of soil survey, soil classification and land evaluation.

FURTHER READING

Cruickshank, J.G., 1972, *Soil Geography,* Newton Abbot: David & Charles.

Davies, D.B., Eagle, D.J. and Finney, J.B., 1972, *Soil Management,* Ipswich: Farming Press Ltd.

Ellis, S. and Mellor, A., 1995, *Soils and Environment,* London: Routledge.

Nortcliff, S., 1984, *Down to Earth. An Introduction to Soils,* Leicester: Leicester Museums Publication No. 52.

Smithson, P., Addison, K. and Atkinson, K., 2002, *Fundamentals of the Physical Environment,* 3rd edn, London: Routledge.

Thomas, D.S.G. and Middleton, N.J., 1994, *Desertification: Exploding the Myth,* Chichester: John Wiley.

Trudgill, S.T., 1989, 'Soil types. A field identification guide', *Field Studies,* 7, 337–63.

Wild, A., 1993, *Soils and the Environment,* Cambridge: CUP.

Journals

There are many journals that report research on soil management. The following is a list of some, but by no means all, of the most helpful sources of research information:

Canadian Journal of Soil Science
Catena
Communications in Soil Science and Plant Analysis
Earth Surface Processes and Landforms
European Journal of Soil Science
Geoderma
Journal of Arid Environments
Journal of Soil Science
Journal of Soil and Water Conservation
Land Degradation and Rehabilitation
Plant and Soil
Progress in Physical Geography
Soil Science Society of America Proceedings
Soil Technology
Soil & Tillage Research
Soil Use and Management

2 Soil erosion and conservation

The dust is gold that bears the harvest;
Save the soil that grows our bread;
Let not wind and rain remove it;
Guard with care for years ahead.
Quoted from S.G. Brade-Birks (1944: 189)

Introduction

This chapter explores the issues of soil erosion and conservation. The chapter commences with a consideration of patterns and processes of water and wind erosion. The issue of what can be considered as acceptable rates of soil erosion is discussed, followed by approaches to the quantification of soil erosion rates. The factors that control erosion risk are then discussed and the chapter concludes with a review of approaches we can use to conserve soils.

2.1 SOIL EROSION

2.1.1 Patterns of soil erosion

Water, wind or ice cause the erosion, transportation and eventual redeposition of soils. Where water is the eroding agent, we refer to this as water erosion. This can cause distinct landforms, which we can divide into erosional and depositional phases.

We can classify water erosion into several types. In the erosional phase, we include splash, sheet, rill and gully erosion. The initial impact of raindrops can break down soil aggregates into primary particles, by the translation of kinetic energy from the drops to the soil aggregates. This process is known as slaking. Plate 2.1 shows the effect of a simulated raindrop hitting a moist sandy surface. The kinetic energy of the impacting raindrop causes the formation of an impact crater, around which the water rises as a rebound corona. The process occurs very quickly; Plate 2.1 was shot at 1/2000 of a second. Larger raindrops (diameter about 4 mm) are most erosive. Drops larger than about 6 mm diameter are unstable and tend to break up in airflows. Because of the influence of gravity, more soil particles are splashed downslope than upslope, and are splashed a greater distance downslope.

Plate 2.1 A simulated raindrop hitting a moist sandy surface. The kinetic energy of the impacting raindrop causes the formation of an impact crater, around which the water rebounds as a corona. The process occurs very quickly, Plate 2.1 was shot at 1/2000 of a second (photo M.A. Fullen).

Plate 2.2 The effect of a stone protecting soft silty sediments from splash erosion in the Tabernas Badlands of southeast Spain (photo M.A. Fullen).

The cumulative effect is a net downslope transfer of detached soil. This seemingly innocuous process can lead to substantial transportation of sediment. For instance, Plate 2.2 shows the effect of a stone protecting soft silty sediments from otherwise rapid splash erosion in the Tabernas Badlands of southeast Spain.

Plate 2.3 High sediment concentrations in the Yellow River of China. It is estimated that the river transports some 1800 million tonnes of sediment per year to the Yellow Sea. Much of this sediment is entrained when the river flows through the Loess Plateau, an area of erodible soils derived from wind-blown silt (photo M.A. Fullen).

When rainfall intensity exceeds the infiltration capacity of the soil, runoff occurs and this leads to the erosion of soil. Initially, the process starts with a sheet of water of fairly uniform thickness flowing over the surface. This causes quite uniform erosion beneath it, known as sheet erosion. However, this state is unstable, as flowing water has a natural tendency to concentrate in the lower parts of the surface and then incise into the soil. Where these channels or rills are shallow, the process is known as rill erosion. However, if rill erosion continues, it leads to the development of gullies.

The distinction between rill and gully erosion is problematic. Early guidelines in the 1930s from the US Soil Conservation Service defined a gully as a feature a prairie dog could not jump across! Later definitions described a gully as a feature which would need to be mechanically infilled for agricultural activities to proceed. Others argue that rills are features incised into the topsoil (the A horizon), whereas gullies incise into the subsoil or parent material (the B or C horizons).

Eroded soil is eventually redeposited. When this occurs on slopes, the deposited material is described as colluvium. Colluvial deposits are particularly common on concave sections of slopes or against obstructions, such as walls and hedges, as decreasing overland flow velocities mean the runoff water can no longer transport sediment. Sediments can also enter water courses. For instance, Plate 2.3 shows high sediment concentrations in the Yellow River of China. The Yellow River flows through the Loess Plateau, which is composed of a thick deposit of fine silt. Silt-sized particles (2–60 μm) with little clay to bind them together have little resistance to erosion and are thus highly erodible. Sediments that are reworked as part of the fluvial system of streams and rivers are described as alluvium. Alluvium is reworked either as overbank deposits by streams or by channel migration within the floodplain. Sediments derived from farmland soils are often rich in agro-chemicals, such as phosphate, nitrate and pesticides. These can cause pollution of the water and damage to aquatic ecosystems (Sections 5.4 and 5.5).

Wind can also erode, transport and redeposit soil. Processes dependent on wind action are often described as aeolian processes, named after Aeolus, the Greek god of the winds. Loss of material by wind erosion is described as deflation. Wind erosion can transport fine sediments considerable distances. For instance, scientists at the Mauna Loa Observatory in Hawaii can detect the onset of the Chinese spring planting season by increased dust fallout, even though Hawaii is some 5000 km east of China (Parrington *et al.*, 1983).

The severity of soil erosion varies markedly. Langbein and Schumm (1958) proposed a model relating water erosion to rainfall. In very dry climates, there is usually little water erosion, but a slight increase in rainfall amount often causes a large increase in erosion rates. The occasional rainstorms tend to be very intense convectional storms, which have considerable energy to cause erosion; that is, they are very erosive. Also, there is little protection from vegetation, which is sparse in semi-arid environments. In more humid climates, the vegetation cover increases and protects the soil surface from erosion. In very wet climates, erosion rates then increase again, because of high rainfall erosivity.

The Langbein and Schumm model has been critically assessed in many parts of the world. Certainly, in the semi-arid tropics, erosion rates are very high. It is estimated that some 6000 million tonnes of soil per year are washed off the croplands of India and that some 1600 million tonnes per year are transported by the River Ganges to the Bay of Bengal. The Mediterranean environment has long been noted for soil erosion, because of its semi-arid climate and the long human occupancy of the landscape, which has promoted deforestation.

A useful and simple way to express rainfall erosivity is the Fournier Index (Fournier, 1960), which takes the form:

$$\text{Fournier Index} = p^2/P \qquad (2.1)$$

where p is the highest mean monthly precipitation (in mm) and P is the mean annual precipitation (in mm).

Climates with a highly seasonal rainfall tend to have high Fournier Index values. Numerous studies indicate positive correlations between the Fournier Index and measured erosion rates. Using climatic data, it is possible to calculate the Fournier Index for a given station or region. Rudloff (1981) is a useful source of such data, reporting climatic data from selected meteorological stations throughout the world.

There are areas of the world where erosion rates should theoretically be low, according to the Langbein and Schumm model or Fournier Index, but actually are high. This is largely because of human activities, particularly vegetation removal. For instance, deforestation of tropical rainforests is leading to very high erosion rates. The soils of the humid tropics and subtropics are usually highly leached, acidic and nutrient deficient. These are defined as Oxisols and Ultisols in the US Taxonomy (Section 1.2). These soils contain little organic matter, unlike the soils of more temperate environments, and removal of vegetation can lead to further irreversible losses of soil organic matter. Soil organic matter stabilizes soil structure and its loss therefore further increases erosion risk (Sections 2.4 and 8.2.3).

Temperate continental interiors should have a natural or climax vegetation cover of grassland. These areas include the Prairies of North America, the Steppes of Russia and the Ukraine and the Pampas of Argentina. Under these conditions, the soils are deep, black and organic Chernozems or 'black earths' in the Russian and FAO/UNESCO soil classification systems (Mollisols in the US Taxonomy). Over the last 100–130 years many of these continental grasslands have been converted to arable use, particularly for cereal production. The soil organic content has fallen and the soils have become more erodible and this has increased the risk of severe soil erosion. For instance, it has been estimated that some 4000 million tonnes of soil per year have been eroded from the continental USA since the 1930s. This would fill a freight train long enough to encircle the equator 24 times! There is also increasing evidence of soil erosion in northern and western Europe. In the Langbein and Schumm model, these areas should experience little erosion. However, the damage to soil structure imposed by increasingly mechanized and intensive agriculture in Europe, particularly since the 1940s, is believed to be a major factor increasing erosion rates.

Erosion directly related to cultivation is often termed tillage erosion or tillage translocation (Govers *et al.*, 1994; Van Oost *et al.*, 2003). For instance, mouldboard ploughing on slopes results in downslope displacement of soil, which occurs when the soil is lifted by the plough perpendicular to the slope, but falls back vertically (and therefore downslope) under the influence of gravity. The strongest effect results from sideways displacement of soil during ploughing along the contour, but it can also occur with cultivations up and down slope. Montgomery *et al.* (1999) measured movements of 110–113 kg m^{-1} per tillage operation, with degradation of soil by losses from slope convexities and accumulation in concave sites: this rate of movement is an order of magnitude greater than that resulting from soil creep and splash under cultivation. The problem can be eliminated by zero or minimal tillage (Section 7.2). Concern has also been expressed over the amount of soil lost by levelling operations in preparing sites, for instance for construction of greenhouses along the Mediterranean coast (Poesen and Hooke, 1997). Furthermore, the amount of soil removed adhering to crops and agricultural machines can equal or even exceed the amount eroded by wind or water (Poesen *et al.*, 2001).

There have been several attempts to quantify global erosion, particularly by Lester R. Brown of the Worldwatch Institute in Washington DC (Brown, 1984, 1991). He described soil erosion as 'a quiet crisis, one that is not widely perceived. Unlike earthquakes, volcanic eruptions or other natural disasters, this human-made disaster is unfolding gradually' (Brown, 1984: 162). Brown estimated that the maximum rate of soil formation is 2 tonnes of soil per hectare of land per year, but that the average soil erosion rate is about 20 t $ha^{-1} yr^{-1}$. The calculated global excess of soil erosion over formation is 25,730 million tonnes, roughly equivalent to the amount of topsoil in Australia's wheatlands. These estimates must be treated very cautiously, but they are indicative of the scale of the problem.

2.1.2 What is an acceptable rate of soil erosion?

A definition of what is 'acceptable' as a rate of soil erosion has been debated by soil scientists for many decades. We know that soil erosion is a natural process, a mechanism by which uplands are worn down or denuded over the geological eons of time. 'Natural' or 'geologic' erosion is an integral component of the process of soil formation. Some spectacular geological features, such as the Grand Canyon in the USA, testify to the effectiveness of natural denudation processes. We need to separate this background level of natural soil erosion from 'accelerated' erosion, where human activities are increasing erosion rates much above their background levels. The most common activity is vegetation removal, exposing soils to wind or water erosion.

Distinction between 'natural' and 'accelerated' erosion is complex and often uncertain. To illustrate the point, we can refer to 'Tu Lin' (the 'Soil Forest') in Yunnan Province, China (Plate 2.4). Erosion has caused deep incision into soft Tertiary sediments. The climate is monsoonal, with highly erosive summer rains. Human activity has also played a role in promoting erosion, as the area has been occupied for several thousand years. Intense land pressure has resulted in areas marginal to Tu Lin

Plate 2.4 Tu Lin (the 'Soil Forest') in Yunnan Province, China. Highly erosive summer monsoonal rains have incised into soft Tertiary sediments. Human agency has also played a role in promoting erosion. For instance, in the top-centre, note the cultivation of melons right at the edge of the gullied area (photo M.A. Fullen).

being cultivated, further accelerating erosion rates. Thus, both natural and accelerated erosion rates are high.

In an attempt to define what we mean by 'acceptable' soil erosion, many soil scientists use the argument that the soil system should be in a state of dynamic equilibrium; that is, the rate of soil loss by erosion should not exceed the rate of soil formation. It is analogous to a bath of water, where the rate of water input from the taps equals the rate of water loss down the drain. The actual particles of water in the bath are in a constant state of motion, entering and leaving the system, but the mass of water in the bath remains about constant. The concept of dynamic equilibrium is important in our understanding of many environmental systems.

We know very little about rates of soil formation, but we know that it is a slow process. Soil formation occurs essentially by the weathering of material at the soil–parent material interface or by deposition of sediment on slopes. The parent material is usually rock, but often includes soft sediments, such as glacial, alluvial, estuarine or aeolian deposits. Weathering is essentially a process of disintegration or decomposition, which covers a complex range of physical, chemical and biological processes. Pedologists agree that it occurs at a slow rate, perhaps taking 1000 years to weather 10 cm of material at the B/C horizon interface. Tillage can speed up this process, by scraping shallow bedrock and increasing soil aeration and microbial activity. This can increase weathering to a maximum of about 10 cm in 100 years; that is about 1 mm yr^{-1}. This is equivalent to about 2 t $ha^{-1} yr^{-1}$. Many have argued this gives us some guide or 'yardstick' as to what is an acceptable rate of erosion. Theoretically, if erosion and soil formation occurs at about 2 t $ha^{-1} yr^{-1}$, then the soil system will be in a state of dynamic equilibrium. Such erosion rates have been labelled 'tolerable' or 'T' values and are used in planning soil conservation strategies.

Many have criticized the 2 t $ha^{-1} yr^{-1}$ 'T' value, suggesting it is too high (Schertz, 1983). For instance, it assumes an unlimited supply of weatherable material. Also, pedogenesis is a much more complex process than simply weathering at the base of the soil. Soils mature through time, usually incorporating organic matter into the topsoil (A horizon). Erosion usually removes topsoil, often with serious implications for soil fertility. Topsoils are essentially the 'seats' of most biological activity within soils, containing most of the soil fauna, the soil organic fraction and most of the soil nutrients, both natural and applied. Therefore, erosion involves more than the loss of inorganic components of the soil system. Often the most fertile material is lost and any new soil formed at the base of the profile is much less fertile. In cases of severe erosion, the A horizon may be completely stripped away, deposited downslope and then less fertile subsoil (B and/or C horizon) is deposited on top. This 'soil profile inversion' markedly diminishes soil fertility.

In critically assessing 'T' values we need to recognize the complex environmental effects of erosion. The erosion rate may be tolerable from the viewpoint of maintaining soil fertility in the affected field. However, the eroded sediments may promote several problems in streams and rivers. Sediments rich in agro-chemicals may introduce pollution to surface waters (Chapter 5) or they may block water courses and roads. Therefore, we need to distinguish what is 'tolerable' erosion in soil fertility terms on slopes (T_{SL}) and what is tolerable in terms of sedimentation and pollution of water courses (T_{WQ}). Generally, T_{WQ} is much less than T_{SL}.

Clearly we can level many criticisms against the 2 t ha^{-1} yr^{-1} 'T' value, all of which suggest it is too high. However, soil erosion rates can be several hundreds of tonnes per hectare per year and so can represent a very serious depletion and, in extreme cases, the complete removal of the soil resource base. In the long term, even low erosion rates continuing over many years can be damaging.

2.2 METHODS OF QUANTIFYING SOIL LOSSES

Numerous techniques exist for quantifying soil losses. For water erosion a simple technique is to establish runoff plots. These are bounded on three sides (usually using wood, metal or plastic) and runoff and eroded sediment are collected at the downslope end. It is often useful to have an additional small wall or bund on the upslope side, to prevent runoff from upslope entering the plot. The 'runoff plot' approach was established in Germany by Ewald Wollny in the 1880s. The approach measures runoff and erosion rates in precisely defined conditions, with specified conditions of soil type, slope and vegetation cover. However, it has been criticized, mainly because plot boundaries interfere with erosion processes and rates; for example, the plot boundaries can prevent the full development of rills.

Runoff plots were adopted by the US Soil Conservation Service, following its establishment in 1934 (Woodruff, 1987). Standardized 0.01 acre (1 acre = 0.405 ha) plots were constructed in a range of agricultural environments, principally east of the Rockies. For each year a plot was operational, it generated a 'plot year' of data and by the 1960s over 10,000 plot years of data had been collected. The advent of computers enabled scientists at the National Soil Erosion Laboratory in Purdue, Indiana, to collate and analyse the database. This effort was led by the statistician Walter Wischmeier, who proposed the 'Universal Soil Loss Equation' (USLE) in 1960 (Wischmeier, 1960). The equation states that:

$$A = R \cdot K \cdot L \cdot S \cdot C \cdot P \tag{2.2}$$

where A is average annual soil loss (tons per acre, now expressed in tonnes per hectare); R is rainfall and runoff factor, that is the amount of energy available to cause runoff and erosion; K is soil erodibility factor. This is based on several soil physical properties (soil texture, soil organic matter content, soil structure and infiltration capacity). The resistance of the soil to erosion was assessed on a scale of 0 (low erodibility, high resistance to erosion) to 1.0 (high erodibility, low resistance to erosion). In practice, K values are between 0.03 and 0.7; L is slope length; S is slope steepness; C is cropping and management factor; and P is conservation practice.

The soil conservation engineer would assess each of these factors, using established field manuals, and multiply them together to produce a value for A. For each soil series, a 'T' value was established, which varied according to soil depth. If A was greater than T, some action was necessary to decrease erosion to tolerable levels. Then the conservationist drew up a soil conservation plan in consultation with the farmer. This usually involved altering either the C (cropping) or P (conservation practice) factor. Such assessments are now made using laptop computers with in-built databases. Morgan (1995) comprehensively reviewed the development of the USLE.

The USLE has been subject to many misinterpretations. First, it is not 'universal'. It is essentially an equation applicable to the USA east of the Rockies, particularly the Mid-West. However, it has been used in many other parts of the world and in climatic zones for which it was not designed. Nevertheless, the K factor does provide a standardized protocol to assess soil erodibility and has been widely and successfully adopted. Furthermore, the USLE is not a predictive equation. Comparison of measured and A values calculated from it has indicated accuracies of plus or minus 50 per cent. The USLE was designed to indicate erosion risk and not as an accurate predictive equation. Wischmeier (1976) explicitly recognized the misapplications and misinterpretations of the USLE.

Since the USLE was introduced, many soil erosion equations and models have been developed. These include RUSLE (the revised USLE), followed by WEPP (the Water Erosion Predictive Equation). EPIC (the Erosion Productivity Impact Calculator) attempts to predict the long-term effects of erosion on soil properties, fertility and crop productivity, simulating erosion for up to hundreds of years (Williams *et al.*, 1990). EUROSEM is an attempt to predict erosion rates in European conditions (Rickson, 1994), and LISEM (the LImburg Soil Erosion Model) is a model to predict erosion on loess soils (Jetten and de Roo, 2001); it can be accessed via the worldwide web at: http://www.geog.uu.nl/lisem/ (accessed 17 February 2004).

Models are also used extensively in wind erosion research. The most notable example is the Bagnold Equation, based on extensive work by Major R.A. Bagnold and first published in 1937. During his military service in the British army, based in Libya, Bagnold studied the dynamics of sand dunes. In the equation, sediment transportation by wind is related to wind velocity in the form:

$$Q = 1.5 \times 10^{-9} (V - V_t)^3 \tag{2.3}$$

where Q is total load of sediment carried (t m^{-1}h^{-1}); V is velocity at measuring height (m s^{-1}); and V_t is fluid threshold velocity for sand movement (m s^{-1}).

V_t is the wind velocity necessary to initiate effective sand transportation. Once that velocity is exceeded, then sediment transport increases as the cubic power of wind velocity above the fluid threshold velocity. Thus, slight increases in wind velocity above V_t cause marked increases in wind erosivity. The Bagnold Equation has formed the basis of further predictive equations, especially those developed in North America (Argabright, 1991). Morgan (1995) presented a comprehensive review of the rationale and development of several wind and water erosion models.

Considerable progress has also been made in understanding the mechanisms of wind erosion (Lyles and Tatarko, 1986). Finer particles, especially silts, can be transported in suspension (flowing above the land surface); coarser material is usually transported by saltation, a bouncing mechanism, with particles often imparting their kinetic energy to other surface particles on landing and thus initiating further motion. Saltating particles tend to be coarser than particles in suspension and are usually sand-sized.

There are many ways to quantify and monitor erosion patterns and processes in the field and laboratory (De Ploey and Gabriels, 1980). A simple field technique is to insert pins vertically into the soil. These allow accurate measurements of changing surface levels to show how much erosion (surface lowering) or accumulation (surface raising) has occurred. Simple traps can be established in the field to collect sediment. These are often referred to as 'Gerlach troughs' after the Polish geomorphologist who developed them. In the field, much attention is devoted to assessing the dynamics of sediment movement. This often involves 'tagging' soil particles with a tracer and using the movement of the tracer to indicate soil movement. Many different types of tracers have been used, including painted stones and soil particles, fluorescent dyes, radioactive isotopes and magnetic materials. The fallout of isotopes from nuclear explosions or accidents, such as Chernobyl, has also been used to quantify erosion rates. The most widely used are isotopes of Caesium (Cs), such as Cs^{137} and Cs^{134}. These are positively charged and, as they fall to Earth, they become attached to clays and organic materials, which tend to be negatively charged. As a site is eroded, it becomes depleted in the isotope and so, comparing the radioisotope content of eroded soils with non-eroded soils, such as found in woodland, enables approximate quantification of erosion rates (Walling and Quine, 1991). Another method is to investigate the distribution of fallout isotopes in colluvial deposits. Loughran (1989) provided a useful summary of field methods to quantify erosion rates.

Soil scientists have particularly favoured simulating erosion processes in the laboratory. This usually involves exposing soil samples in trays to simulated rainfall. The stability of soil aggregates under simulated rainfall has often been used as a measure of soil erodibility (Grieve, 1979). A strength of this approach is that we can control precipitation amount, duration and intensity and study system responses (e.g., runoff, time to runoff, sediment concentration and erosion rate) in varying environmental conditions (e.g., slope, soil, vegetation type and cover). By exerting a fairly uniform environmental stress on soil aggregates, which approximates to conditions in natural rainfall, erodibility assessments under simulated rainfall can indicate differences in soil aggregate stability. However, there are difficulties with this approach, such as how realistically laboratory studies simulate field conditions and whether we can extrapolate from laboratory to field conditions (Foster *et al.*, 2000). At the time of writing advanced laboratory simulation studies are in progress in the experimental laboratories of the Universities of Leuven (Belgium) and Toronto.

2.3 FACTORS CONTROLLING SOIL EROSION RISK

Soil erosion is the product of many complex and interacting factors. For instance, in a laboratory study to assess erosion risk on 55 soil samples from the US Cornbelt, Wischmeier and Mannering (1969) found 22 soil and surface properties were necessary to explain 95 per cent of soil loss variance. However, Morgan (1995) offered a useful qualitative simplification, stating that soil erosion results from the dynamic interaction of:

1| the energy of the water or wind in causing erosion (erosivity);

2| the inherent resistance of the soil system to detachment and transport (erodibility);

3| the protection factor of vegetation.

Many studies have attempted to relate rainfall erosivity to erosion rates. The erosivity of rainfall is a function of its intensity and duration, and the mass and velocity of raindrops. The fundamental equation of kinetic energy used to calculate rainfall erosivity is:

$$\text{Kinetic energy } (\text{J m}^{-2}) = 0.5\, mv^2 \qquad (2.4)$$

where m is mass of raindrops (g); and v is velocity of raindrops (m s^{-1}).

Hence, a small increase in the velocity of raindrops produces a large increase in kinetic energy. There have been many attempts to convert rainfall intensity, usually expressed in mm h^{-1}, to kinetic energy per millimetre of rain ($\text{J m}^{-2}\text{mm}^{-1}$). Numerous empirical equations have been developed to relate rainfall intensity and kinetic energy. An equation often used in the USA is:

$$KE = 11.87 + 8.73 \log_{10}(I) \qquad (2.5)$$

where KE is kinetic energy ($\text{J m}^{-2}\text{mm}^{-1}$); and I is rainfall intensity (mm h^{-1}) (source: Wischmeier and Smith, 1958).

The energy of water flowing over the soil surface also affects erosion. The erosivity of running water is related to its velocity, volume, turbulence and shear stress. The predictive equations rely heavily on accepted equations in hydraulics (e.g., the Chezy Equation and the Reynolds Number). In recent years, the role of the shear stress exerted by water on the soil surface has been emphasized. For example, Govers (1985) suggested that shear stress values over $4\,\text{cm s}^{-1}$ are needed for rill incision.

Slope angle, length and shape profoundly affect the erosivity of running water. Generally, as slope angle increases, so does erosion risk because runoff velocity and energy increase. Usually, erodible soils on slope angles greater than about 10° are particularly susceptible to rill erosion. Rills are hydraulically efficient systems for transporting soil so they promote high erosion rates. Increased slope angle also increases the efficacy of splash erosion processes (Section 2.2). On longer slopes, runoff has more time to accelerate and thus achieve greater erosivity. Slope shape is also important, as runoff tends to accelerate rapidly over convex slope segments, achieving higher velocities and thus greater erosivity, but decelerate over concave slope sections, often leading to sediment deposition. Thus, the typical convex–concave morphology of slopes in humid temperate environments makes them particularly vulnerable to water erosion.

Soil erodibility is influenced by many factors, principally texture (Section 1.2) and soil organic content. The relationship between soil particle size distribution and erodibility can be described as a non-linear polynomial relationship. Clays are strongly cohesive because of electrostatic attraction between layer silicates. Consequently clay particles ($<2\,\mu\text{m}$) tend to resist detachment, but are easily transported once they are entrained because they are very small. Coarse materials, defined as particles >2 mm diameter (i.e., gravel, pebbles, cobbles and boulders) are not cohesive (unless cemented) but resist transportation because of their size. The most erodible material tends to be of intermediate size, that is silts and sands, which are not cohesive and are small enough to be transported by the flow rates characteristic of rills. Coarse silts and fine sands, that is particles with a diameter of approximately 50–150 μm, are most erodible. Therefore, sandy loams, loamy sands and sands are the most erodible soil textures while clays and clay loams are the least erodible.

The effect of soil organic matter content on soil erodibility can be described by a polynomial relationship. Numerous field and laboratory studies have shown that soils with low organic matter contents are more erodible than more organic soils. Generally, soils with less than 2 per cent organic matter by weight are highly erodible. As organic matter increases above the 2 per cent critical threshold, soil erodibility decreases to a minimum at about 10 per cent organic matter, which is a typical value for a deciduous forest topsoil in a humid temperate environment, such as the British Isles. An important characteristic of soil organic matter is that it binds mineral soil particles together, forming chemically complex organic bridges between them. Thus, the soil becomes more resistant to the erosive forces of wind and water. The organic fraction is analogous to a 'sticky glue'. This is particularly the case with organic residues secreted by soil fauna. Earthworms in particular break down plant residues in their gut to a range of polysaccharides, which have a gum-like consistency and are very effective in binding soil particles. These processes are responsible for creating soil aggregates, which are often too large to be transported by wind or water. Soils with more than approximately 20 per cent organic matter tend to be more erodible, as there is less clay to form aggregates with the organic matter, and the organic material is very light and therefore easily transported. Organic particles have densities typically about $0.8\,g\,cm^{-3}$ compared with over $2.5\,g\,cm^{-3}$ for most mineral particles. Low density is the main reason why very organic soils, such as the peaty soils in the Fens of East Anglia, are susceptible to wind erosion.

Vegetation protects the soil from erosive forces. Many studies around the world agree that, when plant cover exceeds about 30 per cent, it is very effective in protecting the soil from erosion (Quinn *et al.*, 1980). Vegetation has a number of effects. First, its surfaces dissipate raindrop energy and prevent slaking. This is particularly true for short vegetation. A fall of about 8–9 m is required for drops to achieve their terminal velocity, so it is possible for raindrops falling from tree canopies to regain their erosivity. Second, infiltrating water often follows vegetation root systems. Thus, roots encourage high infiltration rates and so decrease runoff rates. Third, vegetation also acts as a 'brake' on runoff velocity, thus decreasing erosivity. Fourth, vegetation acts as a coarse filter, removing much of the sediment transported in runoff. The protective effects of vegetation were well summarized by Bollinne (1978), who said 'vegetation reduces, brakes and filters the runoff and in this way reduces its erosive effects'.

2.4 SOIL CONSERVATION

Hordes of gullies now remind us
We should build our land to stay,
And, departing, leave behind us
Fields that have not washed away;
When our boys assume the mortgage
On the land that's had our toil,
They'll not have to ask the question
'Here's the farm, but
Where's the soil?'
Quoted from S.G. Brade-Birks (1944: 185)

Society can take action to try to minimize soil erosion and decrease it towards the natural geological rate. Soil conservation may be 'active' or 'passive'. Active soil conservation is taking positive measures to decrease erosion rates, such as terracing or contour farming. 'Passive' conservation is avoiding certain actions, such as ploughing on steep slopes or overgrazing erodible soil. Passive conservation is just as valid and indeed is usually much cheaper than active conservation. Prevention is better than cure! We also need to distinguish between soil conservation and sediment control. Soil conservation is taking action to prevent erosion and keep soil *in situ*. Sediment control attempts to deal with soil that has already been eroded and transported. This includes technologies to keep sediments within fields or to deal with sediment that has entered water courses.

Conservation strategies can be complex and varied. However, a useful start is to consider the proposal by Morgan (1995); that is, to conserve soil we must decrease erosivity, decrease erodibility or improve protection, or any combination of these. Soil conservation has long been employed by human societies, testified to by such features as the terraced fields of Southeast Asia and the Mediterranean basin. For instance, the Banaue rice terraces of the Philippines are believed to be over 2000 years old. However, modern soil conservation technologies have been considerably improved by the US Soil Conservation Service.

The US Mid-West suffered extreme soil erosion in the 1930s, an era often referred to as the 'dirty thirties'. By this time, many of the organic soils (variously classified as Mollisols, Chernozems or Prairie Soils) of the Mid-West had been cultivated for over 80 years. Soil organic contents were falling, thus increasing soil erodibility. Many erodible soils were brought into cultivation by the increased grain prices associated with the First World War. Then, in the 1930s, there was a severe drought. Limited archaeological evidence from native North American peoples shows that drought is a regular and recurrent feature of the climate of the continental interior of North America. This combination of circumstances led to severe erosion, especially wind erosion, in the Great Plains States (Thomas and Middleton, 1994). The worst problems were in Kansas, Texas, Nebraska and especially Oklahoma. A journalist, Robert E. Geiger, labelled the area 'The Dust Bowl' (Hurt, 1981). The social upheaval associated with these events was graphically portrayed in the novel *The Grapes of Wrath* by John Steinbeck, which tells the tragic tale of a family from Oklahoma and their struggle for survival (Steinbeck, 1939). One song writer described the experience of being in a Dust Bowl dust storm as 'It fell across our city like a curtain of black rolled down. We thought it was our judgement, we thought it was our doom' (Woody Guthrie, quoted in Worster, 1979: 10).

In response to the problems of the Dust Bowl an agricultural engineer, Hugh Hammond Bennett, led a campaign to promote soil conservation. Bennett was almost evangelical in his campaigning among farmers and politicians. During one presentation to the US Congress, a dust storm blew into Washington DC. Hammond declared, pointing out of the window, 'there, gentlemen, goes the state of Oklahoma!'. Congress then allocated the funding for the foundation of the US Soil Conservation Service in 1934. Bennett was a prolific writer and his seminal book *Soil Conservation* guided much soil conservation work in subsequent decades (Bennett, 1939).

Various soil conservation techniques have been proposed, and in soil conservation projects we can adopt a specific technique or combination of techniques. The choice depends on many factors, including environmental conditions, such as the climate, topography and soil type. We must also consider social, economic and political circumstances. A major factor is the cost and the availability of resources for conservation work. In the following paragraphs we will review specific soil conservation measures.

Windbreaks can be very effective in decreasing wind velocity and thus erosivity. Windbreaks are aligned perpendicular to the direction of the most frequent erosive winds. Usually, they are effective for 10–20 times their height downwind. Dense windbreaks brake wind velocity very markedly, but wind velocities soon increase in their lee. Better effects are achieved with more permeable windbreaks; wind velocity is not decreased so much, but their effectiveness in the lee is greater. For maximum benefit, windbreaks with a porosity value of about 50 per cent are recommended. Having a mix of heights is also beneficial (Morgan, 1995).

Hedgerows can be very effective windbreaks and their large-scale removal in the British countryside since the 1940s is believed to have contributed to increased wind erosion on arable soils, particularly on the Fens of East Anglia. It is estimated that between 1946 and 1963 an average of 22,000 km of hedgerows were removed per year. Hedgerows also protect against water erosion, as they divide slopes into shorter sections. Their removal has allowed erosive runoff to operate over effectively longer slopes, thus increasing erosion risk. Reinstating all hedgerows is not feasible, as modern farm machinery cannot operate efficiently on fields as small as those that dominated the British countryside before the Second World War. However, it is important to identify 'key hedgerows', which protect areas exposed to predominant wind directions and convex slope sections. Their retention or establishment allows the separation of large catchment fields from lower convex–concave slopes. The absence of key hedgerows integrates fields into geomorphological systems that are more vulnerable to water erosion.

Slope management is a key component of soil conservation strategies. Simply leaving steeper erodible soils with a vegetation cover is a cheap, but effective, form of passive conservation. The agricultural management system of 'set-aside' may have potential. 'Set-aside' involves taking areas out of crop production and leaving them with a permanent vegetative cover. The system was borrowed from the USA by the European Union as an instrument to decrease grain surpluses. Careful targeting of 'set-aside' on steeper erodible slopes could be a very effective means of achieving agricultural, fiscal and environmental objectives.

Terracing is the most spectacular form of soil conservation. The procedure involves dividing slopes into a series of steps, cultivating the flat or flatter sections and protecting the 'riser' wall by vegetation or masonry walls. The flat terrace fields act as effective stores of water and thus increase crop yields. In Southeast Asia terraces are extensively used to grow rice, which is the staple food crop. To retain water, small earth walls or 'bunds' are built on the lower side of the terrace field. However, terracing poses several problems. First, the riser takes up about 10 per cent of land, though this is usually compensated by increased crop yield associated with a better water economy. Second, their construction and maintenance are costly in terms of human resources; most of the world's areas of extensive terraces were constructed over many generations. Third, it is often difficult to operate machines efficiently on terraces.

Another form of slope management is *controlled colluviation*. This is particularly applicable in semi-arid countries with a distinct rainy season. Lines of stones are laid out perpendicular to the slope. When the seasonal rains arrive, soil is eroded and sediment is deposited on the upslope side of stones. Over a few rainy seasons, a fine silty moisture-retentive colluvial soil develops. Moreover, in semi-arid climates there is a tendency for soils to become saline (Section 3.3), and this seasonal flushing with water can desalinize the colluvium. Thus, these zones of controlled colluviation have relatively fertile soils, suitable for crops. The technique is simple and cheap, and so can be afforded in poorer countries.

Contour farming involves orientating agricultural operations (e.g., ploughing, drilling, seedbed preparation and harvesting) along the contour rather than up-and-down the slope. The technique is particularly applicable in areas dependent on extensive mechanized agriculture on fairly uniform, gently to moderately sloping sites. Therefore, it has been particularly adopted in the large-scale farming systems of the Prairies of North America and the Steppes of Russia and the Ukraine. The complexity of slopes tends to limit the applicability of contour cultivation in northern Europe, except for some fairly uniform extensive cereal growing areas, such as Lincolnshire in eastern England. However, slopes must not be too steep ($>15°$), as water can accumulate in furrows and eventually breach crop rows and ridges between furrows. Rapid breaching of contour ridges and cascading of water can cause even higher soil erosion rates than the more common up-and-down slope cultivation. Moreover, farming machinery cannot operate safely or efficiently along the contour on steep slopes.

A relatively simple soil conservation technique is *strip cropping*. In this, alternate strips of land are arranged perpendicular to the relevant erosive agent, wind or water. The crops themselves perform the protective effects of vegetation, decreasing the velocity of the erosive agent and trapping sediments. Temporary grassland (leys) can form part of the strip cropping system. They contribute to increased soil organic matter and hence lower soil erodibility. Usually strips are 15–45 m wide, becoming narrower as erosion risk increases. Hence, on a long, steep slope, strips would tend to be narrow and be more numerous within the system.

Rotation is a well-established agronomic technique, by which different crops are grown in an established sequence. Since crops have quite precise nutritional requirements, a mix of crops avoids depletion of any specific nutrient. A temporary grass ley is often an integral component of a rotational system. Allowing the soil to 'rest' allows a natural recovery, particularly when legumes (plants that can fix atmospheric nitrogen in the soil system) form part of the grass mix. Clovers are very effective legumes. The post-Second World War development and mass production of chemical fertilizers allowed continuous arable production, without temporary leys. These NPK (nitrogen, phosphorus and potassium) fertilizers provided ample supplies of the macronutrients needed for crop production, but often did not increase inputs of organic matter. On many soils, this allowed soil organic contents to decrease so that soil erodibility increased. The increased incidence of erosion on arable soils in much of North America and western Europe has been attributed to continuous arable cultivation. Returning to more traditional rotation techniques with temporary leys could increase soil organic matter, improve soil structure and decrease overall erosion rates.

Addition of *soil organic matter* can improve soil structure and decrease erodibility. The most common organic material is farmyard manure (FYM) and there are many commercial organic fertilizers. 'Green manures' are crops that grow quickly, producing much biomass, but that rapidly decompose to increase organic matter in the soil. Lupins and mustard are useful green manures. In developing countries, human waste is used. This is usually transported and applied at night, hence the material is known as 'night soil'. The use of organic manures in organic farming also contributes to increased soil organic contents (Arden-Clarke and Hodges, 1987).

It is usually beneficial to apply a mix of organic materials. Some will decompose quickly and give a rapid, but often shortlived, increase in organic matter. Others, such as straw, will decompose slowly and give a longer-term progressive increase in soil organic matter. It is useful to consider the 'isohumic factor', that is, the amount of organic material which remains after decomposition (mineralization and humification) processes. This can range from 85 per cent for peat moss to 20 per cent for plant foliage (Morgan, 1995). Trudgill (1988) presented a useful review of organic matter decomposition processes.

Mulching is the application of vegetative or other material to the soil surface, to reduce evaporation and simulate the protective effects of a vegetation cover. Often residues from the previous crop are applied, such as straw on wheat fields. Many studies around the world have shown mulching to be very effective for soil conservation. It is particularly applicable in tropical and subtropical environments. Although effective also in temperate environments, mulches here can decrease the length of the growing season. One of the main limitations to crop production in temperate climates is the short growing season. Soils have to be sufficiently warm to promote active crop growth. This usually takes place with a soil temperature above about 6°C, and a complete mulch cover can prevent effective early soil warming in spring.

Hydromulching is a form of mulching that is particularly applicable for engineered slopes. The mulch consists of a mix of many materials, such as fibre, straw, paper and shredded wood, which is sprayed on a surface with a seed mix. It protects the soil surface while the seeds germinate, grow and

Table 2.1 Infiltration rates (mm $hour^{-1}$) on sandy agricultural soils in Shropshire, UK

Soil condition	Maximum	Minimum	Mean	Standard deviation	Number of measurements
Permanent grassland	807	147	343	192	23
Bare arable	78.6	5.2	30.2	17.9	88
Plough pan	5.7	1.6	3.3	1.3	32
Tractor wheeling	0.37	0.016	0.13	0.07	126

Source: Fullen (1985).

establish a protective vegetation cover. Hydromulching is expensive and applicable mainly to engineered slopes in high-value projects. Typical applications include slope stabilization for road cuttings and construction sites.

Geotextiles are cloths used to protect the soil surface. Usually they consist of biodegradable material, such as jute, which has a coarse mesh and lasts for a relatively short time, usually less than two years. However, this is long enough for seed mixtures to establish protective plant communities. Geotextiles can also be formulated to release nutrients to the soil system as they decompose. There are also modern non-biodegradable geotextiles, which are used as permanent fixtures, for instance in channel bank stabilization.

Compaction can markedly increase soil erosion. It decreases the size and interconnectivity of soil pores and thus impedes infiltration of water, so that more runoff is generated on slopes, with attendant increases in erosion risk. Table 2.1 shows the very low infiltration rates measured beneath a tractor wheeling and in a soil with a compacted subsoil zone (plough pan), compared with permanent grassland and bare arable soils, in Shropshire, UK. Use of increasingly heavy farm machinery is exacerbating compaction problems and thereby increasing erosion. Compaction by animals, especially sheep and cattle, also poses problems (Mulholland and Fullen, 1991). Cattle can produce a fairly small compact hoof imprint in soils. On wet soils, this causes the loss of soil structure, thus increasing the likelihood of the soil generating erosive runoff. This process of 'poaching' is reviewed in Section 7.6.3.

The management of soil structure can make important contributions to decreasing the risk of soil erosion. Such management can also have many other objectives, such as increasing crop growth or infiltration rates. Some of the important techniques are reviewed in subsequent sections. This includes 'subsoiling' (breaking up subsoil compaction) (Section 4.4.7), the use of crop residue management systems such as zero or minimum cultivation techniques (Section 7.2) and techniques to diminish soil compaction (Section 7.6).

Chemical soil conditioners can be added to soil systems to improve structural stability and decrease soil erodibility. They usually bind particles together, thus improving aggregation and so increasing infiltration rates. These conditioners were developed particularly in the 1950s and 1960s, with such chemicals as 'Krilium', 'Flotal' and 'Glotal'. These were followed by a whole range of ionic and non-ionic conditioners. Field and laboratory experiments showed the conditioners were effective, but a major problem was cost. Therefore, their use has largely been restricted to high-value crops or to specialist engineering applications (e.g., stabilizing road cuttings and engineered slopes, oil-well heads and temporary helipads and airfields). Their cost has been deceased by using by-products from industrial or water treatment processes (e.g., ferric-aluminium polyhydroxide sulphate, used for the flocculation and consequent settlement of sediments in drinking water). At the time of writing, there is a new generation of low-cost conditioners, which may increase their applicability for more general agricultural use (Brandsma *et al.*, 1999a). On sandy soils conditioners work well in wet conditions, but the open structure they confer can lead to poor crop growth in dry conditions because of excessive evaporation of soil water.

2.4.1 Soil conservation and socio-economic issues

Soil conservation is not simply a technical problem. Besides engineering issues, successful soil conservation strategies consist of a complex amalgam of agronomic, social, economic and political considerations at a variety of scales (national, regional and local). Soil conservation in the Chinese Province of Yunnan illustrates these points (Fullen *et al.*, 1999). In the face of greatly accelerated erosion, rapid, effective and low-cost solutions are necessary for long-term sustainable agriculture in Yunnan. Efforts to combat erosion are making some progress. The Yunnan Soil Conservation Service, at its headquarters in the provincial capital of Kunming, attempts to formulate soil conservation programmes. These are usually planned at a local scale, with agricultural advisors and soil conservation technicians implementing conservation in the 127 provincial counties. Yunnan Agricultural University also plays a crucial role, which includes providing training to agricultural students, most of whom become provincial agricultural advisors following graduation. The University also undertakes field experiments and relays research results to provincial agricultural advisors in the *Journal of Yunnan Agricultural University*.

The local-scale planned approach to soil conservation can be illustrated using Dongchuan and Xundian Counties as case studies. Instead of adopting a broad plan, specific areas are targeted for conservation. For instance, in Dongchuan, a conservation plan is devised for selected highland catchments in the foothills of the Himalayas and land use zoned on the basis of ecological principles. Upland areas above 1800 m altitude are planted with pines, particularly Yunnan pine (*Pinus yunnanensis*), and lower slopes are stabilized with Eucalyptus trees (*Eucalyptus* spp.). Middle sections are planted with flax (*Linum usitatissimum*), which is harvested by the local population to make rope. The flax harvest is limited to a yield which is sustainable in the long term. This is done in consultation with and with the consent of the local population, thus encouraging them to gain an economic stake and interest in conservation activities while ensuring agricultural development. The provincial government then funds engineering structures, such as diversion channels, to control and regulate storm torrents. Fruit orchards are grown on suitable stabilized slopes and some of the funds from fruit sales are used to pay for conservation work. A similar approach is adopted in nearby Xundian County, where provincial government finance assists in the afforestation of eroded lands with chestnut trees. The local population benefits from the chestnut harvest and pays 8 per cent of income in tax. This revenue is then used to further agricultural and social development within the county.

The work in Dongchuan and Xundian illustrates some important principles for soil conservation. Soil conservationists have often been accused of regarding conservation simply as an engineering problem that can be remedied using technical solutions. Engineering is a crucial component of a conservation plan, but economic and social factors must also be considered. It is essential that an effective dialogue develops between the conservationists and local people. The involvement, support and participatory agreement of the local population are crucial. Coupled with economic incentives, rewards and benefits by the implementation of conservation strategies, both agricultural development and effective conservation are achievable. Thus, soil conservation and economic development can work together as mutually beneficial aims.

In a European context, we can consider recent developments in The Netherlands. Soil conservation organizational structures are developing in the Province of South Limburg, where a coordinated soil conservation project has been conducted since 1991. This involves active collaboration between government (provincial and municipality), agricultural advisory services, three university research institutes and local farmers. Demonstration projects and information dissemination are also important components of the programme (Boardman *et al.*, 1994). This involves the collaboration of extension workers (i.e., agricultural scientists who explain and demonstrate agricultural research to farmers). The Limburg *Erosienormeringsprojekt* may well act as a future model for soil conservation policy in northern Europe (Fullen, 2003).

Taking illustrative examples from very contrasting physical and social environments, we can draw some common lessons. Successful soil conservation requires the participation of many stakeholders. This includes government at all levels (national, provincial, municipal), farmers, scientists, university researchers and teachers, extension workers, agricultural advisors and technicians. Thus, effective soil conservation is essentially a team effort.

Summary

Soil erosion, by both wind and water, takes many forms, introducing problems both to the zones of export and import. How much erosion is tolerable is debated, but a value of $2\,t\,ha^{-1}\,yr^{-1}$ can be used as an approximate 'yardstick'. There are many field and laboratory techniques available to us, by which we can try to quantify and understand soil erosion rates and processes. The erosion that occurs is strongly controlled by the energy exerted on the soil system, the resistance of the system and the protection afforded by vegetation. Many techniques can be used to conserve soils, each with their own relative advantages and disadvantages. However, any successful soil conservation plan is a mix of technical and social objectives. Developing effective and viable soil conservation strategies is one of the most pressing soil management problems we face in the early twenty-first century. These strategies need to be both cost-effective and socially acceptable. However, given the increasing world population and a soil resource base that is diminishing both in terms of quantity and quality (Chapter 1), effective soil conservation is imperative.

FURTHER READING

Kirkby, M.J. and Morgan, R.P.C. (eds), 1980, *Soil Erosion,* Chichester: John Wiley.

Morgan, R.P.C., 1995, *Soil Erosion & Conservation*, 2nd edn, London: Longman.

3 Desertification, salinization and amelioration of arid soils

As whirlwinds in the south pass through, so it cometh from the desert, from a terrible land.

Isaiah, Chapter 21, Verse 1

Introduction

This chapter examines the concept of 'desertification'. First, there is an attempt to define the concept; then the extent, severity and causes of desertification are reviewed. Chapter 3 also examines the related processes of soil salinization and alkalinization. The feasibility and practicality of the reclamation of desertified land are examined, with a review of some of the major techniques used to reclaim both desertified and salinized land. Desertification has been a focus of world attention since the first international conference on desertification was held in Nairobi, Kenya, in 1977, attended by 95 governments. This interest eventually led to the United Nations Convention to Combat Desertification (UNCCD). Of 41 international conventions dealing with the environment, this is the only one that deals specifically with land resource issues.

3.1 THE DESERTIFICATION CONCEPT

The term 'desertification' was coined by the French geographer André Aubrèville (1949). At a superficial level, the definition is quite simple, that is the expansion of desert-like conditions. However, there has been much debate as to its precise meaning and there are over 100 definitions (Thomas and Middleton, 1994). A few illustrate the difficulty of agreeing a precise definition.

Grove (1977) defined desertification as 'the spread of desert conditions for whatever reasons, desert being land with sparse vegetation and very low productivity associated with aridity, the degradation being persistent or in extreme cases irreversible'. A similar definition is offered by Zhu Zhenda *et al.* (1988) that is 'a process of environmental change, through which desert-like landscapes appear on the original non-desertified land, due to excessive human activities and the influences of natural conditions in arid and semi-arid zones during the historical period'. In 1977 The United Nations Environment Programme (UNEP) defined desertification as 'the impoverishment of arid, semi-arid and subhumid ecosystems by the impact of man's activities' (UNEP, 1977).

The first two definitions allow desertification to be either a natural or an anthropogenic process, but the last stresses the importance of human activities. This debate has continued and has intrigued scientists trying to understand and disentangle the relative importance and interaction of the two sets of causes. Indeed, UNEP has been much criticized for placing too much stress on anthropogenic causes, rather than natural environmental causes.

There are many reasons why defining desertification precisely is difficult. A factor in our poor understanding of arid environments is the diversity of arid landscapes. Many arid landscapes are sculptured by aeolian processes. However, water is an important geomorphological agent, with infrequent high-intensity storms causing much erosion of the poorly vegetated surfaces. These produce gullies, known as *wadis* in Arabic and *arroyos* or *barrancos* in Spanish. We tend to think of deserts as sandy lands dominated by sand dunes. These erg, or *shamo* in Chinese, deserts form a significant minority of deserts, estimated at about 40 per cent of the total area. However, there are also stony, 'reg' or 'gobi' deserts, and 'hamada', or bare rock deserts. Many arid lands have dried-up lake beds or 'playas' and saltpans or 'salinas'. Arid environments are also subject to physical weathering processes, with the usual large diurnal temperature change promoting thermal expansion and contraction of rock, leading to progressive disintegration or 'exfoliation'. Arid soils, known as Aridisols in the US Taxonomy (Section 1.2), are also subject to considerable chemical alteration, usually by evaporation of soil water leading to chemical enrichment of the upper soil (Gerrard, 1992).

Databases on the extent of desertification are generally poor. Initial estimates were based on questionnaires to national governments, but responses were very subjective and there was evidence of political bias, with some governments exaggerating the problem in the belief it would increase overseas aid. The GLASOD estimates made a contribution (Chapter 1), but these remain 'best estimates' (Oldeman *et al.*, 1990). A major problem is developing reliable indices of desertification. Most estimates have used vegetation as an indicator, as its extent can be fairly easily and reliably estimated using remote sensing techniques. However, vegetation is not necessarily a good indicator of desertification, and increasingly emphasis is being placed on the quality and productivity of arid soils (Thomas, 1993). But assessment of soils is dependent on the costly and time-consuming process of soil survey.

There are several reasons why vegetation is not necessarily a good indicator of desertification patterns. First, it is highly dynamic in time. The unreliability of seasonal rain and drought are fundamental attributes of the arid climate, so the extent and health of vegetation depend on recent rainfall and do not necessarily indicate long-term trends. Furthermore, there is the phenomenon of 'green desertification', in which there is an increase in total biomass but generally it has little value. An example is the incursion of thornscrub into East Africa, especially in Kenya, replacing much savannah grassland. The dynamics of desertification on the Egyptian–Israeli border have also attracted considerable interest, as the border has moved several times following international political agreements. In the Egyptian Sinai Desert, Bedouin tribesmen graze their camels on the arid rangelands, producing grasses that are palatable but of low biomass. In the Israeli Negev Desert, the Bedouin are prevented from using the border areas for security reasons, tending to be employed in urban occupations in Beer-Sheva. Consequently, arid grasses have become more extensive and have greater biomass, but are not as useful for grazing. Indeed, within two years of border change, the new border could be identified by satellite imagery, with the Negev appearing 'greener' (Pearce, 1992). Therefore, one must ask which is more 'desertified': the more vegetated Negev or the better grazing lands of the Sinai? Inevitably, such questions become enmeshed in political regional issues, which are important characteristics of the desertification debate and will be discussed further in Section 3.2.2.

3.2 CAUSES OF DESERTIFICATION

There is a multitude of causes of desert and desert-like conditions on the planet. Some are natural, some anthropogenic and some result from interactions between the two. According to the Intergovernmental Panel on Climate Change (IPCC) 'unfortunately, the relative importance of climatic and anthropogenic factors in causing desertification remains unresolved' (IPCC, 2001: 517).

3.2.1 Natural causes of deserts

Certainly, there are natural environmental processes at work which keep approximately 25 per cent of the Earth's surface in a very arid state. We know deserts form without human agency, because they developed and waned long before the advent of human society. For instance, many of the rocks in Britain dating from the Devonian and Triassic periods formed when the landmass was very arid. In each period, the landmass experienced desert conditions; the climate of Egypt today would be a suitable analogue. Desert sand dunes accumulated, which eventually lithified to form sandstones. Salts were washed into desert lakes and these lakes evaporated, leaving salt deposits as residues. Torrential desert storms washed coarse debris from the wasting Caledonian Mountains into desert basins. Today, we see these as deposits of coarse debris and pebbles interbedded in the sandstones. A good example is the Triassic Bunter Pebble Beds of the West Midlands of England. If we want to comprehend the environment at the time, we need to consider a comparable modern environment or analogue, such as Death Valley, California.

We can identify several natural causes of aridity resulting from processes associated with the global climatic system. The primary circulation features of the tropics are dominated by features known as the 'Hadley cells', named after the Royal Navy Admiral Sir George Hadley, who noted aspects of these general circulations in 1735. Because of intense heating by the overhead sun, air generally rises in the Equatorial Zone, producing a zone of low atmospheric pressure, typically about 1010 millibars (mb). On ascent, the humid air condenses to form clouds and produces rain; hence the high humidity and rainfall associated with the Equatorial climate. In the upper atmosphere (upper troposphere) there are poleward transfers of air known as the 'counter trade jet'. This air descends at about 25–35° north and south of the Equator. It produces the high pressure of the subtropical high-pressure belts, typically about 1020 mb. Since the descending air is dry, there is little precipitation and it warms more quickly on descent than does humid air. Typically, humid air cools by about 6°C per 1000 m of ascent and warms by a similar amount on descent. However, dry air warms/cools by about 10°C per 1000 m. Temperature changes resulting from operation of the 'Gas Laws', which govern the relationships between air temperature, pressure and volume, are known as adiabatic lapse rates. Because oceans predominate in the southern hemisphere, where 81 per cent of the surface of the hemisphere is water, the subtropical high-pressure belt is a relatively simple global zone. There is slightly higher pressure over Australia and southern Africa compared with the adjacent oceans. However, the larger extent of land masses in the northern hemisphere makes the pattern more complex and there are several extensive high-pressure cells or 'anticyclones'. They include the 'Azores' or 'Bermuda' high in the North Atlantic and the 'Hawaiian' high in the North Pacific.

Winds always blow from high to low pressure, and therefore they blow from the subtropical high-pressure belts towards both the poles and the Equator. These winds are deflected to the right in the northern hemisphere and to the left in the southern hemisphere by the spin of the Earth known as the 'Coriolis force' (Section 9.1.2) (Barry and Chorley, 1998). They are known as the 'Trade Winds', because of their crucial importance for the sailing ships that travelled across the Atlantic from Europe to America. Air masses generally take on the characteristics of their source areas. Thus, these winds are typically 'tropical continental' (denoted as Ct in the Bergeron system of airmass classification); that is, they are typically warm and dry winds (Hare, 1966; Barry and Chorley, 1998). Thus, the Trade Winds export hot arid conditions from the tropical landmasses. They can be quite strong, such as the Harmattan wind of West Africa, and can produce violent dust storms, such as the 'Haboob winds' of the Sudan. The dust storms assist the extension of desert conditions, especially in Africa. The Trade Winds converge on the Equatorial Zone as the 'Intertropical Convergence Zone' or ITCZ, a low-pressure zone of atmospheric uplift. This circulation (ITCZ, counter-trades jet and Trade Winds) form the essential components of the Hadley cells (McIlveen, 1986).

The Hadley cells are dynamic. Generally, for any given latitude, the northern hemisphere is warmer than the southern, because there is more land in the northern hemisphere and land heats up more quickly than water. Also, the large Antarctic Ice Sheet generally cools the southern hemisphere. Moreover, the shape of South America is such that warm Equatorial ocean currents travelling from east to west across the central Atlantic are diverted into the North Atlantic and the Caribbean. Thus, the 'thermal Equator' is generally north of the geographical Equator and migrates seasonally with the overhead sun between about 20°N and 10°S. The ITCZ generally lags about 4–6 weeks behind the overhead sun.

In combination, these processes produce arid conditions in the subtropical high-pressure belts separated by humid conditions in the Equatorial Zone. Thus, in the extreme north and south of the tropics, there are 'desert' or 'arid' climates. In some of these areas there is virtually no precipitation, and these are referred to as 'hyper-arid' zones. Between the arid areas and the humid Equatorial Zone, there are various humidity regimes, known as semi-arid and subhumid. These zones are often called the 'savannah climatic zone', in association with the dry savannah grasslands, which are the dominant climatically controlled or 'climax' vegetation. Thus, as we move north or south from the Equator, precipitation initially decreases and is increasingly associated with a distinct summer 'rainy season', which tends to become progressively shorter away from the Equator. The rainfall also becomes less reliable and annual variability in precipitation amounts can exceed 40 per cent. We can classify these zones as humid, subhumid, semi-arid, arid and hyper-arid, the fundamental pattern repeating both north and south of the Equator.

A further complexity is the monsoonal airflow, which refers to seasonal reversals in airflow (McIlveen, 1986) and is taken from the term *mausim*, meaning 'season' in Arabic. Land warms up and cools down much more slowly than the sea. This is known as the principle of 'continentality'. Thus, maritime climates tend to lack extremes of temperature. However, with increasing distance from the sea, climates tend to have greater temperature extremes and become continental. This effect is particularly pronounced in the Eurasian landmass. A combination of high latitude, the cold Kamchatka Current off the east coast of the Asian continent and continentality produces the extreme cold associated with the Siberian anticyclone in winter, which has pressures up to 1070 mb. As a consequence, cold, dry, polar continental (Pc) air blows out from central Asia to the coast in winter (about October to March). This 'winter monsoon' produces a distinct cold winter dry season over much of Asia. In Britain, we associate these conditions with very cold dry spells, such as the winters of 1947, 1963, 1979 and 1982. Airflow reverses in summer, with continentality contributing to warming and the development of low pressure, usually centred on two systems, Mongolia and north India. Moist tropical maritime air (Tm) tracks from the Pacific and Indian Oceans in towards these low-pressure systems. This 'summer monsoon' brings very humid conditions to the Asian coast. However, the combined effects of distance and mountain barriers impede deep inland penetration of wet summer monsoon airflows, so that some parts of central Asia receive little or no summer monsoon rain and are hyper-arid. For instance, some parts of the Taklimakan Desert of northwest China receive less than 6 mm of precipitation per year. Less developed monsoonal airflows also affect Australia and South Africa. The effect is well explained by Hare (1966: 146–7): 'the monsoon is a specific flow pattern imposed on the general circulation by the maldistribution of land and sea'. Thus, the combined effects of the Hadley cells and the monsoon contribute to arid climatic zones.

Cold ocean currents also contribute to aridity. As warm moist air streams cross cold water, the water vapour in them can condense as fog, rather than move inland. Thus, cold ocean currents can contribute to foggy coastal conditions and prevent the movement or advection of warm moist airstreams inland. For instance, the cold currents produced by northward-moving extensions of the cold Antarctic current on the western edges of southern continents impede moisture penetration and this contributes to aridity. In this way, the Peru or Humbolt Current contributes to the aridity of the western coastal strip of South America, the West Australia Current contributes to aridity in western Australia and the Benguala Current accentuates the aridity of southwest Africa. Indeed, many species of plants and animals are ecologically adapted to extracting moisture from the frequent fogs of the coast of the Kalahari and other southern deserts. In the northern hemisphere, deep upwelling produces the Canaries Current, which contributes to aridity in North Africa, and the California Current, which enhances aridity in the US southwest and northwest Mexico.

Mountain barriers impede the penetration of moisture-laden airstreams inland. On ascending, these airstreams cool, water vapour condenses as cloud and precipitation is increased. This form of precipitation is called orographic precipitation, although the process usually enhances and is not the sole cause of precipitation. A further effect is that, because of differential cooling, warming by adiabatic processes causes the lee of mountain ranges to be warmer than the range front. Ascending moist air cools at the saturated air adiabatic lapse rate (about 6°C per 1000 m), but the descending air warms at the dry air adiabatic lapse rate (about 10°C per 1000 m). This warmed wind is known as a föhn wind', named after the winds that cross the Alps between Germany and Italy (Barry and Chorley, 1998). The nordföhn brings cold north European air to Italy and the südföhn brings warm air to Germany, often promoting rapid snowmelt and flooding in the Rhine basin, especially in spring. These winds are in response to pressure differentials on the northern and southern flanks of the Alps. Perhaps the best known föhn wind is the Chinook, which advects warm dry air to the Great Plains of North America, the name meaning 'snoweater' to the Blackfoot native Americans. The most rapid recorded air temperature change was associated with the Chinook. On 22 January 1943 temperatures rose in the town of Spearfish, South Dakota, from −20°C to +7°C in two minutes, a change so rapid that windows shattered! Thus, mountain barriers can produce very steep gradients of climatic change, with the Rockies, Andes and Himalayas contributing to aridity in the USA, Chile and central Asia, respectively.

The combination of these climatic processes (Hadley cells, monsoon, cold ocean currents and föhn winds) produces the complex pattern of global aridity (Figure 3.1). Table 3.1 shows one of the UNEP classifications of aridity conditions. This one is for Africa and is based on mean annual precipitation. However, consideration of absolute rainfall amounts is not so helpful. Using the UNEP classification for Africa would mean that most of Europe is semi-arid! What is crucial is the water balance; that is, the balance between precipitation input and the output of water vapour from the land surface back to the atmosphere. This includes evaporation of water from land surfaces and inland water bodies and transpiration by plants. These processes are often grouped together as 'evapotranspiration'. We need to distinguish between actual and potential rates. The actual rates are the true rates at which these processes are occurring, usually expressed in millimetres per day. Potential rates are the maximum rates that can occur in the prevailing meteorological conditions. In arid areas potential evaporation

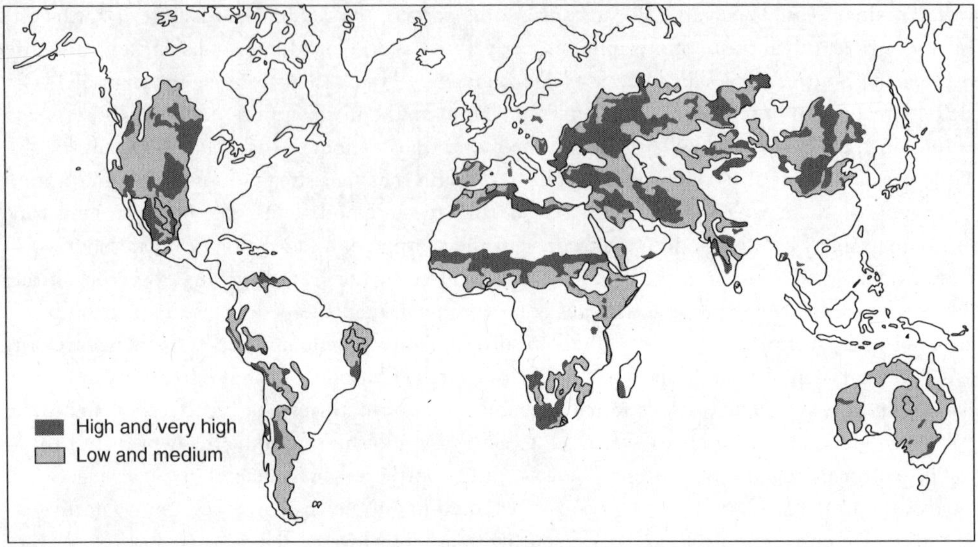

Figure 3.1 The global pattern of aridity
Source: Thomas and Middleton (1994).

Table 3.1 The UNEP classification of aridity conditions in Africa

Aridity status	Mean annual precipitation (mm)
Desert/hyper-arid	<100
Arid	100–400
Semi-arid	400–800
Dry subhumid	800–1200
Moist subhumid	1200–1500
Humid	>1500

rates are usually high, because of a combination of high temperatures, solar radiation, high wind velocities and low air humidity. However, actual rates are usually much less, because part of the water is not available. A useful analogy is a desert lake. Water evaporates from the lake at the potential rate; however, as the lake dries up, the potential rate remains similar but the actual rate progressively decreases. Considering the water balance is a useful approach in investigations of hydrological and pedological processes.

The water balance is a better basis for classifying aridity. There are many indices of aridity, but a frequently used one is the Aridity Index of UNEP, which compares precipitation (*P*) with potential evapotranspiration (*PET*). Table 3.2 shows the UNEP worldwide estimate of the extent of dryland zones with different aridity index values (Thomas and Middleton, 1994). The drylands susceptible to desertification include the dry-subhumid, semi-arid and arid lands. The hyper-arid lands are so dry that they cannot become more arid.

We must also recognize the existence of high-latitude deserts. Generally, polar regions experience high pressure, with little associated atmospheric moisture. Moreover, cold air can retain little water

Table 3.2 The UNEP estimate of the extent of dryland zones

	Aridity index (*P/PET*)	Million ha	Per cent world land area
1 Dry subhumid	0.50-<0.65	1294.7	9.94
2 Semi-arid	0.20-<0.50	2305.3	17.72
3 Arid	0.05-<0.20	1569.1	12.06
4 Hyper-arid	<0.05	978.2	7.52
Susceptible drylands (= 1 + 2 + 3)	0.50-<0.20	5169.1	39.72

Source: Thomas and Middleton (1994: 110).

vapour. Thus, parts of the continental interiors of Antarctica and Greenland can be defined as arid. Although moisture may be present as snow and ice, it has often been in store for many years. However, the potential for development of these areas is limited by both the harsh environment and international agreements to retain them as protected areas.

3.2.2 Desertification and human activities

Human activities undoubtedly contribute to the expansion of desert-like conditions. UNEP estimates current rates of desertification to be about 21 million ha yr^{-1}, mainly by expansion of desert-like conditions into semi-arid and dry-subhumid lands. About 6 million ha yr^{-1} are estimated to be irretrievably degraded (Pearce, 1992). UNEP (1992) has produced a desertification map of the world.

Human-induced desertification has many causes. Overgrazing by animals is a major cause, as it leads to excessive vegetation removal. Often, animal manures are used for fuel and their removal depletes nutrient inputs and cuts an essential link in the biogeochemical cycling of nutrients. Clearly, grazing intensities should accord with the carrying capacity of the land. However, this varies through time, and stocking densities that can be supported in a wet rainy season may be excessive in a drier rainy season. Given the inherent variability of rainfall in drylands, attempts to define sustainable cattle stocking densities are fraught with problems. Moreover, in some semi-arid societies wealth is evaluated in terms of animal numbers, so there are social barriers to decreasing grazing densities.

Woodlands within drylands are tempting sources of firewood, and deforestation and fuelwood removal are another major cause of desertification. Overexploitation of water resources is another contributory factor, not only decreasing water availability but also decreasing water quality, usually causing it to become more saline (Section 3.3). Many of the political problems of drylands are conflicts over water resources. Furthermore, many arid countries are politically unstable and often engaged in military conflict. For instance, civil strive is endemic in Ethiopia, Eritrea, Mozambique, Somalia and the Sudan, which undermines much of the social fabric and decreases ability for development. The post-independence enforcement of borders in West Africa contributed to desertification. Before independence, transhumance was common; that is, nomadic tribes migrated with the seasonal rains to graze their animals, with little regard to political boundaries. However, post-independence enforcement of national boundaries impeded this transhumance and many turned to dryland arable farming, partly in an attempt to produce cash crops such as groundnuts. Dryland farming relies on the highly unpredictable rains, which often fail, leaving bare fields exposed to desertification.

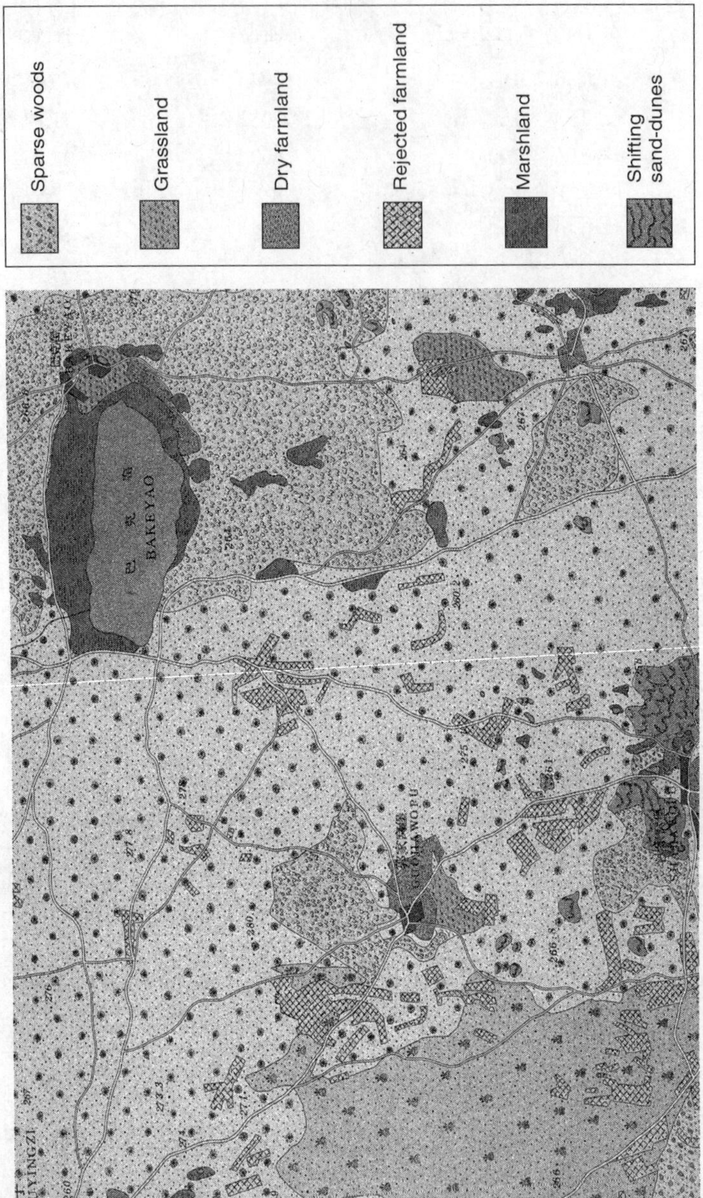

Figure 3.2 The pattern of land use in a small area of Inner Mongolia (Daqinggou, Horqin Steppe), China, in 1958
Source: Institute of Desert Research of the Chinese Academy of Sciences (IDRAS) (Zhu Zhenda, 1958).

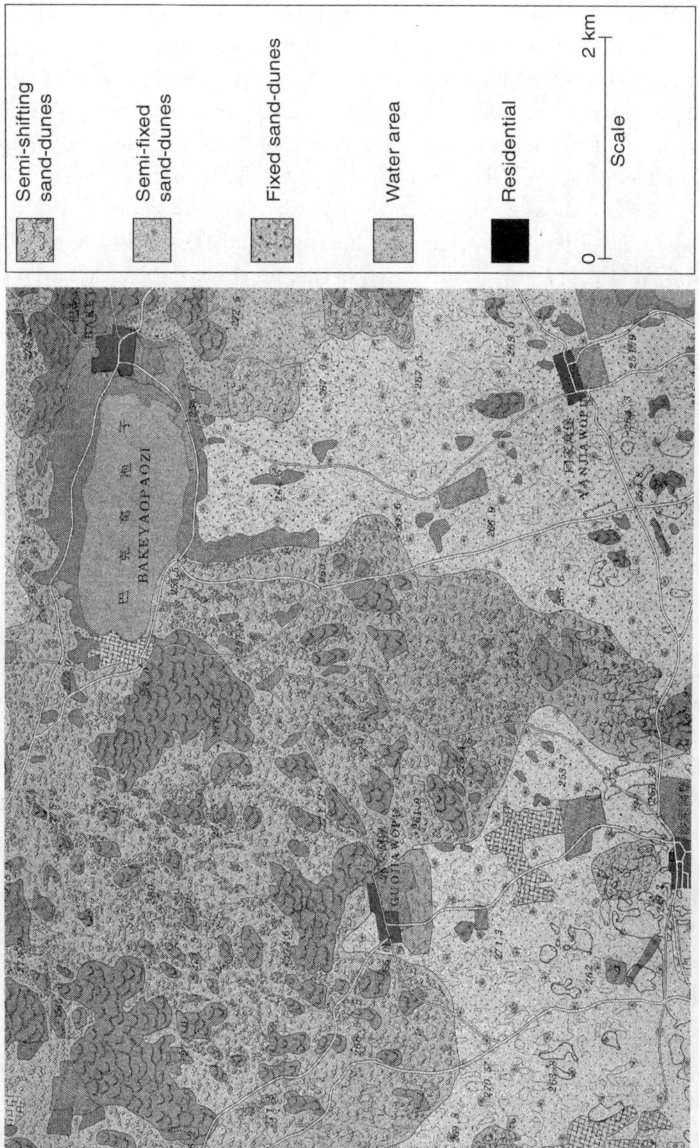

Figure 3.3 The pattern of land use in the same area of Inner Mongolia in 1981. Note how the extent of desert had expanded considerably, but the pattern is very complex

Source: Institute of Desert Research of the Chinese Academy of Sciences (IDRAS) (Zhu Zhenda, 1981).

Sub-Saharan Africa has been particularly intensively studied in terms of desertification processes. The Sahel, from the Arabic for 'border', is the semi-arid to subhumid belt that stretches some 4800 km from east to west on the southern margin of the Sahara Desert. The width of this belt, i.e., from the desert margin to the subhumid zone, is 480–800 km. Annual precipitation decreases northwards from about 600 mm to 100 mm. It includes ten countries from Senegal in the west to Somalia in the east, many of which are subject to civil strife and national and international conflicts. Generally, the southern edge of the Sahara has advanced southwards by about 150 km in the last 100 years (Biswas, 1990) and about 15 years ago the rate was about 5.5 km yr^{-1} (Walsh *et al.*, 1988). An estimated 65 million ha of productive land are threatened by the Sahelian advance (Goudie, 1990).

Detailed studies of patterns of desertification tend to support the argument for anthropogenic causes, particularly overgrazing. The popular concept, often promulgated by the media, is that of an advancing wall of sand. This may be true locally, as in parts of the Kalahari Desert, but desertification processes are usually more subtle. In detail it tends to be a process of the development of 'rashes' (Goudie, 1994). Desert-like conditions develop in small 'blisters' or 'rashes' away from the desert margin and these progressively enlarge and merge. Often they are areas that have been overgrazed. The small whirlwinds or 'dust devils' associated with atmospheric heating and convection contribute to the spread, by ripping and damaging the fragile steppe grasslands. For instance, Figure 3.2 shows the pattern of land use in a small area of Inner Mongolia, China, in 1958. By 1981 the extent of desert had expanded considerably, but the pattern is very complex (Figure 3.3).

Future climatic change will affect desertification patterns. Scientific assessments of the nature and rate of climatic change pose many challenges. We know that climates are dynamic and always subject to change. However, it is clear that the world is generally warming, probably because of increased 'greenhouse' gases in the atmosphere (Chapter 9). However, assessing whether the climate is changing in arid areas, and the causes of that change, pose major challenges. First, we do not know for certain what the climatic change would have been without human interference. Second, it is difficult to differentiate short-term climatic changes from long-term progressive or 'secular' natural change. This problem is compounded by the general lack of climatic data in arid zones. If possible, climatologists use averages of meteorological data over 30 years to describe the 'climate', that is the 'average' weather. In arid zones, such data are generally unavailable or climatic stations are sparse, thus limiting detailed knowledge of the desert climate. To some extent, this situation is improving with the increased use of automatic weather stations (AWS) in drylands, filling in gaps in our global climatic database. However, when climatic data are input into global circulation models (GCMs), most agree that global warming will increase aridity, especially in subtropical continental interiors (Goudie, 1994). It is likely that climatic change induced by increased 'greenhouse' gases will exacerbate desertification by altering spatial and temporal patterns of temperature, rainfall, solar radiation and winds. Likely effects in arid areas include increased temperatures, reduced rainfall, lower soil moisture levels, sparser vegetation and tree cover, decreased soil organic matter and increased wind erosion. In turn, this will cause decreases in both agricultural and livestock yields, leading to increased risk of famine (IPCC, 2001: 519).

Differentiating desertification resulting from temporary climatic oscillations and natural secular change is difficult. For example, the recent evidence of increased aridity in the Great Plains of North America, or the apparent extension of the North African subtropical high-pressure belt, which is increasing aridity in the Mediterranean basin, especially Greece, could be brief local climatic oscillations or part of a pattern of secular natural change. Currently, much research is in progress in Almeria, southeast Spain, which is the most arid part of Europe. The research programme of the *Estacion Experimental de Zonas Aridas* is attempting to establish the causes and likely effects of desertification in Mediterranean Europe (Puigdefrábregas and Menizabal, 1998).

3.3 THE PROBLEM OF SALINIZATION

Problems of salt contamination of soil (salinization) are closely related to desertification (Szabolcs, 1992). Salinization is especially associated with semi-arid to subhumid lands with shallow groundwater or subject to irrigation. The water balance is critical to salinization. Where the *P/PET* ratio is <0.75, soils are very prone to salinization (Lal *et al.*, 1989). Salinization is a significant problem in 28 countries, with a total area of 955 million ha (Rhoades, 1990). The map of salt-affected soils (Figure 3.4) shows the problem to be very extensive and to include even temperate semi-arid environments, such as parts of the Canadian Prairies. Approximately 15 per cent of the world's farmlands are irrigated and, of these, about 10 per cent are sufficiently affected by salt to limit crop production (Rhoades, 1990).

Salt contamination of soil produces multiple problems. With the exception of plant species that are ecologically adapted to saline conditions, i.e., the halophytes, most crops are damaged by salt. Moisture in plant tissues tends to contain higher concentrations of dissolved salts than the soil environment, which allows plants to abstract water from soils by osmosis, drawing in water across semi-permeable plant cell membranes. However, when the salt concentration is greater in the soil, the osmotic potential is reversed and water is drawn out of the plant, producing 'physiological drought'. Generally, crops show little sign of this type of salt stress if salt concentrations are less than 500 mg l^{-1}. However, at 500–1000 mg l^{-1} salt-sensitive crops are affected; at 1000–2000 mg l^{-1} many crops are adversely affected; at 2000–5000 mg l^{-1} only salt-tolerant crops can grow successfully. Very few plants are tolerant of salt concentrations greater than 5000 mg l^{-1} or 0.5 per cent (Agnew and Anderson, 1992).

Salt contamination raises soil pH, sometimes as high as pH 9–10, a process known as 'alkalinization'. Such environments are caustic to plants, causing foliar scorch. Moreover, at high pH, nutrient-solubility is often low and some nutrients become unavailable to plants (Section 6.3.4). At high pH, there can be deficiencies in calcium (Ca), magnesium (Mg), zinc (Zn), iron (Fe), manganese (Mn), copper (Cu) and boron (B).

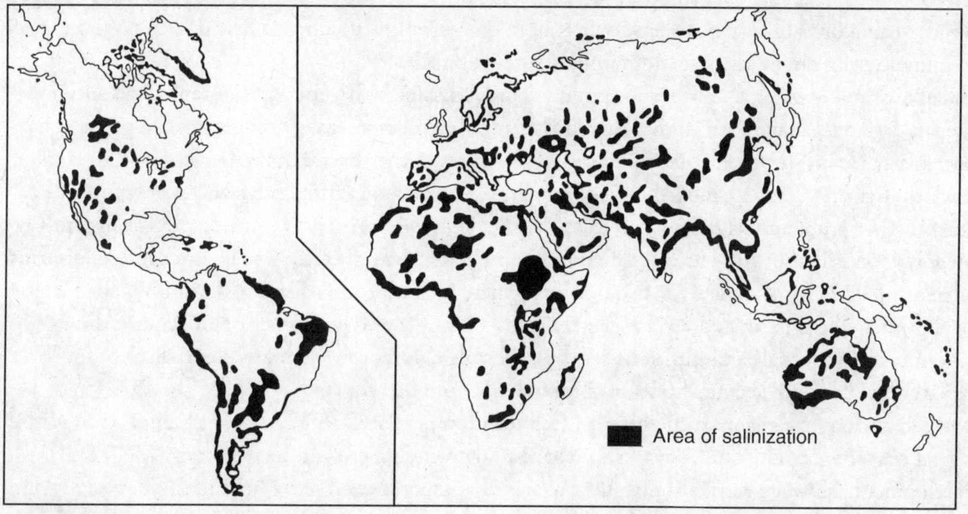

Figure 3.4 The global map of salt-affected soils
Source: Rhoades (1990).

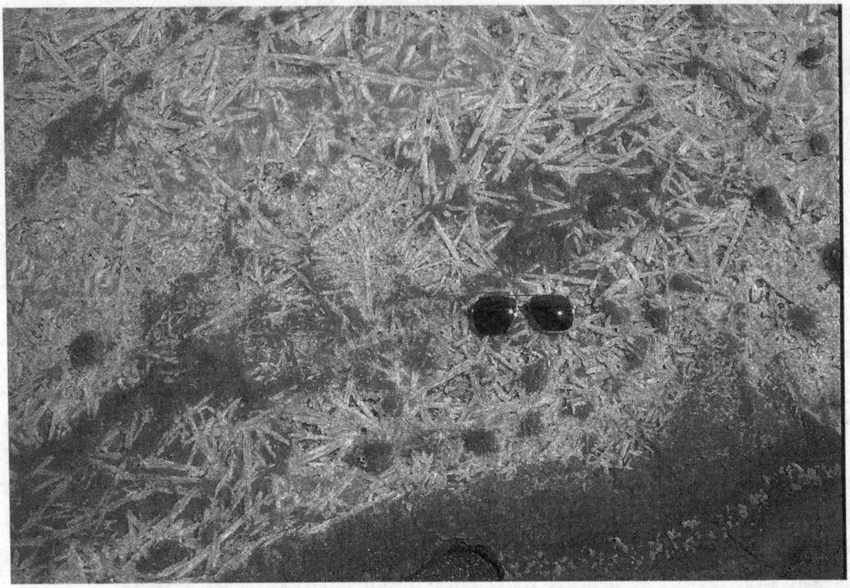

Plate 3.1 Salt crust of a Solonchak on a dried-up lake bed in Yanchi, Ningxia Province, China (photo M.A. Fullen)

The salts are usually chlorides, sulphates and carbonates of sodium (Na), magnesium (Mg) and calcium (Ca). 'Common salt' (sodium chloride, NaCl) occurs most frequently, followed by gypsum (calcium sulphate, $CaSO_4 \cdot 2H_2O$). The sodium salts pose particular problems, as the positively charged sodium ions strongly repel each other, contributing to poor soil structural stability (Nelson and Oades, 1998). The process of sodium salt contamination is a specific form of salinization, known as 'sodification', and can lead to the development of sodic horizons. Salt crystals can grow on the soil surface, producing a shiny white crust; this soil type is known as solonchak (Plate 3.1). Dried-up lake beds or salt flats (salinas) are important sources of salt, which are often even abstracted commercially. Wind erosion on salinas can disperse salt, a problem currently affecting Mexico City. Wind erosion on saline coastal dunes can also contaminate soils with salt.

Shallow groundwater is another important source of salts. With the high potential evaporation rates in drylands, shallow groundwater can rise up by capillary action through the soil pores, particularly in fine-textured soils. The water evaporates at the surface and thus leaves behind salt residues. Irrigation water can also be highly saline. The general salt enrichment of arid topsoils can cause further problems, as the various salts may cement soil horizons (Gerrard, 1992; Ellis and Mellor, 1995). These crusts are classified by their chemical composition. Calcium carbonate forms calcrete, also known as caliche; gypsum forms gypcrete; silica forms silcrete; and iron oxides laterite or plinthite. These indurated layers severely restrict plant root development and impede the movement of water, air and nutrients within arid soils. Moreover, elements such as selenium (Se) can accumulate to toxic concentrations in these salt-enriched topsoils.

Much current expansion in the extent of salinization is related to irrigation schemes. It might seem logical that the introduction of water into sunny environments would be ideal for agricultural development. However, irrigation water can rapidly evaporate and leave behind salt deposits. Many expensive, technologically advanced irrigation schemes are experiencing problems of salinization (Johnson and Lewis, 1995), the most notable being in the Indus valley of Pakistan. Water use efficiency in these environments is often low and it is estimated that approximately 50 per cent of irrigation water is lost by percolation and evaporation (Agnew and Anderson, 1992).

Overabstraction of groundwater resources can lead to salinization. Less dense, relatively pure groundwater often overlies heavier saline groundwater and overabstraction can lead to complete removal of the less saline water and its replacement by deeper more saline water. This problem occurs in many parts of the world. For instance, it is associated with the Ogallala aquifer in the central USA, the El-Hassa aquifer of Saudi Arabia and abstraction of groundwater from the coastal sand dunes of The Netherlands.

Salinization can also result from seawater intrusion into aquifers and transgression of marine waters across low-lying areas. Currently global sea levels are rising at about 1 mm yr^{-1}. This general or 'eustatic' rise is related partly to melting of polar ice caps, but mainly to the thermal expansion of oceans with global warming (Section 9.4). Expensive flood defence schemes, such as in The Netherlands, can prevent transgressions of seawater. However, where these defences are absent, marine transgressions can cause salinization, as in many of the unprotected marshlands and tropical mangrove swamps of the world. Current soil salinity problems in the Senegal Delta of West Africa are related to late Holocene transgressions (Ceuppens and Wopereis, 1999). Drainage of mangrove swamps can also induce salinization, as oxidation of pyrite (FeS) leads to the formation of sulphuric acid (H_2SO_4), which reacts with calcium carbonate ($CaCO_3$) to produce gypsum ($CaSO_4 \cdot 2H_2O$).

Salinity is usually determined quickly and cheaply in the field or laboratory by the surrogate measure of the electrical conductivity of a soil–water mixture. Dissolution of salts increases the solute content of the water extract, so that it conducts an electrical current more easily. Alternatively, filtered soil water extracts can be evaporated in the laboratory, leaving behind the salt residue. This gives a direct measure of dissolved solids, usually expressed in milligrams per litre of water (mg l^{-1}), and provides a sample for further analysis, but is rather time-consuming. Because of the particular problems associated with sodium, much attention is focused on measuring sodium concentrations. The 'sodium-enrichment ratio' (SER) measures the concentration of sodium salts relative to others. Rowell (1994: Chapter 14, 277–302) provides a useful summary of these analytical techniques.

3.4 DESERT RECLAMATION

The desert shall rejoice and blossom as the rose.

Isaiah, Chapter 35, Verse 1

I will make the wilderness a pool of water, and the dry land springs of water.

Isaiah, Chapter 41, Verse 18

Considerable debate surrounds the question of desert reclamation. Some argue for 'making the desert bloom', while others maintain that since deserts are largely natural environments we should leave them to nature. Regional conditions are important. For instance, lack of agricultural land induces strong pressures for reclamation in Israel. Political issues can also be important, leading to expensive reclamation projects, as in the San Joaquin valley of California, the Indus valley of Pakistan and the pivot-irrigation systems of Saudi Arabia. Importing food would be cheaper than these projects.

However, there is considerable potential for both decreasing rates of desertification and for reclaiming some drylands (Lal, 1990). This is especially true in subhumid to semi-arid environments, but there is less potential in arid and hyper-arid environments. Realizing this potential involves both passive management strategies of preventing dryland degradation and more active strategies of reclamation. Both involve considerable degrees of commitment and expenditure of human and financial resources; the latter is especially high for active reclamation. Successful reclamation requires recognition of the complexity of the desertification process, including the multiplicity of causal factors and the socio-economic problems involved. Addressing this complexity requires an integrated approach, giving full consideration to both environmental and socio-economic issues.

Many techniques are available to reclaim desertified land. They usually involve the development of a protective vegetative cover, either by avoiding damaging actions (e.g., overgrazing) or by positive actions to stabilize surfaces (e.g., tree planting, planting shelterbelts, using appropriate irrigation schemes and improving water use efficiency). Goudie (1990) presented a useful overview of the techniques used. The main reclamation techniques are reviewed below, using China as a case study.

3.5 DESERTIFICATION AND DESERT RECLAMATION IN CHINA: A CASE STUDY

Desertification patterns in China illustrate the complexity of the desertification process and the dynamic interactions between causal factors. Approximately 11.3 per cent of the land area is classified as desert, that is 1.1 million km^2. However, there is not just one 'desert', but rather 12 deserts and sandy lands. The distinction between deserts and sandy lands is important, as the Institute of Desert Research of the Chinese Academy of Sciences (IDRAS) believes the sandy lands have been subject to recent desertification and may be amenable to reclamation (Fullen and Mitchell, 1994). They tend to be located in the subhumid east of China, whereas true deserts are in the arid west (Figure 3.5). According to IDRAS, desertification is a major problem, expanding by an average of 1560 $km^2\ yr^{-1}$ over the last 30 years, so that deserts and desertified land together account for 15.5 per cent of the land area of China, that is about 1.49 million km^2.

There are natural reasons for aridity in China. The climate is monsoonal, producing a cold dry winter monsoon. The rainy summer monsoon brings wet conditions to the southeastern coastal areas, where annual rainfall can reach 2000 mm. However, distance from the sea and orographic barriers impede advection of moist airstreams into arid central and western China. Also, the continued tectonic uplift of the Himalayas accentuates aridity in central Asia.

Climatic change in the recent geological past has left important legacies for the current pattern of desert types. Over the last 2.6 million years the climate of central Asia has changed considerably (Walker *et al.*, 1987). In Miocene times, the climate was more humid, as evidenced by the presence of thick, strongly weathered red clays. During the Pleistocene, the region repeatedly experienced intensely cold conditions and vegetation was absent. Strong winds 'winnowed out' fine material, leaving behind a residue of coarse, stony gravel or 'gobi' deserts. Gobi or reg deserts occupy about 41.8 per cent of China's 1.1 million km^2 desert area. Further downwind, sands were deposited, producing sandy deserts dominated by sand dunes. These 'shamo' or erg deserts occupy 58.2 per cent of China's deserts. Further downwind, silts were deposited as loess. The Loess Plateau of China is both the largest deposit of loess in the world, occupying 530,000 km^2, and the thickest, up to 318 m at Jiouzhoutai, near the city of Lanzou. In a very generalized way, the landscape of arid China can be characterized by a progressive change from gobi desert, to shamo desert, to loess, in a transect from northwest to southeast (Fullen and Mitchell, 1994).

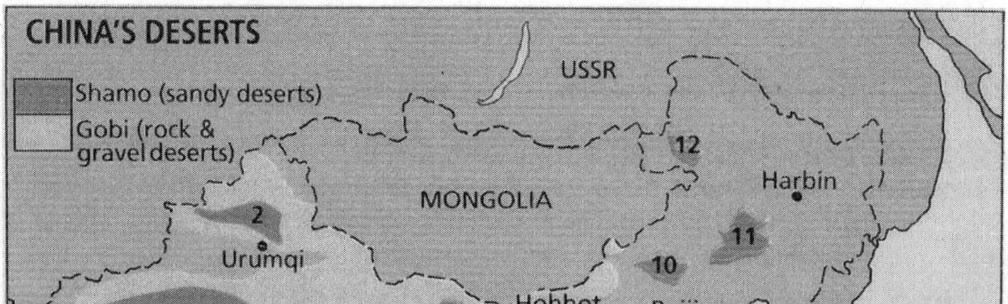

Figure 3.5 The 12 deserts and 'sandy lands' of China. Note that 'true deserts' are in the arid west and 'sandy lands' in the subhumid east of China

Source: Fullen and Mitchell (1991).

Table 3.3 The assessment of the nature and causes of desertification in China

Extent of desertified land and causes	Area (km^2)	Per cent of the total desertified land
Overcultivation on steppe	44,700	25.4
Overgrazing on steppe	49,900	28.3
Undue collection of fuelwood	56,000	31.8
Technogenic factors	1,300	0.7
Misuse of water resources	14,700	8.3
Encroachment of dunes under wind forces	9,400	5.5

Source: Zhu Zhenda *et al.* (1988).

There are also many human reasons for desertification in China. The land has over 3000 years of human occupancy and so the human imprint on the landscape is very profound. Table 3.3 shows the IDRAS assessment of the nature and causes of desertification in China. They include overgrazing, overcultivation, dryland farming, woodland removal and misuse of water resources.

Efforts are under way to combat desertification, coordinated by IDRAS (Yong Zha and Jay Gao, 1997). The Institute has its headquarters at Lanzhou and operates nine field stations in desertified areas. It investigates various techniques of desert reclamation and is thus able to recommend appropriate strategies in reclamation programmes. Some of the initial pioneering work was done at Shapotou Research Station and Yanchi Experimental Station for Co-operative Desertification Control Research, both in Ningxia Autonomous Province. At both stations, desertified land has been reclaimed experimentally by various techniques over different timescales. The Shapotou Station was established in 1957 to find ways of protecting 40 km of the Lanzhou to Baotou railway line from sand burial. In the past, the subhumid area (mean annual precipitation about 300 mm) formed fertile steppe grasslands. Today, largely because of land mismanagement, the area consists of a series of sandy ridges separated by vegetated depressions.

At Shapotou, Yanchi and other field stations of IDRAS, some of the techniques concentrate on dune stabilization, aimed at halting the desert advance and decreasing aeolian damage to rangelands and croplands downwind. These include the use of windbreaks, the establishment of straw or clay checkerboards and managed successions of xerophytes. Other techniques, such as irrigation, land enclosure, extracting buried soils and chemical treatment, attempt to reconvert desertified areas to productive rangelands and croplands. The more important of these techniques are discussed below.

3.5.1 Windbreaks

At Shapotou, Ningxia Province, pine, poplar and willow trees have been planted parallel to the railway line as windbreaks. On a larger scale, there is a very ambitious project to develop a major windbreak across arid north China, known as the 'Great Green Wall'. Biological controls, particularly planned and controlled grazing and improved land management practices, have increased woodland by 22.7 per cent at Yanchi, thus reducing severely desertified land by 10 per cent (Fullen and Mitchell, 1994).

3.5.2 Irrigation

Irrigation not only adds water to reclaimed desert soils but, when silt-laden river water is used, it can improve the physical, chemical and biological properties of reclaimed soils. The Yellow River (Huang He) is a major source of irrigation water, with sediment concentrations among the highest in the world. Severe soil erosion supplies much of these sediments, especially where the river flows through the Loess Plateau. The river deposits an estimated 1600 million t into the Yellow Sea each year and an estimated 63 million t of sediment per year are diverted from the Yellow River onto the land. This has beneficial effects, adding fines, nutrients and organic matter to the desert sands. Other organic inputs to reclaimed areas include sheep manure, added at rates up to 45 t ha^{-1} yr^{-1}, and crop residues. A similar approach is used in the Nile valley of Egypt (Goossens *et al.*, 1994).

At Shapotou, dunes are levelled; initially this work was done manually but bulldozers are now used. Then pumping stations are used to transfer both water and sediment to the levelled sands. These additions result in the rapid accumulation of minero-organic topsoils (Ap horizons) on the buried levelled sands (bCu horizons). Measurements of profiles on sites reclaimed for 0, 4, 9, 15 and 25 years showed a mean accretion of 18 mm yr^{-1} (Fullen *et al.*, 1995). The developing topsoils are capable of supporting arable crops. Soil organic content increased, especially in the first few years of reclamation, generally at a rate of 0.1–0.2 per cent per year. These horizon differences were not evident in soil profiles at Yanchi irrigated with clean groundwater.

At Shapotou, experiments are in progress on 'drip irrigation' systems, whereby water is fed directly to plant roots via buried pipes. These systems have the advantage of decreasing water losses by both evaporation and infiltration and thus improve water use efficiency. However, water from the Yellow River requires filtering before it is passed along the narrow irrigation tubes, an expensive process that starves the soil of desirable silt inputs. Furthermore, algal growths can block pipe outlets.

3.5.3 Straw/clay checkerboards

To provide an environment for indigenous xerophytic plants to colonize and survive, localized surface stabilization is essential. For this technique to succeed, a crucial initial stage is the installation of a 'sand barrier', that is a fence made from woven willow branches or bamboo, to act as a 'windbreak' (Watson, 1990). Wind velocity and erosivity are thus reduced, causing aeolian deposition in the lee. Then checkerboards are constructed behind the sand barrier. Artificial checkerboards of either straw or clay are used to increase surface roughness and reduce wind velocity, thereby decreasing wind erosion, stabilizing the sand and encouraging colonization. Some 4.5–6.0 t ha^{-1} of wheat or rice straw are embedded in the sand, in a 1–3 m grid (i.e., from 1 m^2 to 9 m^2). The straw is inserted vertically 15–20 cm deep, so that it protrudes 10–15 cm above the surface. Where locally available, clay is used to stabilize the surface, because of its greater durability and resistance to erosion.

The checkerboards remain intact for 4–5 years, allowing time for cultivated xerophytic plants to establish. These plants are grown in a nursery and, when sufficiently mature to survive transplantation, they are planted in the centre of the checkerboards. In the Shapotou nursery, 128 plant species are used in desert reclamation (Fullen and Mitchell, 1994). The main species are *Hedysarum scoparium*, *Artemisia ordosica*, *Caragana korshinskii*, *Ephedra przewalskii*, *Haloxylon ammodendron* and *Tamarix* spp. *Salix psammophila*, *Hedysarum fruticosum* spp. *laeve*, *Artemisia sphaerocephala* and *Agriophyllum squarrosum* are used in addition at Yanchi. Within the checkerboard areas, several factors promote surface stability and incipient soil development. These include decreased wind velocities, increased organic-matter inputs, improved moisture retention and accumulation of aeolian dust. Once the shifting sand has been stabilized, soil algae (especially blue-green algae or Cyanobacteria) form the next successional stage (Fearnehough *et al.*, 1998).

Based on results from physical and ecological experiments at Shapotou, IDRAS has devised a procedure for establishing an artificial ecosystem on mobile dunes. The process alters shifting sands, with less than 5 per cent cover, to fixed dunes, with 30–50 per cent cover (Table 3.4).

3.5.4 Land enclosure

As overgrazing and use of trees and shrubs for firewood have severely degraded semi-arid and steppe lands, proper management of desert margins is essential for successful large-scale reclamation (Walls, 1982). At Yanchi, stocking areas enclosed by fences and employing the recommended stocking density of one sheep per 0.87 ha have stabilized and revegetated within five years.

3.5.5 Extracting buried soils

Because of the recent desertification, fertile steppe soils have often been buried by encroaching sand dunes (Plate 3.2). The Ordos grassland steppe was buried about 300 years ago by the Mu Us Sandy

Table 3.4 Phases in the establishment of an artificial ecosystem on mobile dunes at Shapotou

Ecosystem of mobile dunes
Features: bare sand surfaces with vegetative cover <5 per cent, sand shifting
Stabilization phase 1
Establish sand barrier
Establish straw checkerboards
Plant sand-stabilizing species
(*Artemisia*, *Caragana*, *Hedysarum*)
Establish semi-stabilized surface
Stabilization phase 2
Development of topsoil minero-organic crust
Growth of bryophytes
Increase in soil micro-organisms
Regeneration of colonizing plant species
Increase in animal populations
Establishment of artificial ecosystem on dunes
Features: dune movement stopped
Vegetative cover increased to 30–50 per cent

Plate 3.2 Fertile steppe soils buried by encroaching sand dunes in the Mu Us Sandy Land, China (photo M.A. Fullen)

Land. Thus, a fertile calcareous sandy silt loam frequently underlies a cover of desert sand. Sand mobility can be directly attributed to overgrazing on the semi-arid steppes.

Reclamation efforts are in progress by 'mining' the buried soils. Trenches are dug and the buried Ah horizon removed and then mixed with the surface dune sands. On the southern Mu Us, areas reclaimed in this way only a few years before have yielded productive crops of maize, wheat, soya beans and vegetables. Productivity was maintained by additions of sheep manure (about 110 t ha^{-1} yr^{-1}) and irrigation with clean groundwater. Soil samples removed from areas reclaimed 7, 5 and 3 years previously showed that soil organic matter contents have increased at a rate of about 0.1 per cent per year (Mitchell and Fullen, 1994).

3.6 RECLAMATION OF SALINE SOILS

Desalinization of soils initially requires careful evaluation of the extent, depth and severity of salinity. It may prove too difficult and expensive on very saline soils (salt content over 10 per cent of the soil mass) and reclamation attempts are then inadvisable. On highly saline soils (salt content 5–10 per cent), hydrological measures are needed to flush salts away. This involves flooding the soil with water, then allowing it to drain through the soil and leach dissolved salts. Drainage systems (Section 4.4) must be in place to remove the saline water at depth. Often the problem of drainage has received insufficient attention. There is also the problem of disposal of the saline wastewater.

The flooding approach requires definition of the 'leaching requirement', that is the amount of water (usually expressed as millimetres depth) required to desalinize soil to at least the required concentration. As a rule of thumb, to flood a unit depth of soil with the same depth of water will decrease salt content to about 20 per cent of the original concentration. The quality of leaching water is generally less important than the amount. On uneven soil surfaces, such processes are inefficient, as water ponds as depression storage and micro-ridges are not leached. Efficiency is markedly improved if surfaces are pre-levelled. One way to do this is laser-levelling, with a soil scraper computer-adjusted with reference to a laser-beam source. This is one particular aspect of 'precision agriculture'; that is, very precise application of technology to improve agricultural efficiency over a small area, such as a single field.

The need for frequent leaching increases with salt content (Agnew and Anderson, 1992). At a salt concentration of 0.5–1.0 g l^{-1} (0.05–0.1 per cent), it should be every 1–2 years, increasing to once or twice per year at concentrations of 1.0–2.0 g l^{-1} (0.1–0.2 per cent). At over 2.0–3.0 g l^{-1} (0.2–0.3 per cent) there should be several flushings per year.

Where salt contamination is moderate (3 per cent), various techniques can be employed, including physically removing the salt by scraping and decreasing the concentration of the remaining salt by deep ploughing. A cover of sand can diminish the capillary rise of water through the soil. Capillary rise is much more efficient in the small pores of fine-textured soils, but addition of a sand cover produces larger interstitial pores, which do not transmit water so efficiently. It is important to carefully control the water table, usually by deep drainage. A low water table, at least 1–2 m below the ground surface, impedes capillary rise and diminishes the capacity for salinization.

Chemical amendments such as gypsum can be very useful for removing sodicity (El-Swaify *et al.*, 1983). Na^+ ions are strongly attracted to the negatively charged clays. However, addition of

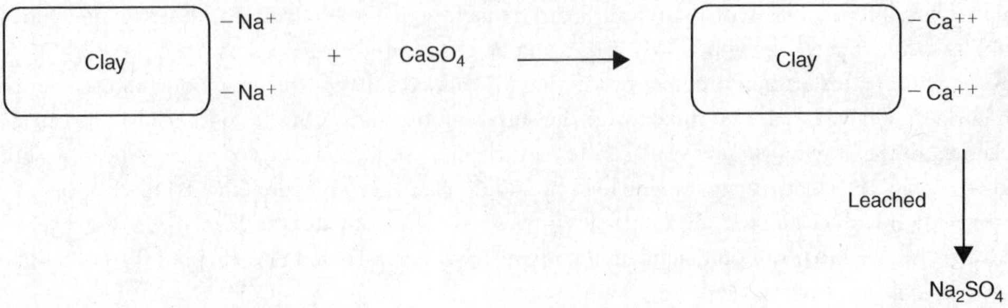

Figure 3.6 Sketch illustrating exchange mechanisms between sodium and calcium in saline soils, leading to enrichment of topsoils with gypsum and leaching of sodium sulphate

Table 3.5 Safe chloride concentrations in irrigation water according to seasonal water need (for sandy textured soils)

Crop tolerance group	Crops in group	Safe chloride concentration (mg l^{-1})	Seasonal irrigation need (mm)
Very sensitive	Peas, dwarf beans, plums, strawberries, blackberries, gooseberries	300	25
		200	50
		100	100
		500	200
Moderately sensitive	Broad beans, lettuce, radish, onion, celery, maize, clover, cocksfoot, apples, pears, raspberries, redcurrants	400	25
		400	50
		300	100
		150	200
Slightly sensitive	Potatoes, cabbage, carrots, cauliflower, wheat, oats, ryegrass, lucerne, blackcurrants, vines	700	25
		700	50
		500	100
		250	200
Least sensitive	Sugar beet, mangolds, red beet, spinach, asparagus, kale, rape, barley	900	25
		900	50
		900	100
		450	200

gypsum sets up exchange mechanisms in which divalent Ca^{2+} replaces the Na^+. The Na then forms soluble sodium sulphate, which can be leached to depth with irrigation water (Figure 3.6).

Crops tolerant of some salinity, such as rape, kale and barley, can be selected once the salt concentration is less than 1000 mg l^{-1} (<0.1 per cent). Table 3.5 shows tolerance groups of crops and the safe chloride concentrations for differing irrigation needs. Currently, there is much research into genetic engineering of salt tolerance, particularly in cereals.

Summary

Deserts form by the action of several natural and anthropogenic mechanisms, often in combination. Human activities can promote desertification, that is the expansion of desert-like conditions in drylands. The human influence on desertification is especially evident in subhumid to semi-arid environments. Salinization is also a major problem in drylands, especially on irrigated land. There is considerable potential to reclaim both desertified and salinized land if the degree of degradation is not too severe. However, reclamation involves the commitment of considerable resources. Reclamation of arid, hyper-arid and highly saline soils is generally too expensive. A broad range of reclamation techniques is available, and can be used singly or in combination. However, a careful management strategy is needed, which is customized to the prevailing environmental and societal conditions.

FURTHER READING

Agnew, C. and Anderson, E., 1992, *Water Resources in the Arid Realm*, London: Routledge.

Ellis, S. and Mellor, A., 1995, *Soils and Environment*, London: Routledge.

Goudie, A.S. (ed.), 1990, *Techniques for Desert Reclamation*, Chichester: John Wiley.

Smithson, P., Addison, K. and Atkinson, K., 2002, *Fundamentals of the Physical Environment*, 3rd edn, London: Routledge.

Thomas, D.S.G. and Middleton, N.J., 1994, *Desertification: Exploding the Myth*, Chichester: John Wiley.

4 Soil water management

The soil was deep, it absorbed and kept the water in the loamy soil.

Plato

Introduction

Water taken up from the soil through roots is essential for the germination and growth of plants. However, too much water in the soil results in anaerobic conditions and prevents respiration by roots (exchange of oxygen and carbon dioxide with the atmosphere). This restricts root development, so that insufficient nutrients are taken up. Soils act as a sponge, retaining water against gravity. Some of the retained water held in larger pores (approximately 0.5–60 μm across) can be sucked out of the soil by plant roots, but the water that is retained in finer pores (<0.5 μm), for example in clay-rich soils, is held too strongly and remains unavailable to plants. When the soil water content available to plants decreases below a certain level, growth can then only be maintained by applying irrigation. Where soil water is so abundant that it fills pores >60 μm, which are normally air-filled, the excess must be removed by artificial field drainage systems. This chapter summarizes the methods used for assessing the need for either irrigation or field drainage, the techniques involved in their application and some environmental problems that may result.

4.1 SOIL POROSITY AND WATER CONTENT

Soils are porous materials consisting of solid particles of various shapes and sizes and intervening voids or pores that may be filled with gases (the soil air) or with water. The solid particles (organic matter and inorganic minerals) are usually arranged into aggregates or peds of various shapes and sizes, and these are separated by large voids (interpedal voids) but also enclose smaller (intrapedal) pores. The total volume of interpedal and intrapedal pores, known as the pore space or porosity of the soil, can be expressed as a proportion of the total volume of the soil. It is calculated by measuring the soil's bulk density (D_b), defined as the mass per unit volume of dry (i.e., water-free) soil, and the mean density of the solid particles (D_p). Then:

$$\text{Porosity} = 1 - (D_b/D_p) \qquad (4.1)$$

This value is usually expressed as a percentage by multiplying by 100. Laboratory methods of measuring bulk and particle density are given by Smith and Thomasson (1974: 42–8). Typical values of D_b range from approximately 0.6 $g\,cm^{-3}$ to 1.8 $g\,cm^{-3}$, though most are in the range 1.0–1.5 $g\,cm^{-3}$. The lowest values generally occur in peat soils and recently cultivated clay-rich topsoils, and the highest in coarser (loamy and sandy) mineral soils compacted when wet by heavy machinery. In mineral soils of the same particle size distribution, D_b is usually 0.1–0.3 $g\,cm^{-3}$ less under permanent pasture than under arable cultivation. Values of D_p for individual soil minerals show a much wider range, but most are around 2.0–2.8 $g\,cm^{-3}$. The value for quartz, which is the most abundant mineral in many soils, is 2.65 $g\,cm^{-3}$, and this is usually used as a default value for mineral soils in the absence of actual measurements.

The proportion of pores filled with water at a given time is measured by weighing the sample moist, then drying to constant weight in an oven at 105°C and reweighing. The weight loss on drying gives the gravimetric water content (θ_g):

$$\theta_g = \text{mass of water/mass of oven-dry soil} \quad (4.2)$$

again multiplied by 100 to express it as a percentage. However, values of volumetric water content (θ_v) are often more useful than θ_g, because they can be used to express the water content as an equivalent depth (usually given in millimetres) spread over the land surface for comparison with rainfall, depth of irrigation water or evaporation, which are also expressed in millimetres depth. Values of θ_v are given as a percentage or as cubic metres per cubic metre. To convert to millimetres, the first is multiplied by 10 and the second by 1000; for example, a value of 25 per cent or 0.25 $m^3\,m^{-3}$ equates to 250 mm per metre depth of soil. θ_v is related to θ_g by the equation:

$$\theta_v = \theta_g \times D_b \quad (4.3)$$

When rain or irrigation water falls on a dry soil, it initially forms films around soil particles and aggregates on or close to the ground surface. The films gradually thicken but, after a certain point, they can hold no more water. Water then moves into smaller pores within aggregates or flows down to create new films and fill pores in the drier aggregates below. This process of downward flow continues until the new water reaches a level (the water table or phreatic surface) below which all the voids between particles are completely water-filled or saturated. Below what is known as a groundwater table, the saturated layer can extend for a considerable distance (tens or hundreds of metres), forming a major aquifer if the rock is porous. Elsewhere the saturated layer may be perched on a slowly permeable layer, which in turn may overlie an unsaturated zone. Further additions of water from the surface then fill the larger voids, previously air-filled in part, at progressively higher levels in the soil, so that the water table rises towards the surface. When it reaches the surface, further rain then ponds or runs over the surface, and the whole soil profile is saturated. The soil water pressure is equal to atmospheric pressure at the phreatic surface, and below that surface it increases progressively above atmospheric pressure.

In unsaturated soil above the water table some of the pore space is air-filled and the pressure of water held in films and smaller pores is less than the atmospheric pressure. The ratio of water-filled porosity to air-filled porosity in unsaturated soil can vary greatly with time and depth below the surface, depending on rainfall, rates of percolation downwards through the profile, evaporation from the surface and uptake by plant roots.

In soil pores containing both air and water, the boundaries between the water films and the soil air form curved menisci generated by surface tension. Some water is also held on the surfaces of soil particles by Van der Waals forces and, in soils containing alumino-silicate clay minerals, some is held between the charged layers of the minerals. Increases in the amount of interlayer water cause the clay to swell and decreases cause shrinkage, processes that affect the size of pores, especially in clay-rich soils.

4.1.1 Soil water contents at different suctions

After saturation or near-saturation of soils by flooding, prolonged heavy rain or excessive irrigation, the larger pores drain under gravity, losing all their water except that held in thin films by surface tension and Van der Waals forces or between the layers of layer silicate clays. The water lost under gravity contributes to a perched or groundwater table below. The downward drainage of larger pores continues at a decreasing rate and, within a few hours or days, ceases at a value of θ_v known as the 'field capacity' (*FC*), which probably has an almost constant value for soils with similar pore space, pore size distribution and pore connectivity. For consistency it is usually quoted as the water content (in volume per cent, $m^3\,m^{-3}$ or mm) after drainage for 48 h. However, exact quantification of *FC* for a given soil type is rarely possible (Cassel and Nielsen, 1986), as pore characteristics vary laterally and over time.

In most British soils, the water content at *FC* defined in this way is in equilibrium with a suction of approximately −5 kPa (−0.05 bar), depending on the particle size distribution, organic matter content and strength of aggregates (Webster and Beckett, 1972). Pores finer than 30–60 μm across retain water against gravity or suctions exceeding −5 kPa, and so remain water-filled at *FC*. Those coarser than this size are air-filled at *FC* and constitute the soil's air capacity. In many countries other than the UK, water contents at slightly greater values of suction (−10 to −33 kPa) are considered equivalent to *FC*. Where winters are cool and wet, as in Britain, most soils return to *FC* in early winter and remain close to that condition until a time in the spring (usually in late March or April) when evapotranspiration exceeds rainfall for at least a few days.

The water held in pores finer than 0.2–0.5 μm, equivalent to suctions greater than about −1500 kPa, cannot be extracted by the roots of most plants. Water held in the suction range −5 to −1500 kPa (−0.05 to −15 bar) is therefore known as the soil's plant-available water capacity (*AWC*). This can also be expressed as a volume percentage of the soil, as cubic metres per cubic metre, or as millimetres depth of water for comparison with rainfall or irrigation. Most plants commence to wilt and grow more slowly or not at all when the *AWC* is partially exhausted; this is mainly because the guard cells of leaf stomata become flaccid, which reduces the transpiration stream (and with it the uptake of nutrients) and limits the intake of CO_2 for photosynthesis. In these circumstances, plants can regain turgor and return to a normal growth rate if θ_v is increased to a value near *FC* by rain or irrigation. However, the value of θ_v at which *AWC* has reached zero, so that the plant is permanently damaged and cannot recover, is known as the permanent wilting point (*PWP*). Generally, *PWP* is approximately equivalent to the water content in equilibrium with a suction of −1500 kPa but, for some plant species, permanent wilting may occur at slightly greater soil water contents. *PWP* is never equivalent to $\theta_v = 0$, as all soils have pores finer than 0.5 μm, which hold water so firmly that it is unavailable to plants. This unavailable water can be a large proportion of the total, especially in clay-rich soils, in which quite large volumes may be held either between the layers of alumino-silicate minerals or within intrapedal pores between closely packed clay particles.

In the laboratory, the total water content of soil samples is usually measured gravimetrically by weighing the field-moist samples then drying them to constant weight in an oven at 105°C. Less precise measurements can also be made in the field, using various pieces of commercially available equipment, such as the neutron probe (Greacen, 1981) and time domain reflectometer (Zegelin *et al.*, 1992). Ochsner *et al.* (2001) described a thermo-time domain reflectometer probe, which can simultaneously measure water content, air-filled porosity and D_b. Moisture content of the surface soil can also be estimated from aircraft or satellites using multispectral remote sensing techniques (Wang *et al.*, 1983).

Laboratory methods for measuring water contents at different suctions are given by Smith and Thomasson (1974: 48–56) and Hall *et al.* (1977: 5–17). Suction tables covered with materials of uniform particle size and composition, such as quartz sand and kaolinitic clay, are used for higher suction values (>−500 kPa), and a specially designed pressure membrane apparatus is often used for lower suctions (Klute, 1986). As these methods are slow, Cornelis *et al.* (2001), Zhuang *et al.* (2001a) and others have suggested ways of calculating the moisture retention curve $\theta_v(h)$ (i.e., the curve showing changes in θ_v with increasing soil water pressure head, *h*) from pedofunctions derived from basic soil properties (particle size distribution, organic matter content, D_b and aggregate size distribution) measured in the laboratory or estimated in the field. For coarse-textured soils (>50 per cent sand) the values estimated from pedofunctions often show good agreement with laboratory measurements, but for finer soils they are usually less than those measured in the laboratory (Pachepsky *et al.*, 2001).

In the field, soil water contents at different suctions <−100 kPa (−1 bar) can be measured with tensiometers. These measure the reduced pressures (negative with respect to atmospheric) in unsaturated soil using a manometer or pressure transducer that is in communication with the soil water via a porous membrane, usually a ceramic cup. The membrane can hold the negative pressure but prevents air from bubbling through.

4.1.2 Factors affecting soil water content at different suctions

Clay content (<2 μm) is the main property influencing water retained at all suctions, but especially at −1500 kPa in subsoils (Figure 4.1). In organic topsoils D_b explains most of the variance in *AWC*,

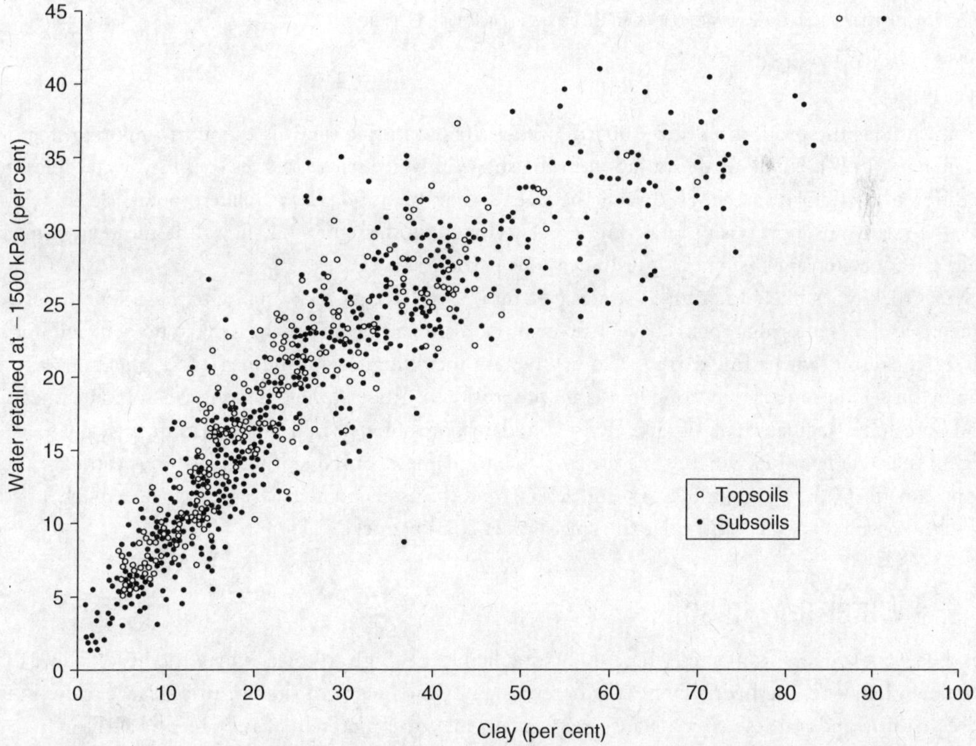

Figure 4.1 Relationship between clay content and water retained at −1500 kPa in topsoil and subsoil samples from England and Wales
Source: Hall *et al.* (1977). Reproduced from National Soil Resources Institute Technical Monograph 9. Copyright Cranfield University, no part of this publication may be reproduced without the express written permission of Cranfield University.

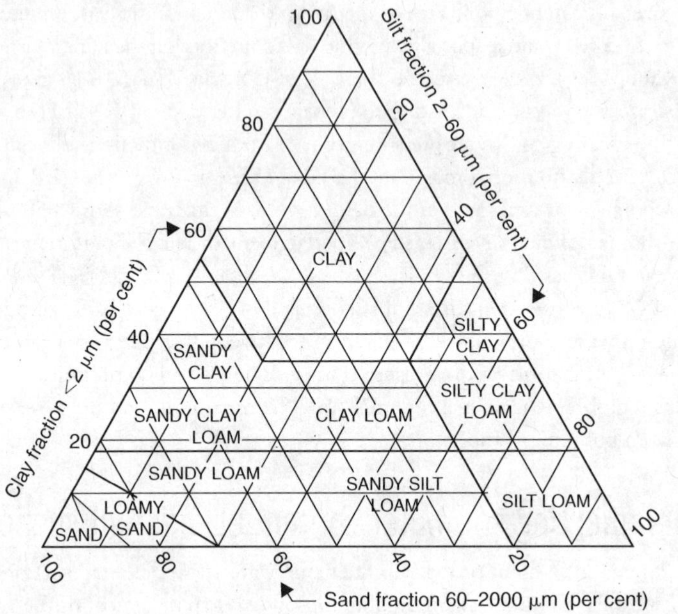

Figure 4.2 Soil particle size classes used in England and Wales

Source: Avery (1980). Reproduced from National Soil Resources Institute Technical Monograph 9. Copyright Cranfield University, no part of this publication may be reproduced without the express written permission of Cranfield University.

but in subsoils the proportion of 2–100 μm particles (silt and fine sand) is the most important factor (Hall *et al.*, 1977). Much of the variation in the shape of water release curves (per cent water released at different suctions over the *AWC* range between −5 and −1500 kPa) is related to particle size class, i.e., the relative proportions of sand, silt and clay (Figure 4.2). Figure 4.3 shows the mean amounts held at various tensions (suctions) for the various particle size classes.

Most rock types likely to occur as stones (>2 mm) in soils (e.g., flint, quartzite, hard limestones, igneous and metamorphic rocks) have little or no porosity, and consequently make no contribution to *AWC*. Even soils with a fine earth (<2 mm) fraction providing an optimum *AWC* (e.g., silt loams) can be drought-prone if they contain a large proportion of these stones. In contrast, soft siltstones and limestones such as chalk often contain abundant pores of a size (around 5 μm in typical English chalk) that retains water which is readily available to plants. Consequently these stones can have the opposite effect (i.e., they can increase soil *AWC*), and thin soils on chalk (rendzinas) are much less drought-prone than soils of similar thickness on less porous rocks.

4.1.3 Water flow in soils

In soils, water moves from sites where the hydraulic head is high to those where the head is low. In the saturated zone, the hydraulic head is determined by the height of the phreatic surface above a given datum, and in these circumstances movement may occur laterally if the phreatic surface slopes relative to the datum (e.g., on a valley side). In the unsaturated zone, the hydraulic head is determined by the height of a porous membrane in contact with the soil (e.g., the ceramic cup of a tensiometer) above the datum, less the negative soil water pressure head resulting from the difference

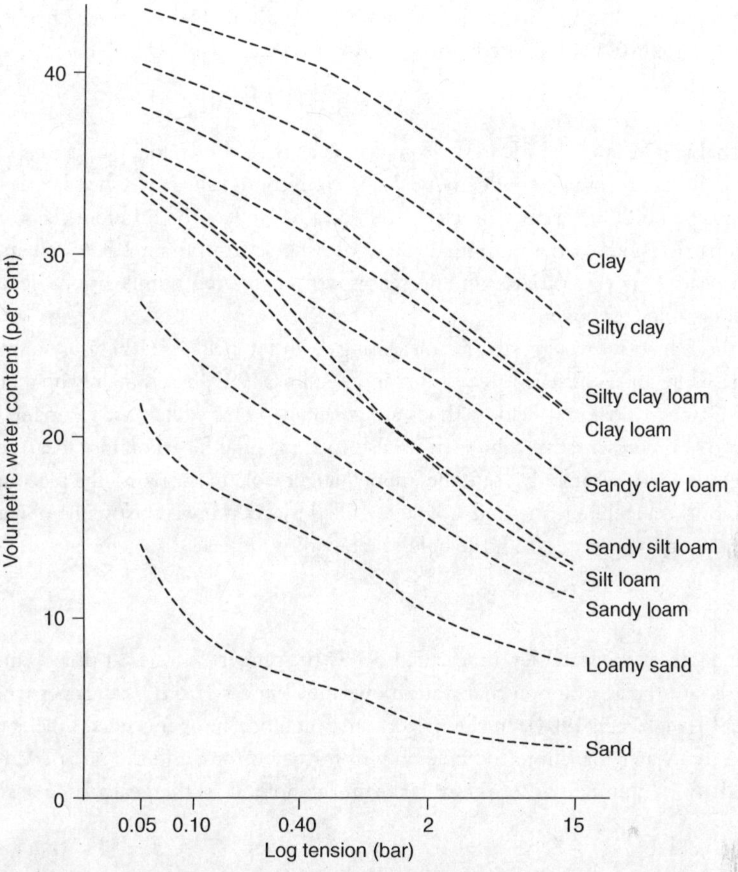

Figure 4.3 Percentages of water released at different suctions over the plant-available water range for the soil particle size classes used in England and Wales

Source: Hall *et al.* (1977). Reproduced from National Soil Resources Institute Technical Monograph 9. Copyright Cranfield University, no part of this publication may be reproduced without the express written permission of Cranfield University.

between the soil water pressure and atmospheric pressure. Negative soil water pressure heads are often referred to as matric potentials (matric suctions if expressed as positive values). Vertical movement can be downwards in response to gravity or upwards because of capillary rise in fine vertical channels or driven by evaporation from the soil surface and transpiration by plants. Upward movement varies with season and occurs mainly near the soil surface. The zone in which it occurs is separated from the underlying zone of purely downward movement by what is known as the zero-flux plane. Movement from sites of high to lower values of hydraulic head is resisted by frictional forces that increase as the channels between solid particles decrease in size and become more tortuous.

Because of solid particles, water cannot move through the entire cross-sectional area of the soil in the direction of flow. The flux density or flux ratio is that proportion of the area occupied neither by solid particles nor (in unsaturated soils) by air-filled pores. In experiments on saturated water flow down columns of sand, Darcy (1856) demonstrated a linear relationship between flux density and the hydraulic gradient or hydraulic head difference causing the flow, such that for his sand the saturated flow (Q) was proportional to the hydraulic gradient (Δh) and the cross-sectional area of the column (A), but inversely proportional to the column's length (L), i.e.,

$$Q = K_s A \Delta h / L \tag{4.4}$$

The proportionality constant K_s in Darcy's Equation is expressed in velocity units (e.g., m day^{-1}) and is termed the saturated hydraulic conductivity. K_s has minimum values of about 1 m day^{-1} in sands, 0.1 m day^{-1} in silts and 0.001 m day^{-1} in clays. These values reflect the different sizes of packing voids between particles; larger than minimal values of K_s arise from fissures between structural aggregates. In peats K_s is very variable and decreases with increasing humification (decomposition of the original organic components).

Various methods have been suggested for measuring K_s in the field and laboratory (Amoozegar and Warrick, 1986; Klute and Dirksen, 1986; Smith and Mullins, 1991) based mainly on a one-dimensional application of Darcy's Equation. Field methods are generally preferred because for many soils it is difficult to remove a representative volume of undisturbed material for laboratory measurements. Two simple field methods are commonly used, the unlined auger-hole method and the piezometric (lined hole) method. In the auger-hole method, a hole is drilled to a depth well below the phreatic surface and, after rapidly pumping water from the hole, the rate of water level rise is measured. Then:

$$K_s = C \times \delta t / \delta z \tag{4.5}$$

where z is the depth of the water level measured below the phreatic surface at time t, and C is a shape coefficient based on the assumption that water flows into the hole from fissures perpendicular to the wall (Boast and Langbartel, 1984). In the piezometric method, an open-ended tube is driven into the auger-hole, so that water flows into the hole only from the bottom, and the rate of water level rise is then measured after pumping part of the water from the hole. K_s is then calculated according to:

$$K_s = (\pi r^2 / C \delta t) \ln(z_1 / z_2) \tag{4.6}$$

in which r is the radius of the tube, z_1 and z_2 are the water levels measured below the phreatic surface between the beginning and end of the time interval δt and C = approximately $5.6r$ for a flat bottomed-hole ($9r$ for a hemispherical bottom).

In unsaturated soils water flow is similar to that in saturated soils except that fewer (only the larger) conducting channels are involved, so that the hydraulic conductivity (K_u) is much less (Richards, 1931). As more and more conducting channels become involved, values for K_u approach those of K_s. Consequently K_s is regarded as characteristic for a given soil type, whereas K_u varies with water content. Nevertheless, it is difficult to measure K_s very precisely; repeat measurements often differ by 50 per cent or more, so values determined by any method are usually given as the mean and standard deviation of numerous measurements.

Other field methods for measuring K_s involve creating a bulb of temporally saturated soil within the unsaturated layer, and measuring the rate at which the bulb accepts water from a cylindrical reservoir inserted into a well. As this rate depends on the hydraulic head (positive water potential) above the bulb, which would decrease as the reservoir empties, a constant head is established in the primary reservoir using a second (movable) reservoir or Mariotte bottle. This is equipped with two exit tubes, one to allow air into the bottle and the other for siphoning water into the primary reservoir, which thereby maintains a constant head on the saturated bulb of soil. This principle is used in the Guelph constant-head well permeameter (Reynolds *et al.*, 1985) and others described by Reynolds and Elrick (1990) and Matula and Kozáková (1997).

Tension disc permeameters (Reynolds and Elrick, 1991) allow water to be supplied at a range of selectable potentials, both negative and positive, to assess the contributions to flow of all sizes of conducting channels. The most widely used type is the CSIRO pattern described by Perroux and White (1988). This consists of a calibrated water reservoir tower and a bubbling tower, in which negative water potentials (usually −10 mm and −50 mm water heads) are determined by the height of water above the bubble entry point and the distance between an air exit tube into the lower part of the reservoir and a basal porous supply membrane.

Another type of field permeameter for measuring K_s, which is cheap and simple to use, is that described by Philip (1993) and Muñoz-Carpena *et al.* (2002). This consists of a clear plastic tube inserted with a tight fit 15 cm into a borehole in the soil, which has been previously drilled with a flat-bottomed auger. A 2-mm mesh is glued to the end of the pipe to prevent erosion of the soil in the bottom of the auger hole when water is added. The tube is then filled with water to a measured height above its base, and the times required to lower the water level to half height and to completely empty the tube are recorded. The gravimetric soil moisture content just beneath the tube is measured before the water is added and again when the tube is empty. Philip (1993) gives the method for calculating K_s by this method.

4.1.4 Computer simulation of water and solute movement through soils

Prediction of the rates of water movement through soils is important for calculating irrigation rates, for the design of drainage systems and for estimating leaching losses of nitrate, other nutrients and pesticides (Chapter 5). One of the simplest yet most effective models of water and solute movement is the Solute Leaching Intermediate Model (SLIM), described by Addiscott and Whitmore (1991).

SLIM divides the soil profile into successive layers of equal thickness, and the soil water into two categories, mobile and immobile (or retained against gravity). In cultivated soils the uppermost layer is equivalent to the topsoil or Ap horizon. The immobile water contents are not decreased by natural or artificial drainage, but can be lost by evaporation. The most strongly retained water (half that held at −1500 kPa suction) is not accessible to anionic solutes, such as nitrate, but can influence the content and movement of non-anionic solutes. Water and solutes entering the surface layer as rainfall or a subsurface layer as percolation water from the layer above are added to the mobile water and mobile solute contents of that layer without displacing them. However, once a layer is saturated, further additions pass through it rapidly by a third category, namely by-pass flow.

Movement of water and solute through a particular layer of the soil operates in five stages. In Stage 1, incoming water/solute is mixed with the existing mobile water/ solute. In Stage 2 a part of the augmented water/solute (α) moves to the next layer down. In Stage 3 half of this α moves down to a further layer. In Stage 4 some of the water/solute remaining in the first layer moves from the mobile to the immobile category. In Stage 5 the other half of α from Stage 3 moves to the next layer down. Stage 3 is performed for all layers, then Stage 4 and finally Stage 5 for all layers. Stages 3, 4 and 5 are performed even if there is no rainfall, because there is always some mobile water in each layer unless evaporation exceeds rainfall. The model has no set time-step, but rainfall/irrigation and evaporation are presented on a daily basis.

In Stage 4 convective movement of solute between the mobile and immobile water components is considered before movement by diffusion. Diffusive movement can then be treated in one of three ways:

- by equalizing solute concentrations between the mobile and immobile water; in practice this does not happen in a finite period;

- by making assumptions about the shape and arrangement of soil aggregates;
- by partially equalizing concentrations between the mobile and immobile water components using a 'hold-back factor' (β) calculated from the size distribution of aggregates. As the topsoil usually has a different structure from the subsoil, different β values are used for each of them.

In the layer containing field drains, a proportion (γ, usually 0.8) of the water and solute entering the layer is diverted into the drains and recorded separately from the seepage from the lowest layer at the base of the profile.

Evaporation from the soil surface (e) is calculated as a proportion of measured open-pan evaporation (e_o). For freshly moistened soil, $e = 0.9e_o$ at any time of the year when rainfall exceeds $0.9e_o$. However, in the winter, when rainfall is $>0.9e_o$, the ratio of e to e_o is based on relationships determined experimentally by Penman (1941). Evaporation is met initially from the topsoil layer, first from its mobile water content, then from its immobile water. Provided the immobile water content does not decrease below a value equivalent to θ (−200 kPa) minus half θ (−1500 kPa), evaporation is restricted to the uppermost soil layer. However, if it does fall below this value, the evaporation losses are shared between the first two layers, then between layers 2 and 3, etc. This gives an exponential decline in evaporation losses down the profile.

The important factor α is determined by the soil's permeability to water but, unlike hydraulic conductivity, is dimensionless. It is calculated either from measured immobile volumetric water contents of the topsoil and subsoil, or from the clay content of the soil, using a regression based on measurements of 12 soils, such that $\alpha = 1.0$ when percentage clay <9.5, and $\alpha = 0.0$ when percentage clay >58.3.

To allow for loss of water by transpiration, SLIM is coupled to a model for crop growth. It can also be linked to models for turnover of organic matter to simulate changes in soil mineral N and therefore in production of nitrate by mineralization.

4.2 EFFECTS OF SOIL WATER CONTENT ON PLANT GROWTH

When θ_v falls below *FC*, because of continuing evaporation from the soil surface or uptake of water by plant roots and transpiration through the leaves, the difference is termed the soil moisture deficit (*SMD*). To maintain a high growth rate, irrigation water is usually recommended when *SMD* reaches a certain value, which varies according to the crop being grown and the stage reached in its growth cycle, the hydraulic conductivity of the soil and the relative humidity of the atmosphere. Because the different horizons within a soil profile may have different particle-, aggregate- and pore-size distributions, air capacity and water-retention properties can vary with depth. To evaluate the effects of these on plant growth, it is therefore necessary to measure them for each horizon, and compare the values with estimated depths of root penetration at different periods of the growing season. The roots of most common plants need pores >100 μm in diameter to spread easily through the soil and respire adequately for good growth. As the air capacity indicates the volume of pores >30–60 μm, it is somewhat greater than the proportion of soil exploitable by roots. The minimum desirable value for air capacity is approximately 10 per cent, giving a root-exploitable volume of at least 5 per cent. A soil horizon with an air capacity <5 per cent is rather impermeable and therefore likely to generate a perched water table. In most soils the total volume of air-filled pores decreases with depth because of deposition of illuvial clay, occurrence of bedrock or other compact or clay-rich material, or the presence of a water table, and this usually limits the ultimate depth of penetration by roots of shortlived plants such as annual crops.

4.2.1 Soil structural quality and droughtiness classes

Air capacity and *AWC* percentage can be used to define simple soil structural quality classes, such as those shown in Figure 4.4. For mineral topsoils (<4 per cent organic C), good structural quality

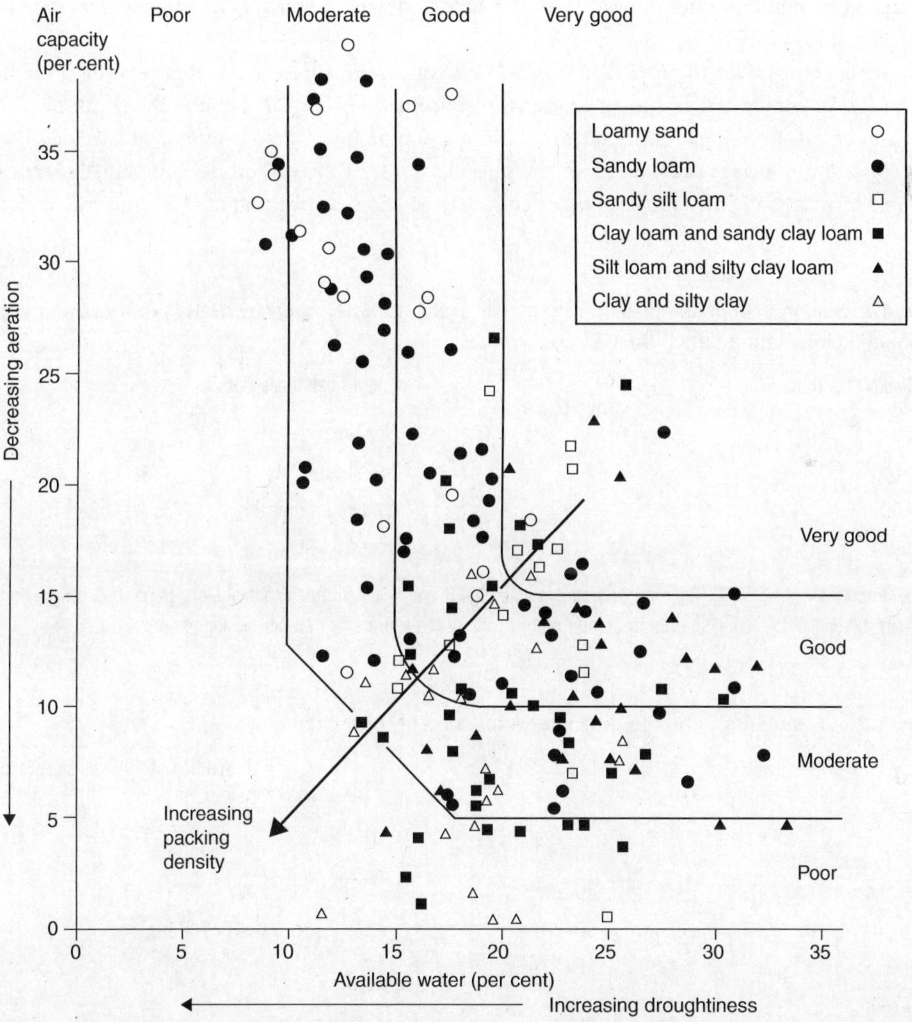

Figure 4.4 Structural quality classes for topsoils of various particle size classes used in England and Wales (see Figure 4.2) according to plant-available water content and air capacity

Source: Hall *et al*. (1977). Reproduced from National Soil Resources Institute Technical Monograph 9.

means >15 per cent air capacity and >20 per cent *AWC*, whereas poor quality means <5 per cent air capacity and <10 per cent *AWC*. In topsoils with >4 per cent organic C, which have greater biological activity and therefore need more oxygen to avoid anaerobic conditions, the upper limit of air capacity for poor quality topsoils should be increased to 10 per cent.

Hall *et al.* (1977) identified four classes of soil dryness or droughtiness based on *AWC* integrated over all horizons in the soil profile to a depth of 1 m (*PAWC*) and values of average potential maximum soil moisture deficit in millimetres (*APMSMD*) for the area (Table 4.1). Maximum *SMD* usually occurs in late summer, and actual values vary in magnitude and timing from year to year, so an average value is calculated for a 30-year period. Values also vary according to the density and type of the vegetation cover, so the potential value is that estimated for a well-established, healthy and therefore uniform grass sward. *APMSMD* is calculated from local meteorological data, including rainfall and evaporation rates derived from sunshine and wind records, and also includes a factor for transpiration losses by grass.

For annual crops both the *PAWC* and *APMSMD* values are adjusted when calculating droughtiness classes. *PAWC* is adjusted according to the crop concerned (Table 4.2) because the depth to which water can be extracted depends on potential rooting depth. Also, there is often a greater density of roots close to the surface than deeper in the subsoil, so less of the subsoil *AWC* is actually accessible to crops. In some soils the actual rooting depth may be less than the potential for the crop concerned

Table 4.1 Soil droughtiness classes from profile available water capacity (*PAWC*) and average potential maximum soil moisture deficit (*APMSMD*)

PAWC–APMSMD (mm)	Soil droughtiness class
>+50	Non-droughty
0 to+50	Slightly droughty
0 to−50	Moderately droughty
>−50	Very droughty

Source: from Hall *et al.* (1977). Reproduced from National Soil Resources Institute Technical Monograph 9. Copyright Cranfield University, no part of this publication may be reproduced without the express written permission of Cranfield University.

Table 4.2 Adjustments to profile available water content for annual crops

Crop	Depth (cm)	Proportion of *AWC* considered available
Winter wheat	0-50	All
	50-120	<−200 kPa
Spring barley	0-50	All
	50-120	<−200 kPa
Potatoes	0-70 only	All
Sugar beet	0-80	All
	80-140	<−200 kPa

Source: from Hall *et al.* (1977). Reproduced from National Soil Resources Institute Technical Monograph 9. Copyright Cranfield University, no part of this publication may be reproduced without the express written permission of Cranfield University.

because of adverse subsoil conditions, such as rock layers, hardpans, severe acidity or poor respiration caused by impeded drainage.

The effects of ground or perched water tables on the supply of water to annual crops are small and are usually ignored in calculations of *PAWC*. This is because:

- permanently shallow groundwater tables usually occur in areas of high rainfall, where APMSMD values are likely to be small, so that the soils are non-droughty in any case;
- in late spring and summer periods when *SMD* is increasing, groundwater tables in drier regions have usually fallen below the rooting depth of most crops; and
- the total volume of pores >30–60 μm (normally air-filled) in a horizon affected by a perched water table is by definition small, and therefore adds little to the *AWC* of that horizon.

APMSMD values need to be adjusted for annual crops to take account of:

- the usually incomplete ground cover in the early stages of growth, which means that transpiration losses of water are less than the full potential; and
- decreased demand for water during the late summer period of ripening. Examples for typical crops grown in lowland areas of Midland and eastern England are given in Table 4.3. These adjustments imply that precise assessments of soil droughtiness for particular crops, soil types and years need to be based on mean monthly, weekly or even daily cumulative values of *PSMD* through the growing season.

Table 4.3 Examples of adjustments to annual potential soil moisture deficit (*APSMD*) for annual crops in eastern and Midland England

Crop	Adjustments	Example (mm)
Winter wheat (full crop cover by mid-April)	End of April *SMD*	16
	Deduct 33% for partial crop cover	−5
	Mid-July *SMD*	115
	Adjusted *APSMD* value	110
Spring barley (full crop cover by mid-May)	Mid-May *SMD*	34
	Deduct 33% for partial crop cover	−11
	Mid-July *SMD*	90
	Adjusted *APSMD* value	79
Sugar beet or potatoes (bare ground until mid-May)	Mid-May *SMD*	34
	Deduct 66% for bare ground	−22
	Full crop cover by July; increase in *SMD* mid-May to July (34–113 mm)	79
	Deduct 33% for partial crop cover before July	−26
	SMD end of August	170
	Adjusted *APSMD* value (170 − 26 − 22)	122

Source: Modified from Hall *et al.* (1977). Reproduced from National Soil Resources Institute Technical Monograph 9.

In Mediterranean countries and others with fairly predictable seasonal wind patterns and rainfall and sunshine values, the effect of *SMD* on crop yield is unlikely to vary much from year to year. However, in countries such as the UK, where the weather varies much more in the short term, the year-to-year variation in *SMD* is often large enough to modify the 30-year average droughtiness by >50 mm, i.e., two classes in the classification of Table 4.1. Without accurate forecasts of seasonal rainfall, it is therefore impossible to predict site-to-site differences in actual crop growth and yield. However, a method of estimating daily changes in *SMD* for individual crops in these areas is given in Section 4.3.2.

4.3 IRRIGATION OF CROPS

4.3.1 Planning an irrigation system

The main purpose of irrigation is to supplement rainfall and provide the additional water that established plants require for optimum growth and yield. It is also used:

- to soften hard aggregates in dry soil so that they can be broken down to create a fine seedbed;
- to ensure the establishment of a crop from seed or after transplanting;
- to wash-in nutrients provided as top dressings of fertilizer;
- to soften dry soil before lifting root crops such as potatoes or sugar beet; and
- to leach salts from saline soils.

However, these uses are not considered in this chapter.

Irrigation equipment suitable for treating a large area of soil such as a field is expensive, both in terms of initial outlay and running costs, so a farmer needs to carefully compare the costs with the cash benefits expected from its use over a period of years. Important considerations include:

- the cost of establishing and maintaining an adequate supply of suitable water;
- the expected rainfall and *APMSMD* values;
- the *PAWC* values of the farm soils;
- the crops to be grown;
- profit from the expected increase in yield including any improvement in timing in relation to market demands;
- the increased management costs (e.g., moving the equipment from field to field);
- the expected life of the equipment, which may be limited if the water used is corrosive.

Some of these aspects are discussed below in general terms; detailed considerations for UK situations were discussed by the Ministry of Agriculture, Fisheries and Food (MAFF) (1982).

When planning long-term irrigation needs, it is necessary to consider a range of soil water contents at which irrigation may become necessary, and the needs of crops with different susceptibilities to water shortage and different responses to irrigation. Hogg (1967) proposed four plans for British situations:

Plan 1. To restore the soil to *FC* when *SMD* reaches 25 mm.

Plan 2. To restore the soil to *FC* when *SMD* reaches 50 mm.

Plan 3. To reduce the *SMD* to 25 mm when it reaches 75 mm.

Plan 4. To reduce the *SMD* to 50 mm when it reaches 125 mm.

Repeated restoration of *FC* from a small or moderate *SMD* value (Plans 1 and 2) is expensive, but is likely to be profitable in most years and on most soils with higher value crops that are more susceptible to moisture stress, such as sugar beet, potatoes, summer cabbage, cauliflower, lettuce, runner beans, marrows, courgettes and black currants. At the other end of the scale, Plans 3 and 4 are cheaper but inadequate for susceptible crops in dry years, and therefore less likely to be profitable. Plan 3 is suitable for the less drought-susceptible crops, such as carrots, brussels sprouts, peas, broad beans, onions and parsnips, and Plan 4 is mainly for tree fruit crops. They are likely to be useful only in dry years and on soils with small *PAWC* values.

Some crops benefit most from irrigation at certain critical growth stages, known as the irrigation response periods, whereas others require adequate water at all growth stages. For example, peas benefit only from irrigation early in the flowering period and at pod swelling, strawberries require adequate water mainly after petal-fall so as to swell the developing fruits, and potatoes respond best to irrigation after the tubers have formed. In contrast, grass, runner beans, beetroot, cauliflowers, lettuce, marrows and courgettes grow best when soil moisture contents are close to *FC* throughout their growing seasons.

For rough planning purposes the average rainfall (*R*) and potential transpiration (*PT*) for the relevant months of the growth period or irrigation response period are calculated from long-term meteorological data, or in England and Wales can be taken from the regional agroclimatic data given by Smith (1976). The difference between them ($PT - R$) is then used to read the irrigation need according to the length of the irrigation response period from graphs given by MAFF (MAFF, 1982: figures 17–20), and another graph (MAFF, 1982: figure 21) gives the frequency of irrigation need.

4.3.2 Calculating actual irrigation needs

For calculating actual day-to-day irrigation requirements during the growing season, *SMD* can be obtained using *in situ* soil moisture measuring equipment, such as tensiometers, neutron moisture meters or time domain reflectometry. However, for most farmers these types of apparatus are expensive, difficult to install (especially to greater rooting depths) and complex to operate. Instead, a simple balance sheet approach is recommended (MAFF, 1982), in which daily *SMD* values are calculated from estimated daily soil water loss by evapotranspiration (*ET*) set against *R* measured daily with a standard rain gauge. *ET* is estimated for individual fields according to the extent of crop cover. When a planned *SMD* is reached according to the balance sheet, the amount of irrigation required under the plan is applied and then deducted from the *SMD* value. Ideally the balance sheet is started at the end of the winter when the soil is still at *FC*; for most parts of Britain this means no later than 1 April.

It is important to avoid overirrigation (exceeding the soil's *FC* value), because the excess water is likely to be lost, either by drainage through the soil profile or by runoff over the soil surface. The latter may cause soil erosion, and both lead to loss of nutrients. In addition, crops growing in soil that is too wet are susceptible to damage by pests and diseases such as those caused by nematodes or fungi. Also, some subtle aspects of crop quality, such as the flavour of grapes and tomatoes, can deteriorate with the continual lush growth that occurs at soil water contents close to *FC*. In practice, with most crops it is best to satisfy all but a small part (a few millimetres) of the calculated water requirement, to allow for any rain that falls immediately after irrigation. In these situations small deficits are unlikely to stress the crop. Where rainfall or irrigation does lead to a water content in excess of *FC*, the excess is written off in the calculation, because *SMD* cannot have a negative value.

4.3.3 Infiltration capacity

An important factor in calculating irrigation rate is the soil's infiltration capacity (i), or the rate at which the surface soil can accept water. If it is exceeded, then water is lost by runoff even before *FC* is reached. In physical terms, infiltration rate depends on the hydraulic conductivity (K_s) of the topsoil and the soil water head gradient at the surface. In practice, it is influenced by various soil characteristics (particle size distribution, organic matter content, abundance of channels or fissures in surface and subsurface horizons that in turn determine K_s, previous soil management (size and strength of surface aggregates, presence of a surface crust (Section 7.3.4) or compacted layers), crop characteristics (percentage leaf cover, angle of leaves) and slope characteristics (angle, profile, aspect), all of which must be taken into account when deciding irrigation rates so as to avoid surface ponding and runoff. Measured infiltration rates are very variable, ranging from $<1\ \mathrm{m\ day^{-1}}$ for crusted surfaces or uncultivated clay soils on slopes, to $>1\ \mathrm{m\ h^{-1}}$ for recently cultivated loamy or sandy soils on level ground.

Infiltration rate can be measured in the field with a simple double ring infiltrometer (Hills, 1970). A constant head permeameter or tension disc permeameter (Section 4.1.3) sitting on the soil surface can also be used (Angulo-Jaramillo *et al.*, 2000), provided any vegetation is trimmed close to ground level. Various methods have also been proposed for calculating it from pedofunctions derived from basic soil properties measured in the laboratory, such as particle size distribution, organic matter content and soil water retention curves (Zhuang *et al.*, 2001b), but these ignore the often significant effects of vegetation and slope characteristics.

The infiltration rate measured initially is usually slightly less than the equilibrium infiltration capacity determined by K_s, which is reached after a period of 15 minutes to several hours. However, even a single equilibrium infiltration capacity measurement cannot be used as the recommended irrigation rate, because the soil is likely to be variable and few irrigation systems apply water evenly. Variation in i can be assessed roughly by measuring infiltration at several sites over the area to be irrigated.

The rate of water movement from the topsoil downwards into underlying horizons is termed the percolation rate. When this is less than the infiltration rate, a saturated topsoil can overlie a non-saturated subsoil. Where such conditions persist for any length of time, the soil has a perched water table. These can arise where pores in the subsoil have been blocked by deposition of illuvial clay or a cementing agent (e.g., calcium carbonate or iron oxide) or have been reduced in size by compaction (e.g., by heavy machinery). Excess water in the saturated layer may lead to surface runoff, but on slopes it may move laterally through upper parts of the soil profile as interflow.

4.3.4 Irrigation of cracking clay soils

In soils with moderate or large amounts of expansible clay minerals, such as smectite or vermiculite, the subvertical shrinkage cracks, which form on drying, provide preferred pathways for downward movement of rain or irrigation water. Cracks often form where there has been earlier disturbance of the soil, for example where mole drains (Section 4.4.5) have been formed. In these circumstances much irrigation water can be lost through the soil's artificial drainage system. Such losses can be minimized by commencing irrigation before the cracks penetrate to mole drain depth (usually 50–70 cm), but the clay-rich soils that require artificial drainage of this type are probably best managed without any irrigation. Heavy equipment for harvesting of late-season crops, such as potatoes and sugar beet, can cause compaction and structural damage if clay-rich soils are close to *FC*, and some cracking in dry years is probably essential for overcoming this problem because it assists development of a structure with sufficient air capacity to dispose of excess winter rain. Irrigation schedules that prevent cracking of these soils may therefore create long-term problems that outweigh any short-term benefits to crop growth.

4.3.5 Modelling irrigation requirement

Bailey and Spackman (1996) proposed a computer-based model known as IRRIGUIDE for predicting *ET* and soil moisture changes as an aid to irrigation scheduling for arable crops. As input data the model requires topsoil thickness, texture or *AWC* of the topsoil and subsoil, the slope of the site, crop species, planting date and a range of meteorological data (*R* for every 12-h period, mean wind speed for every 6-h period, daily maximum and minimum air temperatures, daily sunshine hours and the surface wetness and dew point temperature at 09:00 h). The model then estimates the *PET* of the crop and models canopy and root development, taking account of soil differences and crop development to simulate crop water use in individual fields. Water reaching the soil in excess of *PSMD*, either from rain or excess irrigation, is divided into drainage, stored surplus and runoff according to soil characteristics and the amount of excess. The stored surplus has an upper limit depending on soil type and slope.

IRRIGUIDE was originally tested on potatoes and sugar beet grown on sandy loam soil at Gleadthorpe Experimental Farm, Nottinghamshire (Bailey *et al.*, 1996), where a very strong relationship ($r = 0.99$) was found between measured and modelled soil moisture changes. Subsequently it has been widely used in Britain for a range of arable crops grown on various soil types.

4.3.6 Sources of irrigation water

Suitable sources of irrigation water include lakes, rivers, groundwater from wells and boreholes or the local public water supply. However, supplies are often more limited and consequently more expensive in summer than in winter, so it is usually beneficial to construct a reservoir for storage of supplies abstracted during the winter months. In periods of high *ET* loss in the growing season, when *SMD* may increase by 25 mm every 5–10 days, any supply should be able to deliver up to about 750 m^3 of water per day for every 10 ha of land (e.g., under Plan 1 in Section 4.3.1). If *ET* loss continues at the same rate, applications of this volume would be required every 5–10 days, a period termed the irrigation cycle. A reservoir to supply water at this rate throughout the summer (i.e., with no summer rainfall) would contain about 50,000 m^3. Earth reservoirs are the cheapest way of storing this volume; further details of their design are given by MAFF (1977).

Surface water may not be suitable for irrigation purposes if it is polluted with sewage or runoff from farmyards. This is a particular problem with salad, fruit and vegetable crops, which may be eaten raw. River water containing heavy metals may affect crop growth and poison the population of beneficial bacteria in the soil. However, the fine, chemically inert silt and clay that is often suspended in river water can improve the water retention of sandy soils (Fullen *et al.*, 1995). In some areas water from springs or boreholes may contain dissolved salts, which can also hinder crop growth, degrade soil structure or corrode irrigation systems.

4.3.7 Methods of applying irrigation water

The simplest and cheapest method of applying irrigation water is by flooding the soil surface directly from a supply pipe. However, this never ensures an even distribution of water. Furrow irrigation, in which the water is directed along shallow surface channels between crop rows, is slightly more efficient, but also results in uneven distribution, usually with considerable loss of water by infiltration through the soil beneath the furrows.

Sprinklers and spray booms are designed to provide a more even distribution of water. In order to cover large areas without frequent repositioning, many sprinklers have rotating or reciprocating heads, which have the added advantage that surface runoff is less likely because there is time for infiltration in the brief periods when no water is received. Spray guns distribute the water over large areas, but the time required for application of a given volume per unit area is increased, and the distribution is more likely to be uneven, especially in a strong wind.

Many of the problems of uneven water distribution are overcome by micro-irrigation systems, in which a large area such as a whole field is covered by a semi-permanent grid of small-bore plastic water pipes laid on or just below the soil surface. These systems avoid losses by wind-spray, evaporation, runoff and infiltration in areas of excessive application. They also remove the need for frequent repositioning of equipment, and the system easily can be run at night when there is less evaporation and no risk of damage to wet leaves exposed to intense sunshine. If the distribution lines are buried beneath the soil surface, they are protected from damage by other farm equipment, animals or exposure to the sun. The water is emitted from perforations in the pipes (drip or trickle irrigation) or from microjets, both of which can be positioned close to individual plants or their roots, where the water is actually needed. However, a common problem with such systems is clogging of the emitters, which is best avoided by careful filtering of the irrigation water before it is used.

Micro-irrigation systems are initially more expensive than other types, but the more efficient use of water in terms of avoiding losses and improved crop responses usually means that they are cheaper to run and provide greater increases in crop yields and profit margins. They are probably most useful in orchards, vineyards, ornamental gardens and glasshouses, where the system can be left undisturbed for many years. Micro-irrigation systems have become increasingly popular over the last two to three decades in many areas, such as Mediterranean countries and parts of Australia, India and China, where water is expensive and there is a severe shortage of rain in the growing season. In Britain and the USA, they are used for only a few per cent of the irrigation water applied.

4.3.8 Fertigation

This term is used for the simultaneous application of fertilizer and irrigation water (Bar-Yosef, 1999). In theory, soluble fertilizers can be dissolved in irrigation water applied by flood, furrow, sprinkler or gun irrigation. However, it is achieved most conveniently and efficiently through a micro-irrigation system, which can accurately control both the amounts of nutrients applied and the timing of applications according to crop demand. Large tanks are necessary for preparation and storage of the fertilizer solution and, if carefully located, these can usually provide sufficient head to supply a considerable area. If not, the solution must be pumped through the micro-irrigation lines.

4.4 FIELD DRAINAGE

4.4.1 The need for field drainage

In soils with a high groundwater table or with slowly permeable subsurface horizons, such as those developed on thick clay formations, the large pores that would normally be air-filled are water-filled for long periods. This has various potentially deleterious effects on crop growth, including:

- restricting root respiration (exchange of oxygen and carbon dioxide with the atmosphere), because gaseous diffusion rates are approximately 10^4 times faster in air than in water. This is especially critical in spring when the soil is at or close to *FC* and its temperature is increasing, so that the roots have to compete with increasing microbiological activity for the remaining oxygen in the soil;
- decreasing the uptake of water and nutrients because of poor root development;
- decreasing the availability of nitrogen through denitrification (Fairchild *et al.*, 1999);
- decreasing the rate of humification and mineralization of organic matter so that the nutrients it contains are released more slowly;
- increasing the availability of Fe^{2+}, often to toxic levels;
- increasing wind throw (especially of trees) because of poor root development and weak anchorage;
- decreasing the storage of carbohydrates and proteins in root crops, such as potatoes and sugar beet;
- decreasing the synthesis of the plant growth regulators important for coordinating root and shoot growth;
- encouraging the persistence of root fungal diseases, such as 'take-all' (*Gaeumannomyces graminis* var. *tritici*) in cereals;
- encouraging the spread of soil-borne nematode pests such as potato root eelworm, which require pores that are partially or completely water-filled to move through the soil;
- making the soil slower to warm up, so that spring germination is slow;
- encouraging the growth of competitive weeds, such as cleavers in cereal crops.

Most crop plants can survive soil saturation in the rooting zone for short periods only, usually 2–20 days depending on species and growth stage. Saturation for longer periods can lead to complete crop failure, but even shorter episodes of anaerobic conditions may decrease the growth rate and ultimate yields of most crops. Once established, grasses are among the most tolerant species, which explains why wet soils are suitable mainly for pasture. However, with persistent waterlogging the more productive grass species are progressively replaced by weeds.

Experiments on sites with clay soils suggested overall yield benefits of up to 25 per cent for some crops using the more effective field drainage systems (Trafford, 1972), but the value of drainage is often uncertain, especially as it may lead to environmental problems in peat soils (Section 4.4.11).

4.4.2 Field and arterial systems

In Britain, various methods of improving the drainage and aeration of poorly drained soils, which occupy extensive tracts of the country, have been practised for several centuries. They have two components – a system of subsurface field drains for removing water from higher saturated horizons of cropped soils, and an arterial system of surface water courses (ditches, streams, rivers and estuaries) for collecting the water from the drains and transferring it to the sea, preferably without flooding areas downstream.

Effective arterial water courses are easy to achieve in upper reaches of river valleys, where flow rates are rapid. However, the area of land benefiting from field drainage in these areas is generally small. In contrast, the area of land potentially benefiting from drainage schemes in middle and lower reaches of valleys is much larger, but the arterial channels have diminished gradients and much slower discharge rates, so that field drains cannot operate so efficiently. In flatter low-lying areas, small channels, such as farm ditches and streams, should have a constant cross-section, smooth sides and floors and uniform gradients to provide the best flow velocity. They also require regular maintenance to clear weeds, sediment and other obstructions. However, they are important as wildlife habitats and corridors for movement of wildlife. Wildlife has demands, especially in terms of habitat diversity, which often conflict with efficient transfer of excess water, so a sympathetic compromise is usually required.

Arterial drainage improvements in the lower reaches of valleys are more complex and expensive than in higher reaches, though they usually influence much larger areas of agricultural land. When high water levels in the main outfall channels are shortlived (e.g., at high tide), drainage of surrounding lowlands can often be controlled simply and cheaply by sluice gates, which are opened only when the level in the outfall water course is low. However, if the outfall level is permanently high, control has to be achieved by the more expensive option of pumping to lift water into the outfall channel, and this has to be balanced against the expected benefits in crop production and improved accessibility to land throughout the year. In grassland the improvements in accessibility mainly affect livestock, such as sheep and cattle; in arable areas the benefit is principally in lengthening the period over winter for machinery work days.

Systems of field drains (collectively known as underdrainage) have various purposes:

- in temperate humid countries, such as the UK, they are used to transfer excess winter rain (the rain falling between the return to *FC* date in the autumn and the initiation of *SMD* in spring) from the soil to the arterial drainage system. This helps increase soil strength so as to allow cultivation and other trafficking of the soil without structural damage;
- to ensure adequate aeration of the soil over the rooting depth during the growing season of crops;
- in salt-affected soils of arid regions or areas inundated by the sea, to leach salts away from the rooting zone, where they can cause 'physiological drought' (Section 3.3).

Soil water can enter field drains only if the hydraulic head exceeds atmospheric pressure. Drains are therefore installed below the water table or phreatic surface with the aim of lowering that surface. For most arable crops the system is usually designed to eliminate saturation of the upper 50–200 cm of the soil profile over the whole field. The water table is lower over the field drains than at the midway points between them, so to achieve a minimum of, say, 50 cm aerated soil over the whole field, drain depth must be greater than this; how much greater depends on the drain spacing.

4.4.3 Design of field drainage systems

The design of field drains involves decisions on depth, spacing, diameter, type, use of permeable fill such as gravel over the drains and secondary treatments to increase permeability between the drains, such as moling and subsoiling. It is also influenced by predominant land use and the area and slope angle of the land to be drained. Drain depth determines the maximum depth of aeration under dry conditions when drainflow has ceased. Greater drain depths generally allow wider drain spacing. However, the chosen drain depth must be permanently above the level of outfall into the arterial network.

Drain spacing mainly affects the rates of water table rise and fall. The more closely spaced the drains, the slower the rate of water table rise under wet conditions and the faster the rate of water table fall when rainfall ceases. The time taken for the water table to fall a given distance is inversely proportional to the square of the drain spacing, so that halving the spacing approximately quarters the time.

Subsoils with lower values of hydraulic conductivity (K_s) should have more closely spaced drains (Table 4.4). For example, K_s values <1.0 m day^{-1} require a lateral drain spacing of 10 m or less to prevent the water table rising under a rainfall rate of 10 mm day^{-1}, whereas the same result with K_s values >10 m day^{-1} can be achieved with a drain spacing of 30–40 m. As it is difficult to obtain a representative value of K_s to the required depth across a whole field (Thomasson and Youngs, 1975), drain spacing is usually decided according to what has been found to be effective in the past in similar general conditions of soil, relief, crop and climate.

Equations relating the water table height midway between drains to the drain spacing and hydraulic conductivity of the soil under steady state rainfall conditions, were suggested by Lovell and Youngs

Table 4.4 Minimum values of hydraulic conductivity (K_s, m day^{-1}) required to prevent the water table rising to <50 cm of the surface for >24 h at a rainfall rate of 10 mm day^{-1} with various depths (h) to an impermeable layer

Drain spacing (S, m)	$h = 0$	$h = 0.5$ (m)	$h = 1.0$ (m)	$h = \infty$
10	1.0	0.34	0.27	0.22
15	2.3	0.75	0.55	0.39
20	4.0	1.34	0.94	0.54
25	6.3	2.1	1.42	0.78
30	9.0	3.0	2.0	0.92
40	16.0	5.3	3.5	1.33

Source: Thomasson (1975). Reproduced from National Soil Resources Institute Technical Monograph 7.

(1984). However, because of various physical and mathematical assumptions, they found considerable discrepancies between results from these equations and experimental field measurements. Youngs (1985) suggested instead a single power-law relationship, which agrees better with experimental results and is also simpler than other 'drainage equations':

$$h/0.5S = (q/K_u)^{1/\alpha} \tag{4.7}$$

where S is the spacing needed to give a desired height (h, m) of the water table above drain depth midway between drains, q is the rainfall rate, K_u is the hydraulic conductivity of the unsaturated soil and α has values ranging from 1.384 (where $D/S \quad \infty$) to 2.0 (where $D/S = 0$) (D = vertical distance (m) between the drains and the impermeable layer below). However, equations based on steady state conditions are oversimplifications of real situations, in which the water table moves in response to variations in rainfall and the soil is structurally heterogeneous vertically, horizontally and over time.

4.4.4 Pipe and permeable fill types

Pipes for the initial removal of water from saturated soil (the minor or lateral drains) are made either from earthenware or plastic. Water entry into the earthenware types is mainly through gaps where the short sections are butted together. Plastic drains are continuous and water entry is through numerous regularly spaced perforations. Earthenware pipes are usually 75 mm in diameter, but the more modern plastic pipes are often narrower, which means they reach their flow capacity more easily and then do not work as efficiently. Flow capacity is proportional to pipe diameter raised to the power of 2.67, so a small decrease in diameter results in a large decrease in capacity. The minor or lateral drains discharge either directly into an open ditch or into permeable collector drains (known as main or leader drains) usually laid at approximately 90° to the laterals. These, in turn, feed into a ditch or, where large areas between ditches are involved, into a large-diameter impermeable carrier drain.

In soils with $K_s < 1.0\,\mathrm{m\,day^{-1}}$, a layer of permeable fill (gravel or crushed rock 5–50 mm in size) is often placed in the trench above the pipe to improve flow between water-bearing channels in the subsoil and gaps or perforations in the pipe. The water-bearing channels may be either natural (e.g., root or earthworm channels or desiccation cracks) or artificially generated either by 'moling' (Section 4.4.5) or 'subsoiling' (Section 4.4.7). The permeable fill covers the drainpipe to within 40–50 cm of the surface, so that mole channels subsequently drawn through the subsoil at depths between this level and that of the pipe will intersect the permeable fill and drain through it into the pipe.

In sandy or silty soils, which usually have weak aggregate strength, the water moving through subsoil channels often carries dispersed soil particles and requires filtering to prevent partial blockage of the pipe by deposition of particles. Permeable fill is too coarse for this purpose, and so the plastic pipes intended for use in these dispersible soils are either surrounded with porous matting or factory-coated with synthetic filters.

4.4.5 Mole channels

Mole channels are formed usually at a depth of 50–70 cm by a bullet-shaped plough up to 75 mm in diameter mounted at the lower end of a thin blade or leg, which is pulled through the soil behind a powerful tractor. To prevent the mole plough from following minor undulations in the ground surface and thus producing a channel with sections of reversed slope, the leg is usually mounted on a sliding beam. A sphere of about 100 mm diameter is towed behind the bullet on a short chain to enlarge the channel and smooth its walls. The channels are usually drawn to intersect the gravel-filled drain at approximately 90°, and at a lateral spacing of 2–3 m. As special equipment is required, the work is usually done by specialist contractors rather than by farmers themselves. Moling is usually done in early summer when the topsoil is dry enough to shatter but the clayey subsoil is still sufficiently plastic to form a stable water-transporting channel (Spoor *et al.*, 1982). This combination of conditions usually occurs when the *SMD* is around 50 mm.

Mole channel stability is unlikely in soils with <30 per cent clay (<2 μm). Soils of any texture that are very wet are also unsuitable for moling, as the shear strength is insufficient to maintain channel shape. Mole channels formed under the right soil conditions can, however, survive without collapsing for several years, though they are often redrawn every 3–5 years to ensure that drainage efficiency is maintained. Even in clay-rich soils, well-formed mole channels can become eroded, leading to local collapse and blockage, especially if channel steepness increases the rate of water flow. For this reason moling is usually confined to areas of land with slopes of $<2°$.

4.4.6 Gravel-filled mole channels

In wetter clay-rich soils mole channels are sometimes filled with gravel or stone chips of 12–20 mm diameter, which are inserted through a feeder pipe behind the plough leg as the channels are drawn. This prevents collapse of the channels when the soil is very wet and plastic, or where there are pockets of sand in an otherwise clay-rich soil. Gravel-filled channels often have a longer life than empty channels, but flow rate is reduced and siltation occurs more easily, especially in less clayey soils. Also the cost is about twice that of normal moles.

4.4.7 Subsoiling

Ploughing in very wet conditions and/or with blunt plough shears can produce a compact subsoil zone or 'plough pan'. Plough pans can severely impede the movement of water, air and nutrients within the soil and lead to erosion. Plough pans are often ignored, as they are largely 'invisible' because they lie below the surface and can only be seen if exposed by erosion or cultivation. It is important that farmers regularly examine the extent, depth and severity of subsoil compaction. The penetrometer is a useful instrument to do this, being a probe that can be used to measure soil resistance to insertion. When plough pans are found, they must be broken up by 'subsoiling', which is also an alternative method of increasing subsoil K_s, particularly to improve horizontal water movement to drains.

Subsoiling involves heaving and shattering the subsoil at and above 40–50 cm depth using specially shaped plough-like implements with a tapered 'shoe-shaped' share to heave the soil, mounted behind a powerful tractor. It is essential that the 'shoe' gets beneath the plough pan and breaks it up. Subsoiling that is too shallow can further compact the pan and worsen the problem. The lateral spacing is usually less than that for moling, and the soil should be in a drier, more brittle condition (*SMD* $>$ 100 mm), so that it shatters, rather than deforms plastically. For this reason subsoiling is best done in summer and is not suited to areas with annual rainfall $>$ 750 mm, where moling is preferred. Also, it is suitable mainly for grassland, as arable crops cannot be disturbed in summer. However, some impermeable subsoil horizons (iron pans and fragipans) can be shattered effectively with the soil in a moist state. Like moling, subsoiling usually has to be repeated every few years because the soil gradually settles and closes the fissures. Subsoil surveys and remedial action should be undertaken regularly (every 1–4 years), depending on both the susceptibility of the soil to compaction and the nature of agricultural activities (i.e., more frequently if using heavy farm machinery).

4.4.8 Effect of land use on drainage requirement

Another factor influencing drainage design is predominant land use. Arable land usually requires more effective drainage than grassland because, under grass, the soil organic matter content is maintained or progressively increased so that subsoil aggregates are strong and natural drainage is relatively good, at least within rooting depth. In contrast, with repeated cultivation, arable soils slowly lose organic matter and their structural stability deteriorates, especially when they are compacted by the use of heavy machinery in wet conditions. Also grass can tolerate temporary anaerobic conditions better than most arable crops, and the financial benefits of good drainage are likely to be greater for intensively grown arable crops. However, the extra cost of permeable fill, moling or subsoiling can often be partially offset by wider drain spacing. Further details of field drainage design in regions of temperate climate are considered by Henderson and Farr (1992).

4.4.9 Indicators of drainage requirement

The need for field drainage can be assessed from frequent observations of the depth to waterlogged layers over a period of years. Observations can be made in dipwells (unsealed auger holes drilled to a depth of at least 1.0 m and preferably lined with perforated plastic pipe to prevent collapse) or piezometers (narrow sealed tubes open only at the base) inserted to various depths. To allow for soil variability both dipwells and piezometers should be replicated.

Thomasson (1975) proposed six water regime classes (*WRCs*) defined by numbers of days (not necessarily consecutive) in most (i.e., >50 per cent) years during which time the soil is saturated within certain depths:

WRC I. Soil profile not saturated within 70 cm depth for >30 days in most years.

WRC II. Soil profile is saturated within 70 cm depth for 30–90 days in most years.

WRC III. Soil profile is saturated within 70 cm depth for 90–180 days in most years.

WRC IV. Soil profile is saturated within 70 cm depth for >180 days, but not saturated within 40 cm depth for >180 days in most years.

WRC V. Soil profile is saturated within 70 cm depth for >335 days in most years, and saturated within 40 cm depth for >180 days.

WRC VI. Soil profile is saturated within 40 cm depth for >335 days in most years.

The extent of soil saturation can also be inferred from the occurrence of soil redoximorphic features (Verpraskas, 1992), such as the greyish (iron-depleted) matrix colours, ped faces or mottles produced by repeated reduction of Fe^{3+} by seasonal waterlogging (Blavet *et al.*, 2000). Greyish soil colours are defined using the Munsell Color Chart as having chromas of 2 or less. Well-drained soils with no greyish colours within 70–80 cm depth correspond to *WRC I*, and soils with almost uniformly grey colours to the surface (also often with a peaty surface horizon) correspond to *WRC V* and *WRC VI*. However, between these two extremes it is often difficult to interpret the occurrence of greyish colours. In particular, once greyish colours and ochreous mottles have been formed in subsoil horizons by a prolonged period of seasonal or permanent waterlogging, they are not easily removed by more aerobic conditions. Consequently, they are a poor guide to the present water regime of soils that have already been effectively drained by artificial methods. In cultivated surface (Ap) horizons, ochreous mottles, especially those associated with live roots, indicate anaerobic conditions within the past year or two.

4.4.10 Soil maps as indicators of drainage requirement

Areas likely to require field drainage can often be predicted from general-purpose soil maps, most of which indicate critical factors such as thickness of soil over hard rock or impermeable layers, textural class and structure at various depths and depth to gleyed (greyish mottled) horizons. In many soil classification systems these properties are used as distinguishing criteria for higher soil categories. For example, in the system used in England and Wales (Avery, 1980), soils (except those in recent alluvial or marine sediments) in which greyish mottles occur within 40 cm of the surface are classified in one of two major soil groups, the Surface-water gley soils (perched water table over a slowly permeable horizon) or the Ground-water gley soils (shallow groundwater table). Within the better-drained Brown soils, stagnogleyic subgroups have greyish mottles occurring within 70 cm depth but not within 40 cm depth as a result of a perched water table; and gleyic subgroups have greyish mottles in the same depth range but because of a high winter groundwater table. Maps showing the distribution of these higher soil categories can therefore provide a provisional indication of drainage requirements.

More detailed information relevant to field drainage design is often indicated by the lowest category in most systems of soil classification, usually known as the soil series, which is used to designate map units at scales of 1:25,000–1:50,000. Soil series are often defined more strictly than higher soil categories in terms of properties important for drainage design, including:

- texture and structure (and therefore hydraulic conductivity) of individual horizons;
- approximate depth of permeable upper horizons to slowly permeable subsoil horizons;
- approximate depth to hard bedrock that may prohibit installation of drains or mole channels; and
- mineralogical characteristics that may lead to shrinking and swelling (Clayden and Hollis, 1984).

However, maps showing the distribution of soil series have several limitations for the design of drainage systems. First, the map units are rarely pure, as they often contain small areas of soil series other than the one by which they are named. Second, the map unit boundaries may not be very precisely located. Third, depths to changes of texture and structure can vary quite widely within some soil series, yet are important in drainage design, for instance in distinguishing moleable and non-moleable soils. Consequently, general purpose soil maps are no substitute for a full site investigation at the planning stage of an expensive drainage scheme. However, with suitable experience a drainage engineer can use them to decide on the main aspects of design for small schemes. Once physical measurements, such as pore-size distribution and hydraulic conductivity, have been made for several representative profiles of a particular soil series, and inferences from these have been tested in terms of drainage treatment, routine management and crop response, the results can often be applied to other areas of the same series where measurements are lacking, provided land use history and present management are also similar. In these circumstances, detailed soil maps can provide a cheap alternative to the often difficult and expensive surveys of properties such as K_s.

4.4.11 Environmental effects of artificial soil drainage

In the twentieth century, the availability of government grants for land drainage in the UK led to a rapid increase in the area of drained land. This, in turn, led to concern about possible environmental impacts of field drainage, in particular:

- an increase in frequency and severity of flooding in low-lying areas (Robinson et al., 1985);
- the loss of valuable wetland ecosystems; and
- the possible increase in contamination of surface waters by nutrients and agro-chemicals.

The first two are considered here, and the last is discussed in Chapter 5.

As drainage systems transfer large amounts of water to ditches, streams and rivers, one might expect them to contribute to flooding in the lower reaches of valleys. However, the risk of downstream flooding depends on various rainfall and catchment characteristics (Parkinson and Reid, 1987). Field drainage maintains a lower water table level in the soil profile throughout the year. This limits the amount of water held in higher soil horizons, which have greater hydraulic conductivity, and thus lowers the maximum rate at which water is transferred from the soil to the arterial drainage system. After the installation of a field drainage system, lateral water movement occurs mainly in deeper horizons, where flow rates are generally less. The greater air capacity of drained subsoils also acts as a buffer in the movement of rainwater to streams, because it increases the volume of water that can be stored temporarily in the soil during periods of heavy rain, and thus delays the onset of surface runoff. Although the total annual flow to streams from drained soils is therefore greater than from undrained soils, it is spread over much longer periods of time. The flow is less peaky, and the short periods of high flow rates that lead to flooding during and immediately after heavy rain are in fact minimized.

These effects of land drainage were confirmed by monitoring of surface flow and drainflow over 10 years (1978/79–1987/88) in the hydrologically sealed plots of the Brimstone Farm Drainage Experiment, Oxfordshire (Harris *et al.*, 1993), which then had ten drained and ten undrained plots. With mole drains at 2 m lateral spacing, the mean depth to the water table in winter was increased from 215 mm in undrained land to 435 mm in the drained plots. Total runoff was often less for drained than undrained plots. Over the 10 years the surface flow from undrained plots was 58–76 per cent of the total runoff, but 5–13 per cent for drained plots. Individual winter storms resulted in much shorter periods of total runoff but considerably greater peak flow rates on the undrained than on the drained plots.

From these and other field experimental results it is clear that artificially improved drainage of agricultural land usually prevents rather than causes flooding. The increased frequency of flooding in the middle and lower reaches of many British river valleys, for example in the very wet autumns and early winters of 2000 and 2002, can be attributed mainly to the extensive areas of impermeable cover, such as roofs, roads and car parks, associated with inappropriate development of floodplains for housing and light industry in the late twentieth century. The increase in northern hemisphere cyclonic storm activity and precipitation intensity resulting from recent climatic change may also have increased autumn/winter rainfall (Section 9.2.1), and in the lowest estuarine areas flooding is being exacerbated by the sea-level rise associated with global warming (Section 9.4). Other recent changes in British lowland agriculture (loss of ditches, ponds and hedgerows) may have contributed to flooding by decreasing the amount of water that can be stored in the landscape, which has shortened catchment response times to rainfall events. Farmers should reverse many of these changes in and near flood-prone areas. Flooding in the lower reaches of many rivers can also be minimized by the construction of dams in higher reaches and by installation of drains to carry runoff from roads and car parks directly into aquifers rather than into arterial surface drainage systems.

Wetlands are ecologically valuable because of their contribution to biodiversity, supporting natural plant, insect, reptile and bird communities that depend on saturated soil conditions for much of the year. Land drainage has certainly decreased the area of lowland peat in Britain by accelerating peat wastage (Section 8.7.2). Also, since the mid-twentieth century, much peat and its associated plant and animal communities have been lost from many areas, such as the Somerset Levels, Welsh Borderlands, Humberside lowlands and The Fenland, by commercial cutting for production of horticultural composts. However, this additional demand on wetlands need not continue now that suitable renewable alternatives, such as coir husk of coconuts, wood waste and composted bracken, are becoming available at costs similar to those of peat.

The need for artificial drainage and cultivation of peat and other wetland soils is also questionable. Before drainage these areas were typically under pasture or woodland, which would be more appropriate land uses than arable cropping in the twenty-first century. During and after the First and Second World Wars there was a need to maximize production of arable crops, especially cereals, and this led to extensive conversion of pastures to arable land and the artificial drainage of many wetland areas. However, the need has now passed and there are good reasons for concentrating the production of arable crops on the naturally better-drained soils. The existing drainage systems of peat and other wetland soils should therefore be allowed to deteriorate so that the ecological value of these areas can return.

The unpredictability of drain performance in peat soils and the possible precipitation of iron ochre in pipes and drain outflows provide further reasons for avoiding underdrainage of peats. Ochre is a rusty-coloured accumulation of iron oxides and hydrated oxides, often bound together in a felted mass with filamentous bacteria, which forms when iron is precipitated in the drain water. The Fe in solution is released either directly from peat as it is humified following drainage or is derived from oxidation of pyrite (FeS_2) in marine muds occurring below or within the peat. When precipitated it forms a soft gelatinous mass, which in time may dry and harden into a scale that block pipes, filters and taps. As the oxidation of pyrite also produces sulphuric acid (Section 6.1), drainage of peat and associated pyritic sediments often leads to severe acidification and consequent habitat damage in ditches and other surface water bodies.

Summary

For adequate crop growth, soil water content should be maintained at, or just below, field capacity (the water retained against gravity in pores $<60\ \mu m$ across). When the water content during the growth cycle falls well below field capacity (*FC*), creating a soil moisture deficit (*SMD*), as in dry spring or summer periods when the crop is transpiring rapidly, irrigation must be applied to return the soil to approximately *FC*. However, over-irrigation (water content $>$ *FC*) should be avoided, as it encourages pests and diseases and results in loss of nutrients by leaching or surface runoff. To calculate the irrigation requirement, *SMD* can be determined either by measuring soil moisture content or, more simply, by estimating daily soil water loss by balancing daily rainfall against evapotranspiration judged from the extent of crop cover. In poorly drained soils, where the natural water content semi-permanently exceeds *FC*, leading to poor root development and other problems, the excess water can be removed by installation of a field drainage system with outfalls into ditches and streams. Two types of field drains are now commonly used: perforated plastic pipes and mole channels. To increase the efficiency of these, the hydraulic conductivity of soil surrounding the drains can be increased by inserting gravel above the drain or by disturbing the subsoil (subsoiling). Field drains increase the total annual flow of water to streams and rivers and spread the flow over longer periods of time than is achieved by surface runoff, so that downstream flooding is prevented. However, drainage of ecologically valuable wetlands, such as areas of peat, should be avoided as it creates environmental and infrastructure problems.

FURTHER READING

Castle, D.A., McCunnell, J. and Tring, I.M. (eds), 1984, *Field Drainage Principles and Practices*, London: Batsford Academic and Educational.

Childs, E.C., 1969, *An Introduction to the Physical Basis of Soil Water Phenomena*, London: John Wiley and Sons.

Hanks, R.J., 1992, *Applied Soil Physics: Soil Water and Temperature Applications*, New York: Springer-Verlag.

Kutilek, M. and Nielsen, D.R., 1994, *Soil Hydrology*, Cremlingen-Destedt, Germany: Catena Verlag.

5 Chemical and microbiological pollution of soil and water

The soil must be man's most treasured possession: so he who tends the soil wisely and with care is assuredly the foremost among men.

George Stapleton

Introduction

Stapleton was thinking essentially of soil as a medium for growing crops. However, his remarks apply equally to another role of soil – the extent to which it can absorb, generate or transmit pollutants that may affect the health of humans, animals and plants. Soils may be polluted by artificial organic or inorganic chemicals originating from a range of industries, by excessive amounts of nutrients such as N and P from agriculture, by pathogenic micro-organisms introduced mainly by inputs of infected organic matter, or by dangerous gases from natural sources, which are often concentrated locally by human activities. Small amounts of these materials spread over large areas of soil usually present no health problems, but higher concentrations in limited areas can do so, resulting in pollution incidents. Some pollutants are effectively fixed or degraded to harmless materials in the soil, but others can be transferred to food, drinking water or the atmosphere. Soils therefore play two opposing roles, as either sinks or sources of pollutants, depending on the nature of the pollutant and its reaction with various inorganic and organic (living or dead) soil components. This chapter considers some of the more important pollutants found in soil, their likely sources, their interactions with soil components and some of the health problems they are known to cause.

5.1 WHAT IS POLLUTION?

Pollution is defined primarily in chemical and biological terms as the presence of any element, ion, compound or organism at a concentration increased above the natural background level as a result of human activities. These higher concentrations often pose threats to human, animal or plant health. Where they do not, the elevated concentrations can be described by the less emotive term contamination. The waste products of many human activities, such as human and animal excrements, crop and food residues, ashes from fires and the by-products of various industries, have been deposited on or in the soil for thousands of years. With the development of towns, villages and farmsteads, most waste products were concentrated in small areas rather than dispersed over the entire landscape. However, in recent centuries the expansion of agriculture, the increasing use of agrochemicals and inputs to the atmosphere from power stations, other industrial sites and motor vehicles have all spread some pollutants more widely.

As carriers of pollutants introduced by many human activities, soils can transfer them to food, water, wildlife or the atmosphere. At the same time soils are an important sink for many pollutants, which can be immobilized by precipitation or sorption on soil particles, degraded by soil micro-organisms or taken up by non-food plants. In the UK, pollution limits defining critical concentrations for industrial contaminants in the soil have been set by the Inter-Departmental Committee on the Redevelopment of Contaminated Land (1987).

5.2 ORGANIC FARM WASTES

Animal wastes (dung and urine) have always been incorporated into the soil. Those from wild animals and free-ranging domesticated animals are widely dispersed and, after they fall on the soil surface, they are incorporated mainly by natural mixing processes (pedoturbation) and leaching by rainwater. However, concentrations often occur in small areas, such as farmyards or stables, where domesticated animals are penned for protection.

5.2.1 Farmyard manure and slurry

Farmyard manure (FYM) or stable manure, a solid mixture of dung and urine with the straw or hay used for bedding, is created where animals (mainly cattle and horses) are overwintered in covered areas. It has been used to increase the fertility of agricultural soils for many centuries. Conventionally it is stockpiled and exposed to the weather for 1–2 years to initiate bacterial decomposition before being spread and incorporated into arable land. The N, P and mineral elements (mainly K, Mg, Na and Ca) in the manure are then released for uptake by arable crops during further oxidative decomposition (mineralization) within the soil.

In addition to solid manure, large amounts of liquid manure (slurry) are also produced by cattle and pigs and by washing out the buildings used to house them. Slurry is mineralized more rapidly than FYM and is usually applied directly to land without prior weathering. Some is applied to arable land but most is generated in areas of dairy, beef and pig production and is consequently applied to pastures to avoid transportation costs. In these situations, most is spread on the grass in winter or early spring and is not incorporated into the soil.

In arable farming, manure and slurry are usually spread before ploughing and sowing, and are then incorporated into the topsoil. This means that most of the abundant N, P and other nutrients they contain are being released by mineralization during or even before the period of germination and early growth, when little or none is required by the crops. For example, in temperate regions such as Britain, manure is typically spread in autumn before sowing winter crops such as cereals or oilseed rape, which grow quite slowly for several months during the coldest part of the year. Alternatively, the manure is ploughed in and the land left as bare fallow over winter before crops are sown in spring. In either situation, there is little immediate uptake of the nutrients for crop growth (Figure 5.1), and large amounts, particularly of N, are likely to be lost by leaching to groundwater or through field drains to ditches, streams and other surface water bodies. The problem is greater with slurry than with

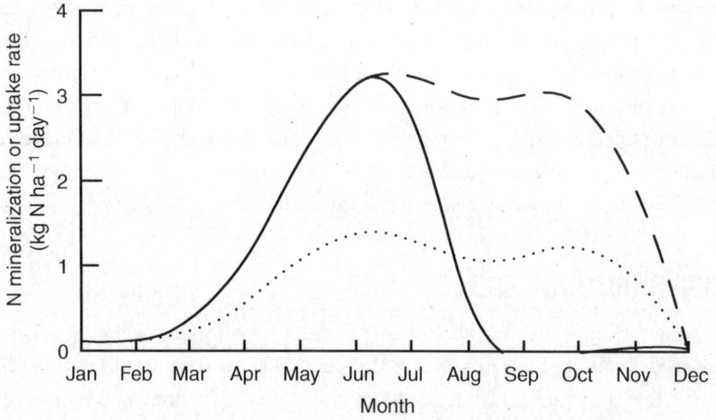

Figure 5.1 Comparison of N mineralization rate (dotted line) and N uptake rates by winter wheat (solid line) and grass (dashed line) in Britain
Source: Powlson (1993).

FYM because it contains a greater proportion of soluble N that is immediately ready for either plant uptake or leaching (Smith *et al.*, 2002). A frequent result is the pollution of either groundwater or surface water by nitrate and smaller amounts of labile (dissolved, finely particulate and adsorbed) phosphorus. Both have occurred extensively in The Netherlands, where large amounts of animal manures have been incorporated into arable soils for many years.

The microbial processes in soil responsible for mineralization of manure, slurry or other soil organic matter are influenced by both soil temperature and moisture content. Mineralization ceases when the soil is frozen, but above 0°C increases in moisture content probably have a greater effect than increases in temperature. Periods when soil moisture content is near to or exceeds field capacity are also essential for leaching of the nitrate produced by mineralization of organic N, and this explains why nitrate losses in humid temperate regions occur principally in winter. Intra-seasonal wetting and drying cycles are also thought to stimulate mineralization of organic N (Powlson *et al.*, 2001b).

In the UK, about 70 million tonnes of manure and slurry are applied annually to about 16 per cent (600,000 ha) of the arable land and 48 per cent (2.3 million ha) of the grassland. Even on grassland, more is applied in autumn and winter than in spring and early summer when it would most benefit growth. This is because few farms have adequate storage capacity, and in many areas the heavy equipment for spreading presents greater trafficability problems in spring than at times when the soil is drier. The large amounts of N, P and K in these organic manures, estimated by Chambers *et al.* (1999) to be countrywide totals of about 340,000, 90,000 and 250,000 t yr^{-1}, respectively, have considerable value as nutrients for production of arable crops and grass. In fact, the British soils given manure applications receive on average 170 kg N ha^{-1}, 30 kg P ha^{-1} and 90 kg K ha^{-1}. If applied in spring, these amounts would be almost sufficient for optimal production of most crops. However, until quite recently, most British farmers have ignored these nutrient contributions and have applied large amounts of artificial fertilizers as well, with the result that unnecessarily large amounts have accumulated in the soil or, in the case of N, have been lost by leaching of nitrate or by emissions to the atmosphere of ammonia and other gases.

Worldwide, livestock excrete annually about 94×10^6 t N, 21×10^6 t P and 67×10^6 t K (Sheldrick *et al.*, 2003). Of this approximately 60 per cent is from cattle, 10 per cent from pigs and 9 per cent from poultry. However, the amounts collected and applied annually to soil as manure (34×10^6 t N, 8.8×10^6 t P and 22.9×10^6 t K) are less than half of these quantities, so large amounts are lost as gaseous emissions to the atmosphere, runoff to rivers and the sea and leaching to groundwater. Most of these losses are from wild animals and domesticated farm animals kept outdoors. Losses are more controllable indoors, where the waste can be collected and then transported to spread over a large area of arable land. However, only a small proportion of the world's total animal population is raised indoors. In underdeveloped countries, farmers can rarely afford the extra cost of erecting and maintaining buildings, and in developed countries the consumer demand for 'naturally raised' animal products has often favoured a move away from indoor units.

5.2.2 Intensive outdoor animal units

Where animals are kept outdoors at high stocking densities, two main problems arise. First, manure and urine inputs to the soil are concentrated in the penned areas, and are often localized in 'hotspots' around sites for feeding and shelter. Many outdoor units are deliberately sited on well-drained soils to prevent the foot problems that arise when animals are treading on soil saturated in wet periods. However, these soils often overlie major aquifers, which consequently become polluted beneath the outdoor units. Second, uptake of N and other nutrients from the soil is less than on arable land because the growth of grass is restricted by intense treading and grazing. Some outdoor pig units, for example, become completely devegetated within a few months (Worthington and Danks, 1992).

In these circumstances inputs in feed greatly exceed N offtakes in meat and other products, so the soil N content increases (Figure 5.2) and losses of N by nitrate leaching and by emissions of gaseous

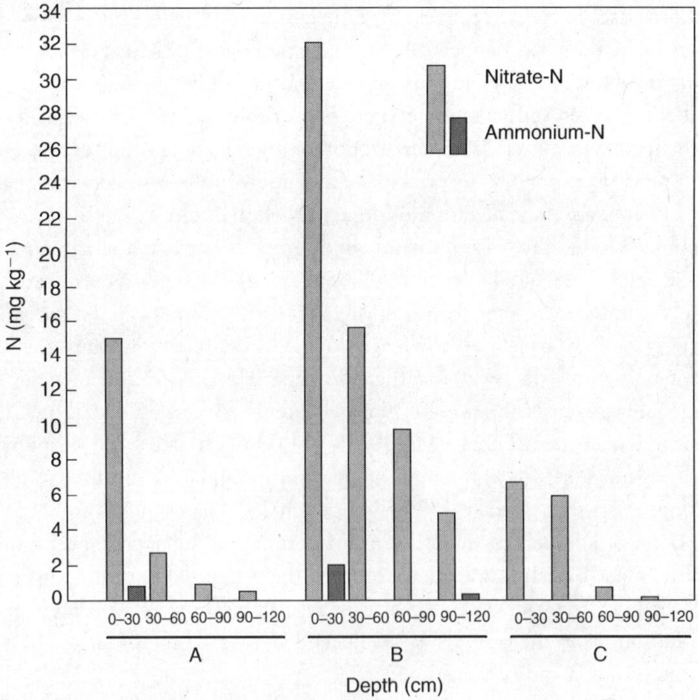

Figure 5.2 Amounts of soil nitrate-N and ammonium-N to 120 cm depth in sandy loam soil following A: newly introduced outdoor pigs following four years of growth of cereals; B: two years outdoor pigs following two years of cereals; and C: two years of cereals following two years of vegetable production
Source: Worthington and Danks (1994).

ammonia and nitrous oxide can exceed 200 kg ha^{-1} yr^{-1} (Table 5.1). The problems associated with outdoor units can be minimized by keeping the animals in small individual enclosures (Eriksen and Kristensen, 2001), and frequently moving them onto fresh well-vegetated areas, preferably where substrata are less permeable. In addition, food rationing for livestock using automated feeding systems can help match the amount of N in food more precisely to that required for protein production, so that less is lost in manure.

5.2.3 Human health problems associated with nitrate in drinking water

Excessive nitrate in drinking water has been associated with two main human diseases, cancer of the stomach/oesophagus and methaemoglobinaemia ('blue-baby syndrome', or inability of the blood to transmit sufficient oxygen in infants up to about six months old). These concerns led to World Health Organization and European Union (EU) regulations that stipulate maximum concentrations of 45 and 50 mg NO_3 l^{-1} (10 and 11.3 mg NO_3-N l^{-1}), respectively, in drinking water. However, the connection between nitrate in water and these health problems remains uncertain. One factor leading to doubt is that fresh vegetables, such as lettuce, turnip and spinach, are often very rich in nitrate, so that people can ingest far more from these than from nitrate-rich water.

More significantly, recent medical research has shown that nitrate, and the bactericidal gas nitric oxide (NO) produced from it in the stomach, are essential for many aspects of good human health (L'Hirondel and L'Hirondel, 2002). They are essential for transmitting information from cell to cell, for regulating blood pressure, for limiting the function of platelets in blood-clotting (thrombosis) and as an anti-infective defence mechanism, in particular to combat bacterial gut infections likely to cause gastro-enteritis, such as *Salmonella enteritidis*, *Shigella sonnei*, *Yersinia enterocolitica* and *Escherichia coli*. Nitric oxide combats gut infections when it is generated in the stomach through the action of the strong stomach acid (pH 1–3) on dissolved nitrite, which is produced in the mouth and throat by bacterial reduction of nitrate (Dykhuizen *et al.*, 1996). Some of the nitrate may come directly from dietary sources (water, vegetables, etc.), but this protective strategy is so important that nitrate in the mouth is more consistently supplied from the blood by secretion through the salivary glands.

Table 5.1 Two-year nitrogen balances for outdoor pig-farming by current commercial practice compared with best management practice to decrease N losses and a typical arable field

N balance	Current commercial practice (25 sows ha^{-1}, cereal stubble)	Best management practice (12 sows ha^{-1}, established grass)	Arable control (fertilized winter cereals)
Feed/fertilizer inputs	1164	558	320
Atmospheric inputs	50	50	50
Leaching losses	192	126	52
Gaseous NH_3 losses	200	96	-
Gaseous N_2O and N_2 losses	10	5	-
Retention in sows	236	116	-
Crop N offtake	-	-	291
N accumulated in the soil	576	265	27

Source: from Williams *et al.* (2000).

Stomach and oesophagal cancer was originally thought to be caused by *N*-nitroso compounds, such as nitrosamine, formed by reactions in the stomach between NO and secondary amines released during digestion of protein. In rats *N*-nitroso compounds are known to cause mutations in DNA that lead to cancer. However, there is no correlation between the nitrate content of drinking water and human stomach cancer (Beresford, 1985), nor is there one between the nitrite content of saliva and people living in areas of high stomach cancer risk (Forman *et al.*, 1985). Also Al-Dabbagh *et al.* (1986) found no increase in incidence of the disease in fertilizer factory workers exposed to high levels of nitrate; in fact these workers showed abnormally low death rates from heart diseases, which were attributed to the suppression by nitrate of platelet function in the formation of blood clots. Further, numerous other epidemiological and case-control studies have found negative rather than positive correlations between gastric cancer and dietary nitrate intake. *N*-nitroso compounds occur in tobacco smoke and many smoked foods (meat, fish and cheese), and are also produced by natural processes of N metabolism in the body, so any link between cancer and *N*-nitroso compounds formed from nitrate in drinking water is very difficult to establish.

A causal relationship between nitrate and blue-baby syndrome has also been questioned recently (Avery, 1999), though any possible problem has been eliminated for decades by feeding infants low-nitrate bottled water. Methaemoglobinaemia is usually defined as a condition in which at least 10 per cent of the haemoglobin in blood cannot carry oxygen or carbon dioxide because it has been oxidized to methaemoglobin by a rapid reaction with nitrite (Kosaka and Tyuma, 1987):

$$4HbO_2 + 4NO_2 + 4H^+ \quad 4Hb^+ + 4NO_3 + O_2 + 2H_2O \qquad (5.1)$$

in which HbO_2 is oxyhaemoglobin and Hb^+ is methaemoglobin. Related processes occur continuously in healthy red blood cells, and are normally reversed by the enzyme methaemoglobin-reductase. At any one time <2 per cent is normally in the methaemoglobin form in adults but, in infants, the enzyme has not reached full activity so that methaemoglobin can accumulate. A threshold of about 10 per cent methaemoglobin causes cyanosis, and levels >60 per cent are fatal unless quickly treated by intravenous injection of methylene blue.

Infant methaemoglobinaemia induced by dietary nitrate is thought to be unlikely, because infants have a very limited ability to reduce salivary nitrate to nitrite. More often the problem results from enteritis, in which diarrhoea induces endogenous nitrite production in the blood. Most investigated cases of methaemoglobinaemia have in fact been shown to result from enteritis caused by infected baby food, and others have occurred in rural parts of the USA and eastern Europe where water supplies were taken from infected private wells close to farmyards or household cesspits. So good hygiene in the preparation of baby foods is probably much more important in avoiding blue-baby syndrome than the nitrate content of the water used. Other inconclusive associations between various health problems and nitrate in water, based on geographical correlations or experiments with animals, are reviewed by L'Hirondel and L'Hirondel (2002).

To meet the EU regulation for nitrate in drinking water, water companies have blended supplies containing >50 mg NO_3 l^{-1} with those containing less. However, this relatively cheap option is becoming increasingly difficult in areas where, for various reasons, the nitrate content of ground- or surface-water supplies is increasing. Also, new, more stringent EU drinking water and groundwater directives are expected.

In the UK, purification of drinking water, principally to remove nitrate and pesticides, costs about £7 per customer annually, but this is forecast to rise by two to three times in the next few years as it becomes increasingly necessary to use methods of chemical or microbiological water treatment, which are more expensive than blending. The necessity for these expensive treatments in terms of human health and the proportion of dietary nitrate in drinking water almost certainly needs more careful evaluation.

5.2.4 Eutrophication of surface waters

In stagnant surface water bodies (ditches, lakes, reservoirs and even parts of the sea close to river mouths) the N and P pollution leads to the problem of eutrophication, in particular the rapid spring and summer growth of Cyanobacteria and aquatic algae (dinoflagellates such as *Pfiesteria*), which form green or blue blooms on the water surface (Kotak *et al.*, 1993; Burkholder and Glasgow, 1997). When these organisms die, they sink and are aerobically decomposed within the water body, depleting it of oxygen and thus leading to the death of fish and other wildlife by oxygen starvation. Carcinogens are sometimes formed when water containing the dead organisms is chlorinated, and other micro-organisms, such as *Prymnesium parvum*, generate toxins, which can directly poison fish and other animals or humans who accidentally drink the water.

5.2.5 Avoiding atmospheric pollution by ammonia, nitrous oxide and odours

Large amounts of the N in slurry and farmyard manure are lost by direct gaseous emissions to the atmosphere. Most is lost as ammonia (NH_3) or nitrous oxide (N_2O). The NH_3 from all these sources contributes to poor air quality by reacting with other atmospheric components to form fine particles of compounds such as ammonium sulphate and ammonium nitrate. When it is redeposited, principally in rain, it also damages many natural plant communities and acidifies soils and surface waters. Plants characteristic of N-deficient environments, such as the mosses, lichens and heather of upland bogs, heathlands and some semi-natural grasslands and woodlands, have considerable conservation value but, when exposed to 'terrestrial eutrophication' by deposition of atmospheric NH_3, they are replaced by a limited range of more common, fast-growing species. In many of these British semi-natural habitats, the current rate of ammonium-N deposition exceeds the critical load for N (i.e., the amount known to cause harmful effects on specified elements of the environment). Manure and slurry contribute about 240 kt NH_3 yr^{-1} to the atmosphere; a further 30 kt comes from spreading of N fertilizers, notably urea, and a further 55 kt from numerous non-agricultural sources, such as sewage treatment, manufacture of fertilizers, landfill sites, vehicles fitted with catalytic exhaust converters, pets and wild animals (Department for Environment, Food and Rural Affairs (DEFRA), 2002).

Although most of the gaseous NH_3 is deposited on soil and vegetation close to areas of emission, ammonia compounds can be transported long distances in the atmosphere, so damage to natural vegetation can occur in very different parts of the country or even different countries well away from the emission sources. Consequently, international agreements are required to decrease the damage from NH_3 emissions. Those to which the UK is already bound are:

- the Protocol to Abate Acidification, Eutrophication and Ground-level Ozone (the Gothenburg Protocol) of the United Nations Economic Commission for Europe (UNECE) Convention on Long-Range Transboundary Air Pollution;

- the European Community (EC) National Emission Ceilings Directive; and
- the EC Directive on Integrated Pollution Prevention and Control (IPPC).

The first two set a target of 297 kt NH_3 yr^{-1} for total UK emissions in 2010 compared with the present total of about 325 kt; the Gothenburg Protocol also bans the use of ammonium carbonate as a fertilizer and limits the use of urea-based fertilizers; and the IPPC directive requires specific measures on housing for pigs and poultry and on manure storage and spreading.

Ross *et al.* (2002) estimated that a typical diary farm can emit 86 kg NH_3-N ha^{-1} yr^{-1} from FYM and 107 kg NH_3-N ha^{-1} yr^{-1} from slurry, which are about 25–50 per cent of the N commonly applied as fertilizer to arable land or pasture. With various abatement strategies, such as direct injection of slurry into the soil, ploughing manure or slurry into soil within a few hours of spreading, covering stockpiles with a rigid cover, minimizing the period of indoor housing of animals in winter and careful control of cattle diet, Ross *et al.* (2002) calculated that the losses could be decreased to 33 and 27 kg NH_3-N ha^{-1} yr^{-1}, respectively. However, some of these strategies are impractical on some farms. N_2O emissions from FYM and slurry are much smaller than those of NH_3 (Stevens and Laughlin, 1997), but are probably significant in the context of accelerated global warming (Section 9.3.8).

Another common environmental problem with application of animal manures to soil is odour. This can be minimized if the manure is liquefied to form a slurry and injected into the soil rather than applied to the surface. However, in well-structured soils, much of the manure may move below rooting depth and contaminate groundwater or surface water bodies (Shipitalo and Gibbs, 2000). In direct drilled or artificially drained soils, it is therefore important to avoid injecting slurry over the lines of field drains. With both surface-applied and injected slurry, it is also better to make numerous applications at low rates ($<35\ m^3\ ha^{-1}$), well spaced in time over the crop growing period, rather than a few applications at high rates (Parkes *et al.*, 1997).

5.2.6 Summary of value of organic manures

Organic manures are therefore a mixed blessing environmentally. They improve soil structure, which is important for improving infiltration, water retention, resistance to erosion and ease of cultivation. However, they are 'leaky' in that large proportions of the nutrients they contain become available at times when crops do not require them and are then likely to pollute ground- and surface waters and increase their chemical oxygen demand. Moreover, the ratio of nutrients in manures is often quite different from that required by many arable crops, so that long-term use of manures can lead to excessive accumulation of some nutrients in the soil. Repeated applications of organic manures increase the microbial biomass of the soil, which should improve nutrient cycling (Section 8.2.2). However, despite this and other physical and chemical indications of improved soil quality, there is no evidence from long-term field experiments that repeated manuring results in a significant increase in crop production compared with inorganic fertilizers (Edmeades, 2003).

Unless they are stored and applied carefully, manures are also a source of atmospheric pollutants, such as NH_3, N_2O and odours. The effects of organic manures on pollution of the atmosphere and water with N compounds arise from the fact that animals do not convert dietary N into protein very efficiently, with the result that the unwanted N is excreted. Dairy cattle, for example, retain only 16–23 per cent of dietary N, so the most promising approach to minimizing N losses from organic manures is probably that of dietary manipulation, such as altering the balance of rumen-degradable protein in cattle food (Smits *et al.*, 1995).

5.3 SEWAGE SLUDGE AND HEAVY METAL POLLUTION

There are similar problems with the use of sewage sludge as an organic manure, as it contains large amounts of N and P (Table 5.2), which can cause eutrophication. In industrial and densely populated regions, sludges may also have high concentrations of heavy metals (e.g., Cd, Cr, Cu, Hg, Mo, Ni, Pb, Zn), other toxic elements (As, B, Se) and synthetic organic chemicals, such as chlorobenzenes (CBs), polychlorinated biphenyls (PCBs) and polynuclear aromatic hydrocarbons (PAHs) (Section 5.5). The heavy metals are very persistent in the soil and can severely limit the activities of the soil microbial biomass that are important in nutrient cycling (Bååth, 1989). For example, Zn pollution from sludge reduces the populations of N-fixing bacteria (*Rhizobium* spp.) on which leguminous plants such as peas and beans depend for their supply of N. Estimates of the size of the soil microbial biomass (Section 8.1.4) can therefore indicate the extent of soil pollution by heavy metals (Brookes, 1995). As well as harming the microbial biomass, Zn and Cu pollution of soil can also reduce earthworm and fungal populations and directly inhibit the growth of some higher plants, so that the

Table 5.2 Composition of British sewage sludges (values for N and P are given at 25 per cent dry solids, the remainder at 100 per cent dry solids)

Element	Composition
N	7.5 g kg^{-1}
P	2.8–3.9 g kg^{-1}
As	<2–123 mg kg^{-1}
B	15–1000 mg kg^{-1}
Be	1–30 mg kg^{-1}
Bi	<2–557 mg kg^{-1}
Br	4–1049 mg kg^{-1}
Cd	<2–152 mg kg^{-1}
Co	<2–617 mg kg^{-1}
Cr	4–23,195 mg kg^{-1}
Cu	69–6140 mg kg^{-1}
F	60–40,000 mg kg^{-1}
Fe	2480–106,812 mg kg^{-1}
Hg	<2–140 mg kg^{-1}
Mn	55–376 mg kg^{-1}
Mo	<2–154 mg kg^{-1}
Ni	9–932 mg kg^{-1}
Pb	43–2644 mg kg^{-1}
Se	<2–15 mg kg^{-1}
V	7–660 mg kg^{-1}
Zn	279–27,600 mg kg^{-1}

Source: data from Smith (1996).

overall soil biodiversity and productivity are decreased. In grassland soils, Zn, Cd and Pb reduce the populations of Myriapoda (millipedes) and Isopoda (woodlice), thus decreasing the rate of litter breakdown and the microbial activity that depends on the litter (Grelle *et al.*, 2000).

The conflicting benefits and problems of sludge can only be resolved by limiting the amounts of heavy metal and organic pollutants added to land in sludges and by repeatedly monitoring the size and composition of the soil biomass. Amounts of pollutants can be minimized by controls on industrial discharges, by chemically detoxifying sludges before they are applied to the soil or by diluting toxic sludges with non-toxic supplies.

In the UK and many other countries, the producers of sewage sludge are required by law to ensure that concentrations of many metal contaminants (usually Cd, Cr, Cu, Hg, Ni, Pb and Zn) are kept below stated limits. For other elements (As, F, Mo and Se) there are recommended (not legally set) limits. Many countries also have legal or recommended limits on the concentrations of these elements in the soil, usually to a depth of about 15 cm on arable land but often to shallower depths (7.5 cm in the UK) under grass. They also have permitted average annual rates of addition over a stated period (e.g., 10 years). As the availabilities of most elements (except Mo and Se) increase in acidic soils, sludge should not be applied to soils with pH values <5.0, and different limits for some elements (e.g., Cu, Ni and Zn in the UK) apply according to various soil pH ranges >5.0.

Manures from intensive pig- and poultry-rearing units may also cause pollution of soils with Cu and Zn, which are used in dietary supplements to improve the growth rate of these animals (Han *et al.*, 2000). The Cu is initially complexed with soil organic matter (SOM) components and the Zn with ferric oxides and hydrated oxides, but both elements are likely to become bioavailable. The Cu may be released by microbial degradation of the SOM and the Zn by reduction of ferric compounds to more soluble ferrous compounds in seasonally anaerobic soil conditions. European legislation has limited the concentrations of Cu and Zn allowed in animal foods, and this has decreased the amounts in manures but, as these elements are very persistent in soil and can also occur in applications of sewage sludge, it is important to monitor the amounts already in the soil before animal manures are applied. Repeated use of some pesticides can also increase the Cu content of soils, and the same requirement for soil monitoring applies when these are used.

Localized and often very intense pollution of soil by heavy metals, metalloids, radionuclides or organic chemicals can result from mining activities, smelting and contamination with other industrial wastes. There are many examples of damage to plant communities or soil fauna resulting from these sources of pollution. In Europe the number of such contaminated sites is thought to be 1.4 million; in the UK alone there are approximately 100,000 with a total area of about 200,000 ha. Several years ago the cost of complete clean-up was estimated at £10–15 billion for the UK alone.

The amounts of some metals (Cd, Co, Cr, Cu, Hg, Mn, Mo, Ni, Pb, Zn) and metalloid (As, Se) contaminants present on many polluted sites exceed toxicity levels. Others (e.g., As, Cd, Hg, Pb) are not essential and are likely to cause problems even in quite small amounts. Localized Pb contamination occurs where the metal was mined and smelted, but it also occurs in soils adjacent to roads because the Pb previously added to petrol was emitted in vehicle exhausts and deposited from the atmosphere. Concentrations in soils close to busy road intersections in cities are often 20–100 times greater than background levels in remote rural areas. In some agricultural areas, especially long-established vineyards, the soil is contaminated with Cu originating from fungicides. The main radionuclides posing a threat are ^{127}Cs, ^{3}H, ^{90}Sr and the radioactive isotopes of U (Negri and Hinchman, 2000), resulting mainly from thermonuclear testing and accidental release.

5.3.1 'Hard' remediation techniques

Various 'hard' remediation techniques have been used at some of the most strongly contaminated mining and industrial sites, but they are often very expensive. They have included excavation of the contaminated soil and disposal at special deep landfill sites; containment within physical barriers; stabilization or binding of the soils by injection of cement, epoxy resins, asphalt or polyethylene to decrease the solubility of contaminants; wholesale incineration to remove organic pollutants; and fusing the soil into a glassy solid by heating to >2000°C with a strong electric current (Bridges, 1991a). Many of these treatments destroy soil structure, organic matter and biomass, so the decontaminated material can be used only as inert fill until these important properties have been restored.

5.3.2 Phytoremediation

'Soft' remediation techniques are about an order of magnitude cheaper than 'hard' techniques, but are much slower. The most promising is phytoremediation (McGrath *et al.*, 2002), which includes (a) stabilization of minespoil by planting metal-tolerant species, thus preventing dispersal of heavy metals by wind and water erosion, and (b) phytoextraction of the metals by hyperaccumulator species (McGrath, 1998).

Hyperaccumulator plants can take up large quantities of particular metals from the soil and yet remain healthy. For example, *Thlaspi caerulescens* (alpine pennycress) can accumulate more than 30,000 μg Zn g^{-1}, which is about 100 times the concentration tolerated by most other plants (Shen *et al.*, 1997); indeed its roots seem to be stimulated by high concentrations of soil Zn, so that they actually grow towards any Zn-polluted zones. In field experiments, Hammer and Keller (2003) reported greater uptake of Cd than of Zn by *T. caerulescens*, and greater uptake of both metals from an acidic than a calcareous soil. The mechanisms involved seem to be that root exudates mobilize Zn and Cd close to the roots and, after uptake, these metals are stored in epidermal cells in the leaves.

Hyperaccumulator plants cannot be simply reincorporated into the soil, as the accumulated metals are in very mobile forms (Perronnet *et al.*, 2000). The metals are therefore recovered by harvesting and burning above-ground parts of the plants, followed by chemical treatment of the ash, a process termed phytomining. Their dual value as fuel and sources of metals can make some hyperaccumulator plants quite valuable crops, especially if NPK fertilizers are used to stimulate growth. For example, Nicks and Chambers (1995) estimated that the profit from a crop of *Streptanthus polygaloides*, which accumulates Ni, can be more than five times that of a wheat crop grown on uncontaminated land.

Reeves and Baker (2000) list approximately 400 known species of possible accumulator plants able to take up a wide range of metals and metalloids, including As, Cd, Co, Cu, Ni, Pb, Se, Tl and Zn. Some can accumulate radionuclides (Negri and Hinchman, 2000). Accumulation of pollutants in above-ground parts of the plant is important, as these are harvested much more easily than roots. Also, many accumulator plants are perennials, and continuing uptake after removal of roots is possible only by reseeding.

Another form of phytoextraction is phytovolatilization, in which elements selectively extracted by plants, often with microbial assistance, are then volatilized in less toxic forms to the atmosphere through their leaves. In particular, Se can be phytovolatilized as dimethylselenide by field legumes such as *Brassica juncea* (Indian mustard) (De Souza *et al.*, 1999). Aquatic plants, such as *Typha latifolia* and *Scirpus robustus*, have also been used to remove Se from contaminated wetlands (Hansen *et al.*, 1998), and others can remove heavy metals from stream waters (Dushenkov *et al.*, 1995).

Phytoextraction of heavy metals from contaminated soil can sometimes be enhanced by adding reagents that mobilize the pollutant metals and make them more available to the hyperaccumulator plants. Examples of the latter include chelates such as ethylene diaminetetraacetic acid (EDTA) for uptake of Pb by *Brassica juncea* (Blaylock *et al.*, 1997), ammonium salts for ^{137}Cs (Ebbs *et al.*, 2000), citric acid for U (Huang *et al.*, 1998) and ammonium thiocyanate for Au and other noble metals (Anderson *et al.*, 1998). However, there are problems with their use. In particular, the mobilized metals may be leached and contaminate the deeper subsoil or groundwater. Chemically enhanced phytoextraction has therefore been restricted to well-contained sites or rural areas that are not strongly contaminated.

Micro-organisms can also be used to remove heavy metals from soil. For example, various strains of the bacterial species *Thiobacillus ferrooxidans* can oxidize and mobilize the sulphides of Cu, Co, Mn, Mo, Ni, Zn and U in mine waste. They require acid conditions (optimum pH near 2) with a suitably warm and moist aerobic environment supplied with sufficient N, P and K as nutrients.

5.3.3 Other 'soft' remediation techniques

Other 'soft' remediation techniques for contaminated soil involve immobilization of metals by applications of lime, phosphate, zeolites, organic polymers or Al/Fe oxyhydroxides, oxidation/reduction to less toxic forms of the contaminant and extraction with strong acid or alkali. The contaminated soil is usually excavated and mixed on-site with the chemical to form a slurry within a reactor vessel. Only small volumes of soil can be treated in these ways, so the methods are used mainly in areas affected by minor spillages. Important soil properties such as structure and the microbial biomass may be destroyed, so the treated 'soil' is best used as inert fill.

Polyacrylate polymer has been used for immobilization of Cu in soil contaminated with Cu fungicides (De Varennes and Torres, 1999). Al/Fe oxyhydroxides can immobilize Cu, Zn and Cd, and improved soil microbiological activity has been noted at sites treated in this way. Leaching with strong acids (HNO_3, HCl or H_2SO_4) can extract some heavy metals and treatment with alkaline solutions (NaOH or Na_2CO_3) can remove metals held in complexes with organic matter. Afterwards both need to be flushed from the soil with excess water to restore the pH to a level suitable for plant growth. Lime is a more persistent alkali and, as in uncontaminated soils (Section 6.4), must be applied in carefully calculated amounts; otherwise a high pH can cause deficiencies of P, K, Mg and some micronutrients in subsequent crops.

5.4 INORGANIC FERTILIZERS

Over the last century or so the use of fertilizers has become a routine part of most farming systems, largely replacing applications of organic manures such as FYM. In the mid-nineteenth century the Rothamsted Classical Experiments demonstrated the value of small quantities of relatively cheap artificial fertilizers, which contain nutrients in fairly pure water-soluble forms, for growing a range of arable crops. For example, on the Broadbalk Winter Wheat Experiment (1843–present) the largest yields of grain on plots given artificial fertilizers have been consistently equal to those from the plot treated with FYM. The FYM has been applied at the high rate of 35 t ha^{-1} yr^{-1}, whereas for most of this period the artificial fertilizers consisted of 144 kg N ha^{-1} yr^{-1} (originally as ammonium sulphate, subsequently as calcium ammonium nitrate or ammonium nitrate) with smaller amounts of P (as superphosphate or calcium phosphate made more soluble by treatment with sulphuric acid),

K, Mg and Na (all as simple salts). Jenkinson (2001) calculated that, because of the large increase in yield compared with no fertilizer, the cost of applying artificial N at the 144 kg ha^{-1} yr^{-1} rate to a wheat crop is only one-fifth of the resulting increase in crop value. The N applied also increases the protein N content of the grain in the approximate ratio 1:1.5.

The Rothamsted and other field experiments have indicated the various proportions of major nutrients (mainly N, P and K) needed for the economic optimum growth of various crops in different soil and climatic situations. Usually the nutrients are applied as a compound fertilizer containing N, P and K in the requisite ratios, as this is cheaper and less labour-intensive than the multiple application of single nutrients. It also minimizes soil structural damage through repeated trafficking.

Artificial fertilizers have major environmental advantages over organic manures such as FYM. First, they do not introduce pathogenic organisms into the soil, which can contaminate crops and water supplies (Section 5.7). Second, there is no odour problem and there is less loss of N by emission of ammonia to the atmosphere or leaching of nitrate (Powlson *et al.*, 1989). Third, they can be applied as a top dressing to the soil surface at the time of year when they are most needed, i.e., when the soil temperature in cooler climatic zones is increasing in the spring and the crop is beginning to grow rapidly. Broadcast into the growing crop, the fertilizer is dissolved during the next rainfall event, the nutrients are carried down in solution and are then easily taken up by the roots of the crop. The nutrients are therefore available much more rapidly than those in organic manures, which have to be mineralized by relatively slow microbial processes before the nutrients become available to the crop (Section 5.2.1). This means there is much less time during which leaching losses can occur, unless there is heavy rain soon after fertilizer-spreading and the fertilizer is washed quickly down fissures or other soil macropores to depths beyond reach of the crop's roots. Using fertilizer labelled with ^{15}N, Macdonald *et al.* (1989) showed that as little as 1–2 per cent of the fertilizer N applied at the correct rate to winter wheat in spring remains in the soil at harvest. In contrast, large amounts of autumn-applied N (in either organic or inorganic forms) are lost by leaching and denitrification during the autumn and winter before they can be taken up by the crop in spring.

Despite these benefits, artificial fertilizers do cause contamination of ground- and surface waters, mainly by nitrate and phosphorus. In surface waters this leads to eutrophication and decreased biodiversity (Vitousek *et al.*, 1997).

5.4.1 Contamination of water by nitrogen from fertilizers

N can be applied to soil as nitrate (originally sodium nitrate), as an ammonium salt (originally ammonium sulphate), as a combined nitrate/ammonium compound (e.g., ammonium nitrate or calcium ammonium nitrate) or as urea. In the soil, ammonium and urea must be microbially oxidized to nitrate before the N can be taken up by plants. This process (nitrification) can take up to several weeks, depending on temperature. Ammonium is also generated by the slower microbial decomposition of soil organic matter (ammonification). Consequently, as well as nitrate, soils usually contain some ammonium, though the amounts can vary widely. Nitrate and ammonium together form the pool of soil N which is termed 'mineral N'.

A problem with respect to N fertilizers is that application to the soil of more than arable and horticultural crops can take up leaves a surplus of mineral N, which is at risk of leaching (Chaney, 1990). In the last 20 years improvements in practices on commercial arable farms in Britain have reduced this N excess from a maximum of about 70 kg ha^{-1} yr^{-1} to around 25 kg ha^{-1} yr^{-1} (Goulding, 2000), but have not eliminated it. In the Broadbalk Winter Wheat Experiment at Rothamsted, leaching losses of nitrate from plots with known fertilizer treatments can be measured in drainwater collected from specially constructed drain outfalls (Plate 5.1). Where up to 144 kg N ha^{-1} are applied in spring the leaching losses are usually <30 kg ha^{-1} yr^{-1}, and the drainage water usually contains less than the EU Drinking Water Limit of 11.3 mg NO_3-N l^{-1} (Goulding *et al.*, 2000). However, much larger annual losses of nitrate (up to 140 kg NO_3-N ha^{-1} in some years) have been recorded on plots given inorganic fertilizer applications >144 kg N ha^{-1} yr^{-1} (the smallest amount giving near-maximum yield, Figure 5.3) and also those given autumn applications of FYM

Plate 5.1 Trench at the downslope end of the plots of the Broadbalk Winter Wheat Experiment at Rothamsted, showing drain outfalls for plots given known fertilizer applications, which have remained constant since 1843. Drainwaters collected from each outfall enable losses of N and P to be estimated (photo J.A. Catt)

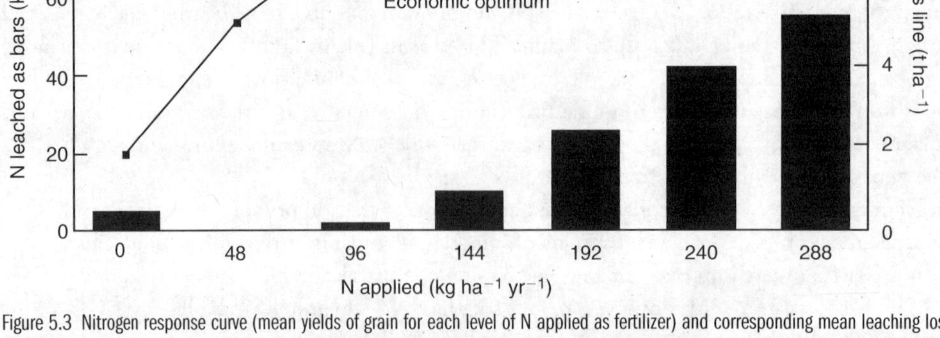

Figure 5.3 Nitrogen response curve (mean yields of grain for each level of N applied as fertilizer) and corresponding mean leaching losses of nitrate-N for plots of the Broadbalk Winter Wheat Experiment at Rothamsted, in which N treatments have remained constant since 1843
Source: Goulding (2000).

(Table 5.3). On all the Broadbalk plots, even the nil plot, which has never received any fertilizer N since 1843, the highest concentrations of nitrate, often exceeding 11.3 mg NO_3-N l^{-1}, occur mainly in the first drainflow of winter after dry summer and autumn periods. In these periods of summer and autumn drought, nitrate accumulates in the soil because large amounts are formed by mineralization of organic matter, but little is taken up because the crop is no longer growing.

Farmers cannot simply decrease N applications too enthusiastically because, for cereals and many other crops, even small reductions of applied N can sometimes change a profit into a loss. The profit on many arable crops comes, in fact, from the last few per cent of the yield response to added N (Figure 5.4). The farmer therefore needs to know crop yields in advance to judge required N inputs exactly, but yields are also influenced by weather, pests and diseases, and these cannot be predicted with any certainty. The calculation is also complicated by inputs to the soil nitrate stock from the atmosphere and from mineralization of soil organic matter even where no organic N is added. The amounts of N deposited from the atmosphere vary with crop type and the extent of local atmospheric pollution with N compounds, but are often >40 kg ha^{-1} yr^{-1} (Goulding *et al.*, 1998a) and can contribute >40 per cent of the nitrate leached (Goulding *et al.*, 1998b).

The nil plot of the Broadbalk Winter Wheat Experiment, which has received no fertilizer input since 1843, still produces approximately 1.5 t grain ha^{-1} yr^{-1}, and the nitrate lost through the plot drain is currently 13 kg N ha^{-1} yr^{-1} (Table 5.2). The nitrate leached and in the crop must come partly from the atmosphere but some results from mineralization of crop residues. In the Rothamsted 'drain-gauges', which are *in situ* lysimeters of previously arable soil maintained since 1871 as bare, unmanured and unfertilized microplots, the amount of nitrate leached annually took 41 years to decrease to half that averaged over the first seven years, i.e., 45 kg NO_3-N ha^{-1} yr^{-1} (Addiscott, 1988). Most of this nitrate would also have come from mineralization of organic matter incorporated into the soil before 1871.

Under pasture the nitrate formed by mineralization of soil organic matter can exceed 300 kg N ha^{-1} yr^{-1} (Bhogal *et al.*, 2001) and, after ploughing up grassland or incorporating crop residues or other fresh organic matter, the amounts are greater than under continuous arable rotations supported by artificial fertilizers. Crop residues with C/N ratios <15 (Table 5.4) mineralize more rapidly than those with less N, especially in tropical conditions. Of the N released in this way (gross mineralization)

Table 5.3 Mean annual leaching losses of nitrate-N for the eight years 1990/1–1997/8 from the Broadbalk winter wheat plots at Rothamsted

Plot treatments	N leached (kg N ha^{-1})
No N	13
48 kg N ha^{-1} as fertilizer	12
96 kg N ha^{-1} as fertilizer	15
144 kg N ha^{-1} as fertilizer	22
192 kg N ha^{-1} as fertilizer	30
240 kg N ha^{-1} as fertilizer	45
288 kg N ha^{-1} as fertilizer	48
265 kg N ha^{-1} as FYM	59
265 kg N ha^{-1} as FYM + 96 kg N ha^{-1} as fertilizer	77

Source: data from Goulding *et al.* (2000).

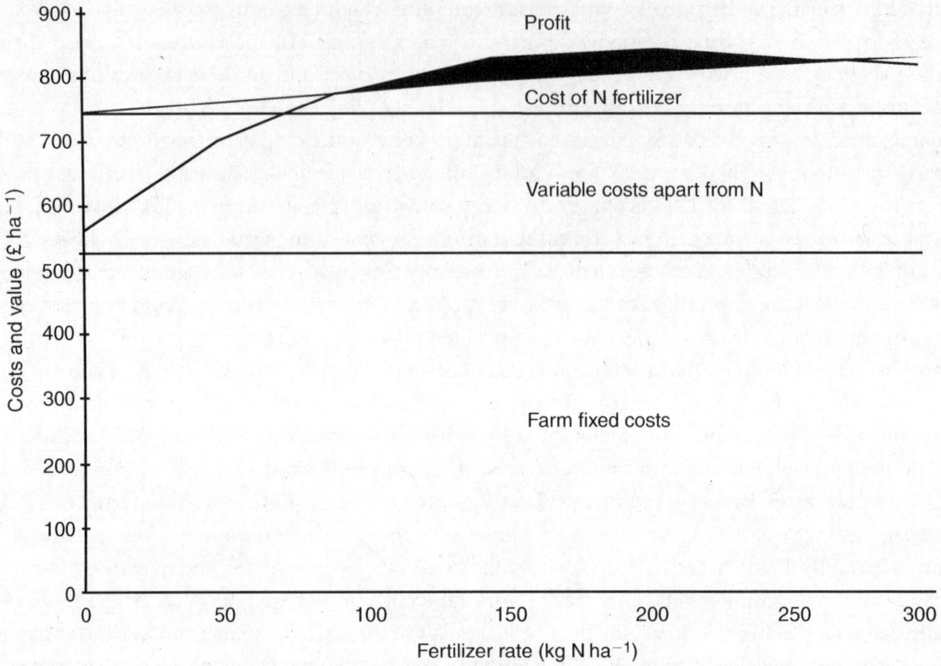

Figure 5.4 Costs and profit/loss from growing winter wheat in the UK in 1997 with different rates of nitrogen fertilizer
Source: Davies (2000).

Table 5.4 Carbon/nitrogen ratios in a range of organic materials

Material	C/N ratio
Soil microbial biomass	2–6
Poultry manure	4–7
Pig and dairy slurry	6–13
Sewage sludge	8–13
Soil humus	7–26
Farmyard manure from pigs	6–11
Farmyard manure from cattle	13–14
Vegetables	11–25
Sugar beet tops	11–27
Potato haulm	15–30
Oilseed rape haulm	20–75
Cereal straw	40–120
Hardwood timber	23–80
Conifer timber	90–625

Source: data from numerous sources.

much is immobilized by growth of organisms, especially the soil microbial biomass, so only net mineralization (gross mineralization minus immobilization) is relevant for nitrate leaching.

Some N is leached from forest soils (Perakis and Hedin, 2002) and arable soils (Murphy *et al.*, 2000) in soluble organic components derived from the partial decomposition of soil organic matter. Most of the N leached from cold mountain forest soils is in this form, but the amounts lost from arable soils in temperate climatic zones are probably less than those in nitrate. Nevertheless, they are still significant and are less subject to seasonal fluctuations than the nitrate formed by complete mineralization of organic matter. Also, once in an aquifer or surface water body, soluble organic matter can be oxidized microbially or even abiotically and thus contribute to nitrate contamination.

5.4.2 Factors affecting nitrate leaching

Measurements of nitrate leaching by various methods at a large number of sites have shown that some soil types are more 'leaky' than others and that some crops leave larger amounts of N in the soil than others. Thin sandy soils directly overlying aquifers and strongly fissured clay-rich soils with efficient subsoil drainage systems lose more nitrate than thick loamy soils, which are more water retentive. With higher infiltration rates, nitrate from top dressings of spring fertilizer or autumn/winter mineralization of soil organic matter moves more quickly to depths beyond the rooting zone of most crops.

However, the extent to which nitrate lost from the soil profile contributes to higher concentrations in aquifers and surface water bodies also depends on the extent to which it is diluted by rain. For example, in the wetter western parts of Britain annual excess rainfall often exceeds 400 mm, but in drier eastern areas it can be $<$100 mm. Although $>$30 kg N ha^{-1} yr^{-1} is lost in many western areas, the average concentration rarely exceeds 8 mg NO_3-N l^{-1}, whereas losses of only 20–30 kg N ha^{-1} yr^{-1} in eastern areas usually result in concentrations greater than the EU limit of 11.3 mg NO_3-N l^{-1} (Lord and Anthony, 2000).

Crops that lead to increased nitrate leaching include potatoes, oilseed rape, sugar beet, legumes such as peas and rotational set-aside (periods of fallow alternating with crops) (Macdonald *et al.*, 1997; Chalmers *et al.*, 2001). Various factors are involved. Potatoes and sugar beet have sparse root systems that invade a small proportion of the soil volume and consequently leave large residues of N in the soil. Leguminous crops are not treated with N fertilizer because they benefit from the activity of N-fixing bacteria (*Rhizobium* spp.) in root nodules; however, their root residues contribute strongly to the soil reserves of organic N, which may then be mineralized at times when subsequent crops need little N. Oilseed rape has a fairly dense root system but, because the crop grows more rapidly than others through the winter, it is usually given some fertilizer N in autumn, and this can be taken up rather inefficiently in cold, wet weather. The unfertilized weeds that grow on fallow land during set-aside take quite large amounts of N from the soil, and thus limit nitrate leaching; however, when they are ploughed in before a new crop is planted, their residues are mineralized and generate nitrate before it is required by the new crop. For the same reason, winter cover crops (e.g., forage rape), which were previously recommended for limiting winter nitrate losses, can increase nitrate losses under subsequent spring-sown or autumn-sown crops (Catt *et al.*, 1998a).

Forests and woodland plantations generally show less potential for nitrate leaching than grassland and arable land. However, N uptake by trees is less (and nitrate leaching therefore greater) in the early stages of establishment, when primary production is still quite small, than in the later period of maximal growth (Hakimata *et al.*, 1997). It is also less at advanced maturity or after harvesting, when decomposition rates increase.

In summary, from a wide-ranging survey of nitrate leaching in various land use systems, Di and Cameron (2002) concluded that the potential for nitrate loss increases in the following order: forest < cut grassland < grazed pastures and arable land < ploughed pasture < market gardens.

5.4.3 Limiting nitrate losses to water

The main techniques used in modern arable farming for limiting losses of nitrate to water supplies are the following.

- Minimizing the disturbance of established grassland and set-aside and other soil cultivations, which stimulate mineralization of organic matter.
- Calculating fertilizer requirements carefully, preferably using a model that allows for measured residues of soil mineral N left over from previous crops and the N likely to be released from previous crop residues, manures and other soil organic matter. Computer-based models of this type were reviewed by Wu and McGechan (1998). For wheat and barley crops in Britain, the total soil N supply in spring should not exceed 180 kg ha^{-1}.
- Avoiding autumn applications of fertilizer. The only arable crop likely to benefit from a little autumn N is winter oilseed rape.
- Incorporating straw and other crop residues with a high C/N ratio. Decomposition of these residues results in immobilization of soil mineral N by the growing microbial population in order to maintain its lower C/N ratio (Table 5.4). This is most effective if the straw is incorporated in spring (Figure 5.5). However, later mineralization of the enlarged microbial biomass can increase nitrate losses (Catt *et al.*, 1998b).
- Careful timing of spring fertilizer applications in relation to anticipated crop growth, which depends on soil temperature, soil moisture content and expected rainfall; splitting the application into two or more dressings a few weeks apart can decrease any losses occurring in subsequent rainfall events.

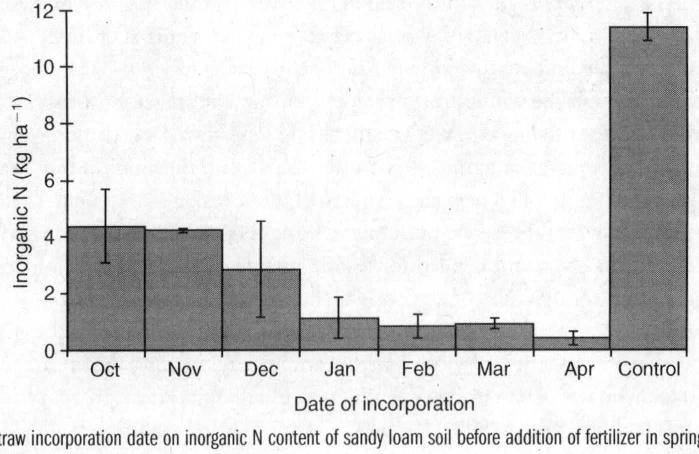

Figure 5.5 Effect of straw incorporation date on inorganic N content of sandy loam soil before addition of fertilizer in spring
Source: Bhogal *et al.* (1997).

- Ensuring good growth through the spring and early summer by eliminating pests and diseases and the competition from weeds.
- Avoiding bare fallow in winter.
- Minimizing the use of organic manures.
- Choosing high-yielding varieties of crops to use available N resources efficiently.
- Applying fertilizers and manures evenly leaving unfertilized 'buffer strips' at the margins of fields adjacent to surface water bodies.
- Avoiding excessive irrigation, preferably using a computer-based model such as IRRIGUIDE (Section 4.3.5).
- Avoiding excessive efficiency of field drainage systems (e.g., not redrawing mole channels too often). This increases the contact time between dissolved nitrate and crop roots, so that more N enters the crop; it also encourages loss of N by denitrification (reduction of nitrate to gases such as N_2O and N_2, which are transferred to the atmosphere).
- Wherever possible using 'leaky' soils for permanent unfertilized pasture rather than arable rotations.
- Using nitrification inhibitors such as dicyandiamide. When sprayed onto soil these compounds inhibit the oxidation of NH_4^+ and NO_2^- to NO_3^- by rendering bacterial enzymes ineffective. Inhibitors have been used successfully to limit nitrate losses under arable crops and grassland (Rogers *et al.*, 1985; Pain *et al.*, 1994).

However, other environmental complications may arise from some of these strategies. For example, reduced drainage efficiency may decrease nitrate losses to water at the expense of increasing nitrous oxide (N_2O) emissions to the atmosphere. As N_2O may contribute to global warming, the problem of water pollution is merely transferred to the atmosphere. Because of these types of interactions it is necessary to adopt a holistic approach to agriculture and the environment, using, for example, an environmental audit system (Lewis *et al.*, 1997) to determine an optimum balance of farming practices.

5.4.4 Nitrate leaching in organic farming

Organic farming is an alternative form of agriculture in which N inputs from artificial fertilizers and organic manures from off-farm sources are replaced mainly by those inherited from leguminous crops, such as grass-clover leys or arable grain legumes (e.g., peas, beans, lupins). Pesticides are also replaced by biological pest-control methods (Section 5.5.3), which decrease pollution of soil, water and food products. The amounts of N accumulated by legumes are usually limited and mineralization mechanisms are difficult to control, so that the yields of subsequent crops are often less than those grown with conventional fertilizers or imported organic manures. Also there is a strong possibility of nitrate leaching if mineralization of the residues occurs at times of rapid water infiltration and slow crop growth. Despite this tendency, comparisons of nitrate leaching from farms under conventional and organic management systems with broadly similar crop sequences and climatic conditions have suggested that N losses from organic systems are generally no greater and sometimes slightly less than from those from conventionally fertilized systems (Stopes *et al.*, 2002).

In both organic and conventional systems, the main nitrate loss occurs after a ley is ploughed out prior to planting an arable crop. Long-term losses therefore might be decreased if the transition is made less frequently by extending either the ley or the arable phase or both. However, the longer the ley the more nitrate is released on ploughing out (Johnston *et al.*, 1994), and organic systems are unlikely to sustain more than about two years of arable cropping. All-arable organic systems, dependent on grain legumes for accumulation of soil N, eliminate the transition from ley to arable and are economically viable on fertile, water-retentive soils (Bulson *et al.*, 1996), but Stopes *et al.* (2002) found that they lead to greater overall nitrate losses than organic ley–arable systems. However, organic ley–arable systems are unlikely to be widely adopted in the near future because of lower yields and the less profitable years when the land is under a ley.

5.4.5 Campaigns to limit nitrate losses

Over the last decade, many countries have introduced schemes to encourage farmers to adopt measures for minimizing losses of nitrate to ground- and surface waters. In England and Wales, following a pilot scheme launched in 1990, the Ministry of Agriculture, Fisheries and Food (now Department for Environment, Food and Rural Affairs or DEFRA) designated 32 Nitrate Sensitive Areas (NSAs) in 1994. Within NSAs farmers were paid to adopt practices that would decrease nitrate leaching, such as restricting the amounts of livestock manure applications, decreasing fertilizer N input rates, planting winter cover crops on land that would otherwise be left bare in winter and converting arable land to low-input grassland. This campaign resulted in an overall 20 per cent decrease in fertilizer use and a somewhat greater decrease in mean nitrate concentrations in drainage waters in the five pilot NSAs that were monitored.

Based on this success, the British campaign was extended in 1996 to 68 Nitrate Vulnerable Zones (NVZs) covering a total of about 600,000 ha in areas where 'leaky' soils occur over important aquifers. Compulsory action programmes for controlling N inputs were introduced in 1998, and farmers in NVZs were offered free advice and specific help on manure management. Since 2002 DEFRA has increased the proportion of land in England subject to NVZ legislation to 55 per cent, and has issued a free CD-ROM supplying a decision support system to help farmers meet their obligations under the NVZ legislation.

5.4.6 Contamination of water by phosphorus

The other main nutrient leading to the environmental problem of surface-water eutrophication is P. This is required by plants and therefore applied in much smaller amounts than N; the mean application rate to arable land in the UK is approximately 23 kg P ha^{-1} (52 kg P_2O_5 ha^{-1}), compared with almost ten times that amount of N, and many fields do not receive P fertilizer every year. It is now applied as either superphosphate or triple superphosphate, in which the P of insoluble rock phosphates, such as hydroxy-apatite [$Ca_{10}(PO_4)_6(OH)_2$], has been made more available to plants by treatment with sulphuric acid or phosphoric acid, respectively. Ectoparasitic mycorrhizae and the fungus *Aspergillus niger* can also extract P from less soluble minerals and make it available to plants (Goenadi *et al.*, 2000).

P is much less mobile than N, as most of it is fixed in acidic soils by sorption on iron/aluminium oxides or layer silicate clay minerals and in weakly alkaline soils by precipitation of calcium phosphate. It is most available to plants in neutral or weakly acidic conditions (pH 6.0–7.0), because release of P by mineralization of soil organic matter is most rapid in these conditions, and any P in solution is loosely sorbed in anionic form on the surfaces of aluminosilicate clay minerals. However, even in neutral soils, the amounts of non-labile P are often two to three orders of magnitude greater than those of labile (plant available) P, and <1 per cent of the labile pool is actually in solution. Removal of the P in solution by plant roots, often assisted by mycorrhizae, results in dissolution of P from the labile but insoluble components or even from more recalcitrant forms, so as to maintain a low concentration of phosphate anions in the soil solution. P availability can also be influenced by soil management; for example, Leinweber *et al.* (1999) found more labile P in grassland soils than in those under arable cultivation, fallow or woodland. Plant-available P is usually determined by extraction with 0.1 M sodium bicarbonate or calcium chloride solution, or by ion exchange resins.

Despite the factors that limit the amounts of P in the soil solution, eutrophication of surface water often occurs because the minimum concentration leading to eutrophication is very low, only 0.02 mg l^{-1} according to the Organisation for Economic Co-operation and Development (OECD, 1982). Drainflow and surface runoff from arable fields into ditches and streams often contain higher concentrations than this. For example, the soluble P in drain water from plots of the Broadbalk Winter Wheat Experiment at Rothamsted, which have received superphosphate applications of 0–35 kg P ha^{-1} annually for over 150 years, almost always exceeds the OECD minimum for eutrophication.

The surface soil of the Broadbalk plots now contains 5–100 mg kg^{-1} bicarbonate-soluble P depending on the P application rate. Heckrath *et al.* (1995) found that the drain waters from plots with up to about 60 mg kg^{-1} bicarbonate-soluble P usually contain <0.15 mg l^{-1}. However, in drain waters from plots with 60–100 mg kg^{-1} bicarbonate-soluble P, they reported linear increases in both soluble and total P to concentrations greatly in excess of 0.15 mg l^{-1} (Figure 5.6). This suggests that the Broadbalk soil can retain amounts of bicarbonate-soluble P up to a maximum of 60 mg kg^{-1} fairly effectively, though rarely so strongly as to prevent drain water exceeding the OECD eutrophication limit. However, where the soil contains >60 mg kg^{-1}, increasing amounts of P are leached in percolating water.

Hesketh and Brookes (2000) suggested that the 60 mg kg^{-1} 'change point' in the relationship between soil bicarbonate-soluble P concentration and the soluble P concentration in drain water can

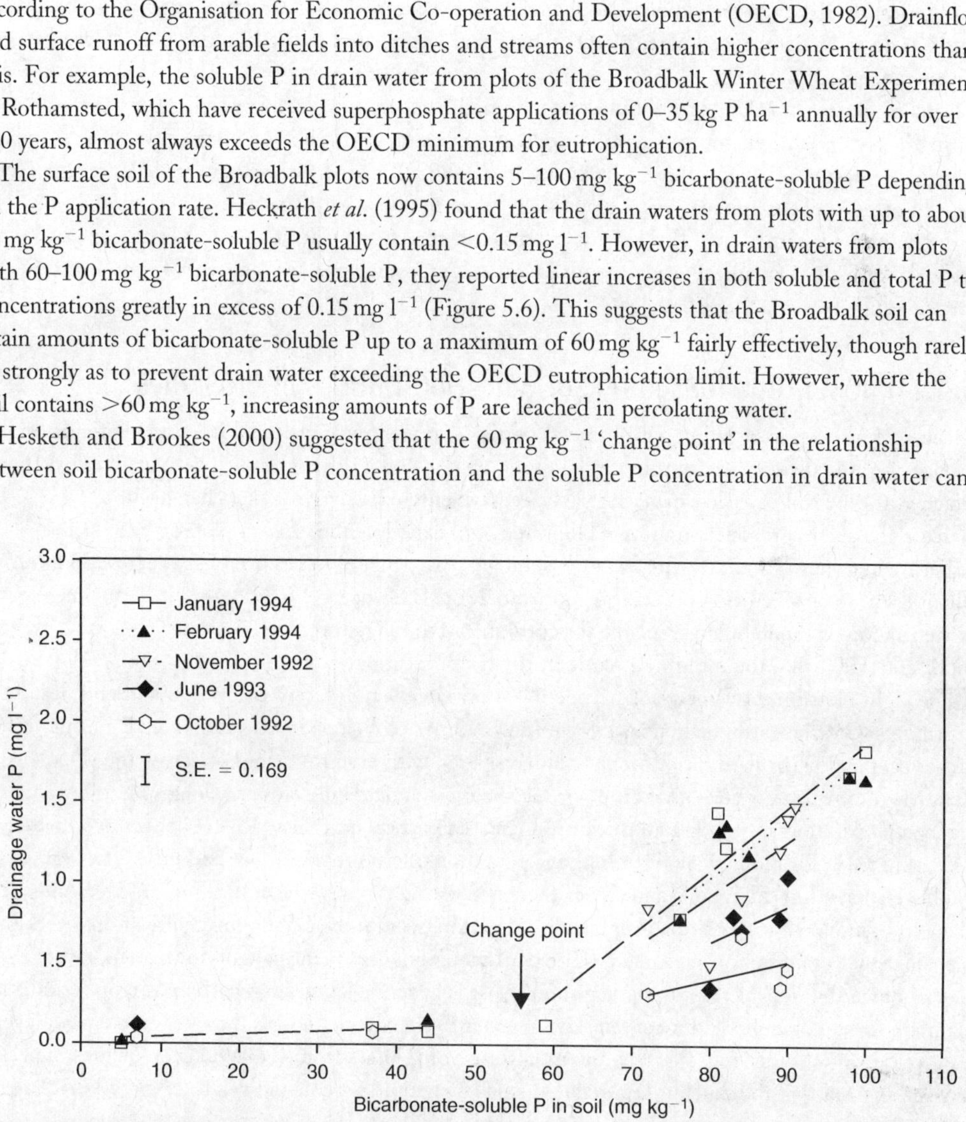

Figure 5.6 Relationship between dissolved reactive P in drainage water from plots of the Broadbalk Winter Wheat Experiment at Rothamsted and bicarbonate-soluble P in the plot soils
Source: Heckrath *et al.* (1995).

be used as an indicator of the risk of P leaching and eutrophication. However, in a range of soils from Britain, the USA and New Zealand with different bicarbonate-soluble P concentrations, pH values and organic matter contents, McDowell *et al.* (2001) found 'change points' ranging from 20 to 112 mg kg^{-1} bicarbonate-soluble P. For some soils of either very high or very low P saturation, they found no 'change point', probably because the P concentrations were either all above or all below the 'change point' for the soil type at the sites concerned. For calcium chloride-soluble P, the rate of increase in P loss above the 'change point' was much greater than for bicarbonate-soluble P, and in another American soil Kleinman *et al.* (2000) reported a 'change point' as low as 0.9 mg kg^{-1} calcium chloride-soluble P. Calcium chloride-soluble P may therefore be a more sensitive soil test for eutrophication risk than bicarbonate-soluble P. In non-calcareous soils, differences in the 'change point' seem to be determined particularly by hydroxyl ions related to the iron and aluminium cations of clay minerals (Blake *et al.*, 2002).

Leaching losses of soluble P are often large immediately after P fertilizer applications, especially during heavy rain and in strongly fissured soils, in which rapid percolation allows little time for plant uptake or sorption of P on soil particles (Catt *et al.*, 1998c). Losses also remain fairly large for several months after fertilizer applications, because P is taken up by crops rather slowly and may not be strongly fixed for some time. The largest P losses can therefore be avoided if fertilizer is applied when dry weather is forecast and thoroughly mixed with the topsoil, so as to improve sorption on soil particles (Djodjic *et al.*, 2002).

5.4.7 Phosphorus losses by surface runoff and soil erosion

Because of its close association with mineral particles, P is often transferred from arable fields to surface-water bodies in surface runoff containing eroded soil. Although much is in insoluble forms attached to fine soil particles, once the particles have entered a river or lake, changes in pH can increase the soluble P concentration in the water. On experimental erosion plots at Woburn Experimental Farm, Bedfordshire, total P (soluble + particulate) losses over a six-year period, during which there were 48 erosion events, ranged from 2.4 to 32.8 kg ha^{-1} (Catt *et al.*, 1998c). The mean concentration of P potentially available for eutrophication ranged from approximately 3.7 to 10 mg l^{-1}, about 200–500 times the minimum required for eutrophication.

The eight plots on the Woburn experimental site were given different cultivation and crop residue treatments. Crop rows aligned up and down the 7–13 per cent slope increased soluble P in runoff by 55 per cent and particulate P by 185 per cent compared with crop rows aligned across the slope. Removal of cereal straw and other crop residues before standard cultivations (ploughing followed by seedbed preparation appropriate to the crop) increased particulate P losses by 69 per cent compared with minimal cultivation (shallow-tining) and retention of crop residues, but had little effect on soluble P losses. On sandy soils and slopes such as those at Woburn the cultivation practices most likely to minimize P losses and eutrophication therefore seem to be (a) minimal cultivation rather than exposing soil to rain by conventional mouldboard ploughing, (b) planting crop rows along rather than up and down the slope, so as to trap any runoff that is generated, and (c) retention of crop residues to maintain topsoil organic matter content and aggregate strength. Although the rare large erosion events lead to the largest individual P losses, the monitoring of P losses from the Woburn plots over a period of years showed that the much more frequent small-magnitude events outweigh the occasional large losses in terms of the amount of P likely to be lost in the long term (Quinton *et al.*, 2001).

P losses in surface runoff from arable fields can be large where heavy dressings of organic manures have recently been applied (Smith *et al.*, 2001). The problem has recently increased because of the cultivation of forage maize for cattle feed, and winter manuring of this crop with generous surface applications of slurry rich in P (Withers and Bailey, 2003). The largest losses occur in runoff under heavy rainstorms soon after slurry application, when P is being released by initial mineralization of the slurry and pore spaces are sealed with slurry. Soil cultivation after slurry application can decrease runoff and release of P from the slurry, and splitting slurry applications decreases the initially high concentrations of soluble P in the first runoff event.

From a water quality viewpoint, the critical question is whether the runoff and eroded soil material reaches a surface water body or not. If the sediment stays within the field and never reaches a ditch or stream, it creates few problems, but cultivation of fields in which eroded soil is likely to reach a water course should be minimized, and areas adjacent to rivers and ditches should be buffered with permanent pasture (Del Campillo *et al.*, 1999).

5.4.8 Modelling phosphorus losses from agricultural land

Heathwaite *et al.* (2003) outlined a catchment model, The Phosphorus Indicators Tool, for loss of P from agricultural land to surface waters in Britain. It incorporates particulate P transfers resulting from soil erosion, leaching of soluble P through drains and the effect of freshly added P in fertilizer and manure applications. The model has a modular structure with three layers, and operates at a spatial resolution of 1 km.

Layer 1 calculates the soluble and particulate P reserves for each land use unit in the catchment, and estimates their potential for loss. Particulate soil P, soluble soil P and P freshly applied in fertilizers and manures have different susceptibilities to loss (Preedy *et al.*, 2001), so the potential losses from these three sources are passed separately to Layer 2.

Layer 2 includes the processes leading to mobilization of P from the three sources. These are driven by soil physical properties (mainly topsoil structure), climate (hydrologically effective rainfall), land use (arable versus grassland) and slope.

Layer 3 defines pathways of P movement from land use units to water courses, and assigns proportions of the various forms of mobilized P likely to reach water bodies or be retained within land use units. Pathways are divided according to soil type into surface runoff, drainflow and percolation to groundwater. All P carried in drainflow is assumed to reach surface water bodies. Surface runoff is subdivided into within-field delivery (the proportion reaching the field boundary), edge of field delivery (transfer across field boundaries) and field to water course delivery (dependent on drainage density and landscape gradient). Layer 3 also includes direct delivery of P from manure on farmyard hard-standings, feed stations and roads.

5.5 PESTICIDES AND INDUSTRIAL ORGANIC CHEMICALS

5.5.1 Industrial sites

On industrial sites, such as oil refineries, gas works and chemical factories, the soil often becomes polluted with a range of organic chemicals, including petroleum hydrocarbons, fuel oils, heavy oils, cutting oils, kerosene (paraffin oil), dry-cleaning fluids and aromatic hydrocarbons. After site closure, these pollutants usually must be removed before the land can be used safely for agriculture or housing. Agricultural land can also be contaminated locally by spillage of fuel oils and pesticides, and more widely it may be affected by the organic pollutants commonly found in sewage sludge, including CBs, PCBs and PAHs (Beck *et al.*, 1995). In industrialized countries, there has also been ubiquitous low-level contamination of soil by PAHs, PCBs, dioxins, furans and other organic compounds originating from combustion of coal and municipal waste (Johnston, 1997).

Spillages of light oil can be removed by digging shallow pits into which it can drain from the surrounding soil, and then repeatedly pumping the pits dry. Heavier oils often stay on the surface and slowly solidify; they can then be scraped up and removed.

Organic contaminants with high vapour pressures, such as light petroleum hydrocarbons, benzene, toluene, CBs and lower molecular weight PCBs and PAHs, are dispersed from soil mainly by volatilization. Their bioavailability to plants, soil invertebrates and micro-organisms of many organic soil contaminants is often decreased by sorption on clay or humic soil components (Chiou, 1990). Soil organic matter is more effective in stabilizing lipophilic than hydrophilic compounds; lipophilicity is estimated from the compound's solubility partition coefficient between 1-octanol and water (K_{ow}). Expansible clay minerals, such as smectite and vermiculite, are more effective in absorbing both lipophilic and hydrophilic organic pollutants than kaolinite, illite and chlorite.

For some organic pollutants, losses by natural passive processes such as volatilization, sorption on organic matter and clay, leaching and degradation by indigenous micro-organisms (Section 5.5.4) are ineffective, and active treatments must then be considered. Soil treatment *in situ* to remove organic pollutants is usually cheaper but slower and less effective than removal for treatment in a reactor vessel. However, as the total volumes of soil affected by spillages are usually quite small, special treatments can often be used that would be prohibitively expensive for large areas. Evaporation of volatile contaminants, such as petroleum hydrocarbons, kerosene, turpentine, perchlorethylene, methanol, benzene and phenol may be accelerated by pumping or sucking air through the soil. Alternatively, they can be removed by injecting steam and extracting the vaporized contaminants through specially installed vacuum drains. Organic solvents, such as acetone and ethyl alcohol, have also been used to remove some of these pollutants, and flushing with water can remove hydrophilic compounds.

More recalcitrant organic contaminants must be removed by incineration or isolated by 'hard' remediation techniques, such as stabilization with cement and use of the hardcore as foundations beneath a road or other sealed surface. However, these treatments are more expensive, and may only transfer the pollution to the atmosphere or, in the longer term, to groundwater.

5.5.2 Organic pollutants in sewage sludge

Sewage sludges are almost always contaminated with organic pollutants from various sources. In sludges from three sites in southern England, Beck *et al.* (1995) reported concentrations of total PAHs ranging from 13 to 89 mg kg^{-1}. PCBs are usually present in much smaller amounts (totals of 1–3 mg kg^{-1}), and CBs are even less abundant (totals of 0.01–0.1 mg kg^{-1}). However, rates of sludge application to soil are usually in the range 10–100 t ha^{-1}, so that an application with a concentration of 10 mg kg^{-1} adds an amount approximately equivalent to the annual dose of total pesticides used on an arable crop.

An important factor is the persistence of these compounds in soil, which influences how frequently sludge can be applied. CBs and PCBs seem to disappear from soil fairly rapidly and are unlikely to build up to levels thought to be dangerous, even with repeated applications of sludge. However, PAHs can persist for many decades in concentrations greatly exceeding published 'target' maxima, such as the 1 mg ΣPAHs kg^{-1} soil recommended in The Netherlands (Beck *et al.*, 1995) or even the 'intervention' value of 200 mg ΣPAH kg^{-1} (Van den Berg *et al.*, 1993), i.e., the minimum concentration at which soil remediation is thought to be necessary. In the UK, the Inter-Department Committee on the Redevelopment of Contaminated Land (1987) set considerably higher 'trigger' (=target) and 'action threshold' (=intervention) values for ΣPAH concentrations, namely 50 mg kg^{-1} and 500 mg kg^{-1}, respectively. Other countries have set values between these two extremes, but generally closer to the Dutch values.

5.5.3 Agricultural use of pesticides

Pesticides are widely used to increase the yield and quality of crops; in fact, it would now be impossible to produce enough food for the world's expanding population without them. The three main types used most extensively are insecticides, herbicides (weedkillers) and fungicides. Some of the early pesticides, such as the insecticides DDT, dieldrin and aldrin, had considerable environmental impact, especially on harmless or beneficial wildlife. As a result, increasingly strict criteria have been introduced and use of many older compounds has been banned. However, in the UK over 300 individual pesticides are currently available, some sold as product mixtures to deal simultaneously with a range of pests. The number of insecticides in particular has increased in order to target specific organisms without causing harm to others, and because insects rapidly develop resistance to new compounds. Of particular environmental importance are the persistence of pesticides in soil, their effects on non-target organisms in and on the soil and their movement to ground- and surface waters.

Most countries now have stringent legal controls and codes of practice for the supply, storage, use and disposal of pesticides. Authorization for their manufacture and use is based on test results indicating that, when correctly used, the compounds have no unacceptable long-term effect on the environment, including non-target organisms. Most such tests monitor short-term effects, because modern pesticides have short half lives in the soil. Most pesticides with half lives longer than 60 days in moist soil at 20°C are not registered for use. However, some with such long half lives have been registered if further tests have shown that they have no harmful effect on beneficial soil microbiological processes such as nitrification.

However, concern is often expressed that frequent repeated use of pesticides at recommended rates can, in time, have deleterious effects on non-target organisms, especially those responsible for soil nutrient cycling. The long-term Boxworth Project (1981–91) of the Agricultural Development and Advisory Service (ADAS) and the follow-up SCARAB Project (1990–6) (SCARAB = Seeking Confirmation About Results At Boxworth) investigated the effects on soil organisms of a range of pesticides applied at recommended and reduced (approximately 50 per cent recommended) rates to various crops at sites with different soil types (Young *et al.*, 2001).

In both projects the main damaging effect resulted from seasonally repeated applications of organophosphorus insecticides, which led to a long-term decline in springtails (soil arthropods). However, the species affected formed only a small proportion (<3 per cent) of the total arthropod species monitored. After six years, repeated fungicide applications briefly depressed soil microbial activity at one site, but insecticides and herbicides had no such effect. No effect on earthworm populations was detected at any of the sites. These results suggest that very few non-target soil organisms are likely to be affected by pesticides used at the manufacturer's recommended rates, and that any problems are unlikely to be widespread. There were some small benefits from the reduced application rates, but these compromised the control of target organisms. For example, low-input herbicides allowed weed numbers to increase up to seven times compared with recommended rates, and this resulted in yield losses, especially in cereals.

The economic and agronomic effects of decreasing pesticide inputs were assessed further in a related ADAS project known as TALISMAN (Towards A Lower Input System Minimising Agrochemicals and Nitrogen), the results of which were also reported by Young *et al.* (2001). Reducing pesticide inputs lowers the cost of crop production, and this can increase gross profit margins by a few per cent, especially with decreased fungicide applications. However, success depends on local knowledge and management skills. Where such skills are absent or limited, the manufacturers' recommended application rates provide the necessary insurance against large decreases in profit margins.

The movement of pesticides through soil to groundwater or field drains is usually assessed in lysimeters (intact monoliths of soil in cylindrical sleeves) at least 1 m deep, or in hydrologically isolated field plots, such as those of the Brimstone Farm Experiment. European Union legislation requires that any groundwater aquifer or surface water body used as a source of drinking water contains $<0.1\ \mu g\ l^{-1}$ of any pesticide as an annual average concentration. Lipophilicity (Section 5.5.1) is a good indicator of a pesticide's susceptibility to leaching through soil. Losses are generally less for strongly lipophilic compounds (e.g., the insecticide permethrin), which are strongly sorbed on soil organic matter, and greater for those that are more soluble in water. However, the longer the period between application and a flow event through the soil profile, the greater the time for sorption on soil organic matter or microbial degradation, and hence the smaller the amount of pesticide leached.

Over the last decade, increasing emphasis has been placed on biological control mechanisms for dealing with insect pests, and this is allowing the total amounts of artificial insecticides used in agriculture to be reduced. For example, the larvae of hoverflies, which feed on aphids, can be encouraged by provision of set-aside areas, such as field margins, in which the wild flowers that are food plants for the adult hoverflies can grow. In Britain, the most suitable wild flowers for encouraging hoverflies are yarrow, white campion and Umbelliferae such as cow parsley and hogweed. When insects are physically disturbed or injured, they emit alarm pheromones, which cause others of the same species to disperse from the pheromone source. Sprays of the otherwise harmless artificial pheromones, such as (E)-β-farnasene emitted by some aphids, can therefore be used to lure insect pests away from high-value crops.

Another approach in biocontrol is to spray the crop with 'plant activator' compounds, such as methyl salicylate, *Cis*-Jasmone and methyl jasmonate. These activate defence mechanisms in the crop itself, such as production of defensive glucosinolates in oilseed rape plants, or attract aphid predators such as the wasp *Aphidius ervi*. At present, the most effective approach to pest control involves integrating a range of chemical and biocontrol methods. In time a range of less harmful biocontrol methods may virtually replace poisonous chemicals for controlling insects and other pests, though insects in particular are surprisingly capable of adaptation to new environments, so that they can rapidly develop resistance to either predators or control chemicals.

Use of pesticides can also be limited by forecasting the arrival of pests, so that applications of appropriate pesticides can be delayed until they are actually needed. The Rothamsted Insect Survey, based on daily identifications of insect populations trapped in a nationwide array of light and suction traps, was originally for this purpose. Recently methods have also been developed for using weather forecasts and past records of fungal diseases in relation to weather patterns to predict fungal attacks on crops.

However, at present there is less hope of decreasing usage of herbicides, because it is more difficult to predict where and when weed populations will increase, and there are few effective non-chemical control methods for weeds apart from bare-fallowing, which means the land is non-productive for a year. Large yield losses can result if weeds are not properly controlled, because crop plants generally compete rather ineffectively with weeds for nutrients from the soil. However, most cheap herbicides are not limited in their effect to individual weed species, and damage to the crop plant can occur if a high dose is applied at critical times in the growth cycle, such as soon after emergence. Farmers therefore usually apply broad-spectrum herbicides in several low doses both before and at various times after crop emergence. This adds considerably to input costs and is also unsympathetic to wildlife, as weeds can provide food for invertebrates and cover for nesting birds. Recently methods have been developed for genetically modifying crop plants so that they have greater herbicide tolerance, and this allows the farmer greater flexibility in the timing of herbicide applications, with fewer applications and therefore lower input costs. It can also allow applications to be delayed until critical invertebrate species, such as spiders and beetles, have reproduced or birds have nested.

The within-field distribution of some weed species affecting cereals, such as cleavers (*Galium aparine*), is influenced by soil moisture, weed patches often occurring repeatedly where the drainage is poor because of locally increased clay content or subsoil compaction. Once such patches have been mapped and geo-referenced, herbicides can be applied to them using a satellite-based global positioning system, leaving other parts of the field untreated. Precision farming of this type can greatly decrease both the cost of herbicide inputs and their detrimental effects on wildlife.

A further possible problem with use of pesticides that is causing increasing public concern is the effect on human health of pesticide residues in food. Exceedingly small amounts of individual pesticides are normally detected in farm food products, though there have been a few examples of accidental contamination at higher rates. However, some believe that even small quantities may accumulate in undegraded or partially degraded form in parts of the human body. For example, lindane (γ-hexachlorocyclohexane), an insecticide often used as a wood preservative, which was previously (though probably incorrectly) linked to the human blood disorder known as aplastic anaemia, may accumulate by attaching to DNA. Also there is concern that, even in small quantities, two or more pesticides entering the human body in food or inhaled from preservatives used domestically (e.g., in carpets) may interact to produce a 'cocktail effect', causing brain damage. Such claims obviously need investigating by careful medical research but, at present, they are unconfirmed.

5.5.4 Phytoremediation for organic pollutants

Phytoremediation is more difficult for soil contaminated with organic than inorganic contaminants, such as heavy metals, and has not been so widely used. Nevertheless, some higher plants, especially grasses, can absorb and metabolize or sequester organic compounds (Cunningham *et al.*, 1996). Examples include accumulation and degradation of PAHs by various prairie grasses, pentachlorphenol by wheatgrass (*Agropyron desertorum*) and the pesticides atrazine and metolachlor by musk thistle (*Carduus nutans*).

Other plants, including most food crops, take up only small quantities of organic soil contaminants. Beck *et al.* (1995) reported that in a wide range of root and foliage food crops PAH concentrations were not significantly greater where the crops had been grown on soils treated for many years with PAH-contaminated sewage sludges than where they had been grown on adjacent untreated soils. Also Wang and Jones (1994) found that carrots took up <1 per cent of the CBs in sludge-treated soil.

The most common mode of plant uptake is in the aqueous phase through the roots, though some plants can also take up volatile organic compounds in the vapour phase through roots or leaves, and roots with large lipid contents can absorb lipophilic organic contaminants. After transportation through the xylem vessels, the compounds are either sequestered in lignin or metabolized by plant enzyme systems similar to those responsible for treatment of toxins in the mammalian liver. Some plant enzymes are released from roots and then react directly with organic soil contaminants in the rhizosphere; for example, peroxidases released by the roots of horseradish (*Armoracia rusticana*) can degrade phenols and analine.

Many organic soil contaminants are also removed by microbial activity. According to Salanitro (2001), indigenous soil bacteria (e.g., *Pseudomonas*, *Acinetobacter*, *Rhodococcus*, *Nocardia*), actinomycetes and fungi (e.g., *Penicillium*, *Candida*) can metabolize fuel oils and other hydrocarbons with molecules up to C_{44}. Hydrocarbons with larger molecules are more recalcitrant to microbial degradation, but usually pose less ecotoxicological risk and are less likely to be leached.

The natural microbial population in healthy soils can develop the ability to degrade complex toxic organic contaminants, such as pesticides, initially to closely related but less toxic compounds and finally to simple materials such as carbon dioxide and water. For example, various species of *Pseudomonas* can transform 3-chlorobenzene and other CBs into chlorophenols and catechols (Spain, 1990). Adaptation to new contaminants usually occurs within a few weeks or months, though it is often slower for aromatic compounds such as PAHs and PCBs than for aliphatic hydrocarbons. In some soils the effect may be enhanced by addition of bulking agents, such as peat, chopped hay, vermiculite or activated carbon (Rhykerd *et al.*, 1998), or power station fly-ash (Wu *et al.*, 2000).

Microbial activity for degradation of organic contaminants is optimal with a soil pH near 7.0, warmth and adequate supplies of water, oxygen and nutrients. Mixing of topsoil with the contaminants by shallow cultivation can increase the supply of oxygen, and a light dressing of artificial fertilizer containing N and P ensures adequate nutrients. However, organic manures should not be added, as these compete with the organic contaminants for oxygen and microbial activity. Strongly contaminated soils are often best excavated and treated in a bioreactor, in which the soil is inoculated with fresh bacteria given adequate supplies of moisture and nutrients in an aerobic environment. Microbial remediation is usually much cheaper than chemical extraction methods.

5.6 GASEOUS POLLUTION OF SOILS

Soil air can be partially or completely replaced by toxic, ionizing or asphyxiating gases, which may cause stress, illness or death in vegetation, animals or humans. Where buildings are sited on gas-contaminated land, they must be adequately ventilated to prevent the gas accumulating, especially in basements.

5.6.1 Methane and hydrogen sulphide

These gases are often produced naturally in anaerobic soils, especially over marshland, sewers, landfill sites, former oil refineries or infilled docklands. CH_4 is an asphyxiant and is also inflammable and explosive in concentrations of 5–15 per cent in air. Emissions from landfill and other CH_4-contaminated sites are usually controlled by barrier and collection systems, which allow the gas to be burned, sometimes for heating purposes. H_2S is toxic or fatal at very low concentrations (a few parts per million), but its characteristic smell even at these concentrations usually provides an adequate warning of its presence.

5.6.2 Radon

The radioactive gas radon (^{222}Rn) is also produced naturally. It forms in rocks, such as granite, organic shales and limestones rich in phosphates, all of which contain radioactive U and Th minerals.

In Britain it probably accounts for about half of the ionizing radiation dose received by the human population. It is a suspected carcinogen, probably responsible for the frequently observed increase in incidence of lung cancer in people living in areas of these rocks, especially those breathing in higher concentrations because they work in confined, poorly ventilated areas, such as mines. Soils derived from these rocks are thought to generate >90 per cent of the Rn entering the surface environment (Nero *et al.*, 1990). In the UK, Rn is thought to be the second most important cause of lung cancer (after smoking). The main areas affected are in Cornwall, Devon, Derbyshire and Northamptonshire. Carbon dioxide often acts as a carrier for ^{222}Rn and, as this is a heavy gas occurring at much higher concentrations in soil than in the atmosphere, the Rn from soils tends to accumulate in basements. Adequate ventilation of rooms below or close to soil level is therefore important in preventing dangerously high concentrations of Rn.

5.7 MICROBIAL CONTAMINATION

Micro-organisms of various types (viruses, bacteria, fungi, mycorrhizas, protozoa and nematodes) are abundant in all soils, of which bacteria and fungi are the largest components by weight. Less than 10 per cent of the bacterial population are responsible for the known biochemical transformations in soil. The remainder have no known functions.

In view of the very wide range of micro-organisms occurring naturally in soil, it is unlikely that types completely new to soil are introduced by any process of soil contamination. However, the balance of different types can be changed locally by inputs of organic matter such as animal remains, dung, farmyard slurry or sewage, which are strongly infected by certain bacteria or viruses. Unusually high concentrations of single species are probably eliminated with time by natural competition with the indigenous biomass but, before this happens, pathogenic organisms may be transferred to food crops, surface water courses or groundwater.

5.7.1 Pathogenic micro-organisms

Some indigenous soil organisms are known plant, animal or human pathogens. Plant pathogens include the fungi *Pythium* spp. (responsible for 'damping-off' in legumes, sugar beet and cucumber, and browning root rot in cereals), *Fusarium oxysporum* ('fusarium wilt' or 'blight' in tomatoes), *Fusarium culmorum* ('ear blight' in cereals), *Gaeumannomyces graminis* var. *tritici* ('take-all' in wheat), and the bacterial infections responsible for 'potato scab' (*Streptomyces scabies*) and 'club-root' in brassicas (*Plasmodiophora brassicae*). Many enter soil from previously infected crop litter, such as straw, and may persist if this remains on the soil surface, as in zero or minimum tillage regimes. Some (damping-off, fusarium wilt and club-root) are more prevalent or persistent in acid soils, whereas others (e.g., potato scab) are favoured by a high pH and can be increased by overliming. In addition to decreasing crop yields, *F. culmorum* infections produce mycotoxins, which can be a serious health risk to humans and animals or affect fermentation when grain is used in brewing.

The effects of most soil-borne plant diseases are minimized by use of fertilizers, which stimulates root growth and increases natural resistance (Huber, 1990). Some forms of *Pythium* are suppressed by certain particulate soil organic matter components (Stone *et al.*, 2001). Modern fungicidal sprays can protect high-value crops from fungal diseases, and for glasshouse crops the problem can be virtually eliminated by steam or chemically sterilizing the limited volume of soil involved. However, the volume of soil in a field is much too large to treat in this way. More promising are a range of biocontrol methods, such as competitive but non-pathogenic species of *Fusarium* for control of 'ear blight' in cereals.

'Take-all' is a very widespread disease in cereals. The fungal hyphae invade the xylem tissue of the roots, preventing water and nutrients from reaching above-ground parts of the plant. This usually happens fairly late in the growth cycle soon after ear emergence, and results in premature ripening, the ears drying before the grains are filled, and the green leaves initially becoming yellow or brown and later grey or black as a result of secondary fungal infections. First cereal crops in a rotation rarely suffer badly, and the worst infections occur in second, third or fourth consecutive cereal crops, when the fungal pathogen has increased in abundance. Non-susceptible crops, such as legumes, potatoes or oilseed rape, diminish the pathogen's abundance in soil, and are often grown every third year as 'break crops' in an otherwise continuous cereal monoculture. However, the pathogen can sometimes persist through a one-year break to infect the first cereal if volunteers from the preceding cereal are common among the break crop.

Some soil fungi of the groups known as Chytridiomycetes (e.g., *Olpidium* spp. in glasshouse crops) and Plasmodiophoromycetes (e.g., *Polymyxa* spp. in field crops) are important vectors of plant virus diseases. Examples of fungally transmitted field virus diseases include Rhizomania in sugar beet and Barley Yellow Mosaic Virus (BYMV) in winter barley (Adams, 1990). They are often spread within fields by movement of the vector spores in water-filled channels on or within the soil, and consequently are more prevalent in wet years or on sites which have been overirrigated.

The commonest human pathogen derived from soil is *Clostridium tetani* (tetanus). Its spores are absorbed through skin lesions and are very persistent and widespread in the soil, where they probably originate from past applications of dung and sewage. Tetanus affects the nervous system, causing convulsions and death in about 50 per cent of cases, though it is well controlled by immunization.

Other human pathogens entering water or food supplies, mainly as a result of soil contamination with manure, farmyard runoff, sewage or abattoir waste, include virulent (verotoxigenic) strains of the bacteria *Escherichia coli*, *Campylobacter* and *Salmonella*, certain enteric viruses and cysts of the protozoa *Cryptosporidium parvum* and *Giardia lamblia*. In Britain *E. coli*, *C. parvum* and other pathogens are often abundant in the numerous earth-banked lagoons used for on-farm storage of animal slurry. However, even where these have been constructed on thin soils over strongly fractured aquifers few pathogens occur in the groundwater, possibly because near-surface fissures become blocked with organic solids from the slurry (Gooddy *et al.*, 2001). *E. coli* O157 occurs in the intestines of about 15 per cent of British cattle, and transmission to humans is usually via manure- or slurry-contaminated drinking or bathing water, or inadequately washed raw vegetables. It is highly resistant to temperature changes and chemical stresses, and can persist in soil for many months (Jones, 1999). Water draining through soil contains sufficient to cause serious infections if used for drinking or food-washing within the first week following a manure or slurry application (Table 5.5); numbers decrease thereafter, but it can still be detected after 2–3 months (Vinten *et al.*, 2002). Cysts of *Cryptosporidium parvum* are also very resistant to degradation in the soil and even to disinfection of water supplies with chlorine. The only effective treatment of water supplies containing *C. parvum* is the rather expensive process of ultrafiltration.

Other pathogenic micro-organisms can be removed from FYM, slurry and sewage sludge by pasteurization, composting, mixing with lime and the industrial process of mesophilic anaerobic digestion. Even extended storage of FYM and slurry can kill some pathogens because of the high temperatures generated by microbial activity. However, the effectiveness of most treatments with

Table 5.5 Applications and losses of *E. coli* in drainflow and surface runoff from three Scottish sites[a]

	Glencorse (gleyed soil on glacial till)	Crichton Royal (gleyed soil on glacial till)	Field No. 3 (well-drained soil on sandy till)
E. coli applied (c.f.u. m^{-2})[b]	2.1×10^8	4.7×10^8	3.0×10^8
Drainflow period (days)	29	38	73
Drainflow (mm)	74	25	259
Surface runoff	1.9–11.9 mm	-	-
Percentage of *E. coli* leached from grassland	4.4–(1.8)	19.6–(1.8)	-
Percentage of *E. coli* leached from arable land	2.1–(1.0)	-	0.1–(0.03)
Percentage of *E. coli* lost by runoff from grassland	Nil	-	-
Percentage of *E. coli* lost by runoff from arable land	0.003	-	-

Source: from Vinten *et al.* (2002).
[a] Figures in parentheses are standard deviations.
[b] c.f.u., colony-forming units.

respect to individual pathogens is unknown at present. As disposal of sewage sludge to farmland is now regarded in the UK and many other countries as environmentally better than incineration or disposal in the sea or to landfill sites, some pretreatment to prevent pathogen transfer from sludge to soil and crops is becoming a necessity.

Summary

Knowledge of soil components, processes and physical structure is necessary to understand whether soil of a particular type acts as a carrier and source or as a sink for a certain pollutant. Most agricultural soils are sources of nitrate and phosphorus, because most crops cannot be grown efficiently without supplementing the amounts of these nutrients occurring naturally, and it is impossible to prevent small amounts from being leached to surface waters, where they cause algal blooms harmful to wildlife. The human health problems resulting from nitrate in drinking water are probably less serious than previously thought. N and P are leached in smaller amounts through use of inorganic fertilizers than organic manures, provided the fertilizers are applied at times and in amounts matched to crop requirements. Soil erosion is another important means by which nutrients, especially P, are transferred from arable soils to surface waters. The N in organic manures is also lost to the atmosphere as gaseous ammonia, which can cause terrestrial eutrophication problems when it is redeposited on N-deficient ecosystems, such as heathlands, bogs and semi-natural grasslands and woodlands.

Soil pollution by heavy metals occurs mainly on mining and industrial sites or where metal-contaminated sewage sludge has been applied to agricultural land. Heavy metals can adversely affect soil organisms, especially the bacteria responsible for nutrient cycling. Metal-polluted soils can be treated in various ways, the most promising being growth of hyperaccumulator plants, which store large amounts of specific metals in their leaves. The metals can then be recovered by harvesting and burning the plants. Organic pollutants from industrial sources, sewage sludge applications or pesticide spillages do not persist in soil as long as heavy metals, as they are degraded by microbial activity or sorbed on clay minerals or soil organic matter.

FURTHER READING

Addiscott, T.M., Whitmore, A.P. and Powlson, D.S., 1991, *Farming, Fertilisers and the Nitrate Problem*, Wallingford: CABI.

Adriano, D.C., Bollag, J-M., Frankenberger, W.T., Jr and Sims, R.C., 1999, *Bioremediation of Contaminated Soils*, Madison WI: ASA, CSSA and SSSA.

L'Hirondel, J. and L'Hirondel, J.-L., 2002, *Nitrate and Man – Toxic, Harmless or Beneficial?*, Wallingford: CABI.

6 Soil acidification

A nation that destroys its soil destroys itself.

Franklin D. Roosevelt

Introduction

Acidification is another common problem resulting from pollution of soil. Slow acidification occurs in humid regions by natural leaching processes, but the rate of acidification has been greatly increased in recent centuries by inputs of sulphur and nitrogen compounds from the atmosphere, arising mainly from industrial and domestic emissions. Accelerated acidification causes numerous agricultural, environmental and ecological problems. Various forms of calcium carbonate and other mildly alkaline materials are used to counteract soil acidity. The final part of this chapter describes methods for calculating the amounts required to meet target levels of decreased acidity suitable for various crops in soils of different acid buffering capacities (abilities to absorb acidity).

6.1 CAUSES OF SOIL ACIDITY

The unaltered parent materials of many soils contain calcium or other carbonates, which dissolve slowly in the soil water to make it weakly alkaline (pH 7.8–8.4) (for definition of pH see Section 6.2.1). Others are neutral or very weakly alkaline (pH 6.5–7.8) because they contain abundant basic cations (mainly K^+, Na^+, Ca^{2+} and Mg^{2+}) sorbed on the surfaces and occurring within primary minerals, such as feldspars, micas, pyroxenes and amphiboles.

Soils initially become acid when water moving downwards through the profile contains compounds that act as donors of H^+ ions (protons) and can dissolve the carbonate or remove the sorbed cations faster than they are replaced by weathering of the primary minerals. Cations sorbed on the surfaces of minerals and organic matter are exchanged for H^+ ions (protons) in the water, but this replacement is reversible if further cations are released from the primary minerals by weathering processes, or are supplied in some other way: e.g., by addition of lime (CaO) to the soil. In releasing basic cations the primary minerals are often weathered to form clay minerals. The amounts of carbonate and minerals containing basic cations are usually quite large, so at this stage soil pH decreases very slowly. It is buffered in the range 5.0–6.5 by the large resources of basic cations, but once these are exhausted the pH can continue to decline below 5.

When all the basic cations are used up, the surfaces of clay minerals and organic matter become saturated with H^+ ions. At a pH of approximately 4, the alumino-silicate lattices of clay minerals become unstable, and aluminium and silica are slowly released into solution. The silica dissolves in the soil water as H_4SiO_4, and is partly or completely lost by leaching. However, the aluminium is sorbed on clay and organic matter surfaces as Al^{3+} ions. Hydrolysis of the sorbed Al^{3+} then produces more H^+ ions, which cause further breakdown of the clay mineral lattices and continued loss of silica. At this stage the pH is buffered around 4 by the slow breakdown of the reserves of alumino-silicate clays. However, loss of clay minerals involves an irreversible decrease in cation exchange capacity, which implies a permanent loss of nutrients and therefore a decrease in soil fertility. Eventually, through loss of silica the only minerals remaining are the aluminium and iron oxides, gibbsite, goethite and haematite, though with further inputs of H^+ ions even these minerals dissolve, buffering the pH at approximately 3.

6.1.1 Sources of protons

The main sources of protons in the soil water are as follows.

Carbon dioxide from the atmosphere and soil air

This dissolves in water to form carbonic acid, which dissociates according to eq. (6.1):

$$CO_2 + H_2O \quad H_2CO_3 \quad H^+ + HCO_3^- \quad H^+ + CO_3^{2-} \tag{6.1}$$

The carbon dioxide content of the atmosphere is sufficient to lower the pH of otherwise pure rain only to 5.67. However, respiration by soil animals and plant roots increases the carbon dioxide content of the soil air, driving the reaction in eq. (6.1) to the right and thus lowering the pH further by producing more H^+.

Inputs of SO_2, SO_3, N_2O, NO, NO_2, NH_3 and HCl from the atmosphere

The SO_x, NO_x (NO and NO_2) and HCl atmospheric pollutants arise mainly from burning of coal and oil, which contain some S, N and Cl. However, some SO_x comes from volcanoes, some NO_x from soils and atmospheric reactions caused by lightning and some HCl from incineration of municipal waste. N_2O is emitted by soils, animals and the leaves of plants, especially trees (Jenkinson, 2001). The principal sources of atmospheric NH_3 are emissions from plant leaves and the microbiological decomposition of manure and urine from farm livestock (Section 5.2.5). Lesser amounts are derived from N fertilizers, sewage sludge, industrial emissions and combustion of coal. Most is in gaseous form, but small amounts are dissolved in rain. NH_3 is acidifying because in the soil water it dissolves to form ammonium ions (NH_4^+), which are then nitrified (oxidized) by chemoautotrophic bacteria (*Nitrosomonas* and *Nitrobacter*) to nitrate ions and in the process release protons:

$$NH_4^+ + 2O_2 \quad NO_3^- + 2H^+ + H_2O \tag{6.2}$$

If the nitrate ions are taken up by plants, the protons are balanced by OH^- ions exuded from the roots, but if the nitrate is leached acidification occurs.

Oxidation of iron pyrite (FeS_2) and other sulphides in soils derived from marine and brackish water clays

The SO_4^{2-} ions in seawater are reduced to H_2S in anaerobic sediments (e.g., estuarine mangrove swamps), and this combines with Fe^{2+} initially to precipitate metastable iron monosulphides and

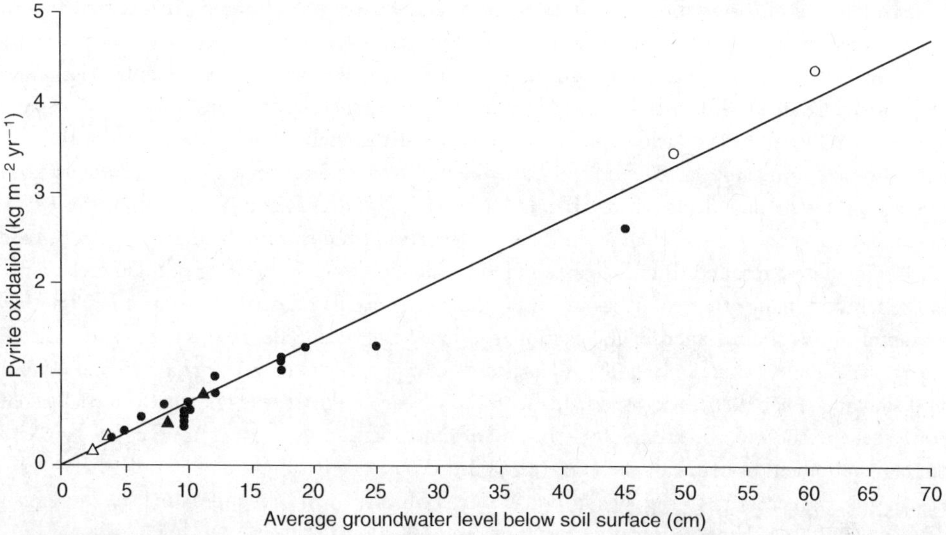

Figure 6.1 Relationship between annual pyrite oxidation rate and mean groundwater depth. Closed circles are model calculations; other symbols are for Indonesian field and laboratory measurements
Source: Bronswijk *et al.* (1995).

then to form pyrite (FeS_2) after deeper burial within the sediment (Wilkin and Barnes, 1997). Re-exposure of the sulphides to the atmosphere by natural lowering of the water table, installation of subsoil drains or excavation of sulphide-rich subsoil leads to the reactions:

$$FeS_2 + 3O_2 + 2H_2O \quad Fe^{2+} + 2SO_4^{2-} + 2H^+ \qquad (6.3)$$

$$2Fe^{2+} + O_2 + 2H_2O \quad 2FeOOH + 2H^+ \qquad (6.4)$$

which can lower the soil pH to <3.5 if no carbonate or basic cations are present. The lower the mean water table level, the greater the rate of pyrite oxidation (Figure 6.1). The resulting soils are often termed acid sulphate soils. FitzPatrick *et al.* (1998) distinguished Actual Acid Sulphate Soils (sulphides have been or are being oxidized) and Potential Acid Sulphate Soils (sulphides not yet oxidized). The oxidation process is greatly accelerated by *Thiobacillus ferrooxidans* and other Fe- and S-oxidizing bacteria (Lundgren and Silver, 1980).

Organic compounds rich in carboxyl and phenolic groups produced by humification of soil organic matter

These compounds are most abundant in woodland soils, podzols and humified peats. Organic acids are also exuded by plant roots.

6.1.2 Human effects on soil acidification

Many of these processes occur naturally, but soil acidification can be accelerated by human activities, such as burning of fossil fuels and wood, increasing the humification of soil organic matter by cultivation, improving subsoil aeration by installation of field drains, application of ammonium fertilizers, removal of basic cations in harvested crops and growth of legumes such as clover. In Britain, where the predominant wind direction is from the west, the combined emissions from power stations, industry and vehicle exhausts result in a strong eastward increase in the acidity of rainfall (Review Group on Acid Rain, 1997), with localized increases near cities, concentrations of heavy industry and motorways. Long runs of acid deposition data, such as the analyses of rain collected in the one-thousandth-acre rain gauge at Rothamsted Experimental Station, southeast England, from 1852 to the present, show gradual decreases in pH and increases in S, NH_3-N and NO_3-N from the 1850s to 1950s, followed by more rapid changes in the same direction from the 1950s to early 1980s (Goulding *et al.*, 1998a). These resulted from increases in industrial and domestic burning of fossil fuels.

In the UK and many other countries, deposition of S has decreased markedly since the 1980s. For example, at Woburn Experimental Station, southeast England, S deposition was approximately 85 kg $ha^{-1}yr^{-1}$ in 1970 but, since 1980, it has decreased to approximately 15 kg $ha^{-1}yr^{-1}$, the pre-1850 level (Goulding *et al.*, 1986), because of the decrease in heavy industry, the change from coal to natural gas as a fuel and the widespread desulphurization of power station flue gases. As 15 kg $ha^{-1}yr^{-1}$ is less than the requirement of many arable crops (e.g., oilseed rape needs approximately 50 kg S $ha^{-1}yr^{-1}$), it is now becoming necessary to supply additional S in fertilizers. In Britain approximately 75 per cent of arable land could now benefit from S fertilization, and 30–40 kg fertilizer S ha^{-1} yr^{-1} is now recommended for oilseed rape, and 15–25 kg for cereals.

Since 1980 NH_3-N and NO_3-N dissolved in rain have also decreased. However, the pH of rain has decreased only slightly over the same period because these inputs have been replaced by increasing amounts of other acidic N compounds, such as gaseous NO_2, HNO_3 and NH_3 and particulate N in dust. Gaseous forms of N are measured by perspex diffusion tubes, in which gases are absorbed on gauze impregnated with absorbent reagents such as triethanolamine for NO_2. The increases in these N compounds have been observed in many parts of the world, and have resulted mainly from increased burning of oil and natural gas for transport, power generation and indoor heating.

Recent measurements at Rothamsted Farm indicate a total N deposition of approximately 44 kg $ha^{-1}yr^{-1}$ to winter cereals, of which about 9 kg is ammonium and nitrate in the rain, 2 kg is in gaseous ammonia, 3 kg is in particulates (dust), 14 kg is in NO_2 and 16 kg is in nitric acid aerosol (Goulding *et al.*, 1998a). Approximately twice these amounts are deposited in areas of deciduous woodland at Rothamsted, because trees can scavenge more from the atmosphere than smaller plants such as grass or cereals. Conifers and other evergreen trees are more effective scavengers than deciduous species. The total amounts at Rothamsted are greater than expected for a rural site because of local conditions, such as the proximity of the M1 motorway and other major roads, the town of Harpenden and several other large cites and the international airport at Luton. However, the soil in many urban and industrial areas is likely to receive even greater inputs of acidifying compounds.

The Rothamsted results also show that amounts of most N compounds in wet and dry deposition are greater in winter than summer, mainly because of increased heating of buildings. However, for NO_2 there is often a small, short-lived decrease from the winter peak over the Christmas/New Year holiday, when many factories and offices close and there is less use of cars for travelling to work (Hargreaves *et al.*, 2000).

6.1.3 The pH buffering capacity of soils

The effects of acidic inputs from various sources are counteracted in soils by the various weathering reactions outlined above, which consume protons. The abundance of carbonates and weatherable alumino-silicate minerals determine most of the soil's pH buffering capacity. This is largest in calcareous (carbonate-containing) soils, but much less in non-calcareous soils because the weathering reactions involving alumino-silicate minerals are very slow. In soils containing few or no easily weathered alumino-silicates (e.g., sandy soils composed mainly of quartz and with little alumino-silicate clay), the buffering capacity is very small, and acidic inputs can rapidly decrease the pH.

The balance between acidic inputs and buffering capacity gives rise to the concept of 'critical load', which is the highest level of acidic (or other polluting) inputs that does not cause chemical changes in the soil leading to harmful effects in sensitive ecosystems (i.e., cannot be buffered for a long period by weathering reactions). A widely used factor for assessing the critical load of acidic inputs is the ratio of $Ca^{2+} + Mg^{2+}$ to Al^{3+} in the soil solution at which significant damage to fine tree roots is observed (Critical Loads Advisory Group, 1994).

By absorbing Al^{3+} ions, organic matter also plays a role in buffering the pH of many soils. Al^{3+} is also sorbed on clay minerals, and distinguishing the two types of sites is difficult, especially in soils formed on volcanic deposits (Andosols). In these soils poorly crystallized alumino-silicate clays (allophane and imogolite) form complexes with organic matter that prevent mineralization of the latter, so that it accumulates rapidly, giving the soils an intense black colour to depth. The large amounts of Al^{3+} held on the abundant and very reactive clay–humus complexes give Andosols a large pH buffering capacity.

Al^{3+} associated with different soil components, such as humus and clay minerals of different crystallinity values, is usually estimated by selective dissolution techniques (Paterson *et al.*, 1993). Extraction with sodium pyrophosphate solution at pH 10 is often used to estimate the Al held in humus complexes, and acid ammonium oxalate is used to measure the total reactive pool of Al associated with organic complexes, poorly crystallized clays such as allophane and imogolite and well-crystallized layer silicates. Al extracted with 0.3 M lanthanum chloride solution at pH 4 has been used as an indicator of lime requirement in acidic organic soils (Hargrove and Thomas, 1984). Extraction with 1 M potassium chloride has been proposed as a method for estimating total exchangeable (i.e., plant-available) Al, and therefore for assessing the lime requirement of mineral soils (Kamprath, 1970), but it is probably unsuitable for organic-rich soils with abundant Al–humus complexes (Ponette *et al.*, 1996).

6.2 DEFINITION AND MEASUREMENT OF PH

6.2.1 Definition of pH

The concentration of H^+ ions in water or soil samples is indicated by the pH value, which is defined as the negative $\log_{10}$ of the concentration of active hydrogen ions (in moles dm^{-3}). The term is derived from the French 'puissance d'hydrogen'. Pure water is ionized or dissociated into protons (H^+) and hydroxyl (OH^-) ions to only a very small extent. At 25°C the ionic product for water (K_w), i.e., $[H^+][OH^-]$, is only 1.0×10^{-14}. As a solution is neutral when $H^+ = OH^-$, the concentrations of H^+ and OH^- ions at neutrality are both approximately 10^{-7} (i.e., pH = 7) though the exact value is actually temperature-dependent. As the pH scale is logarithmic, each unit decrease in value indicates ten times the abundance of H^+ ions (i.e., there are ten times more H^+ ions in a solution of pH 5 than a solution of pH 6). Consequently calculations of measures such as the mean or standard deviation of numerous pH measurements should be based on antilogarithms of the values, not the raw values themselves. The theoretical range of pH is from approximately −1.0 (extremely acid) to approximately +14.7 (extremely alkaline), though values for natural soil and water samples are usually within the range 1–12.

6.2.2 Methods of measuring pH

Filter papers impregnated with indicator dyes such as bromothymol blue or chlorphenol red, which change colour over certain pH ranges, are used in field kits for determining soil pH, but the values obtained are only approximate, usually within 0.5 pH units. In particular they are affected by the ratio of soil:water in the mixture tested and by masking of the colour of the dye by that of the soil itself.

The now-standard glass electrode pH meter is based on the observation that the electrical potential between membranes of low-alumina (low melting point but relatively high electrical conductivity) glass is dependent on the pH of the solution in which they are immersed. The glass electrode is a thin-walled glass bulb containing a solution of known and constant pH and a platinum wire connected to a sensitive thermionic valve potentiometer. The potentiometer is also connected to a calomel electrode containing 1N potassium chloride solution, which is allowed to leak slowly from the electrode through a capillary tube into the solution to be measured.

The pH is measured on mineral soil by immersing the probe in a carefully homogenized 1:2.5 weight:volume suspension of air-dried soil:distilled water or soil:0.01M calcium chloride solution. For organic (e.g., peat) soils the reading is taken in a 1:20 w/v suspension using calcium chloride only. Readings in calcium chloride solution help eliminate short-term changes in pH and electrolyte content resulting from addition of fertilizers, oxidation of organic matter, leaching by rain and other factors, and therefore better indicate the long-term pH value than do readings in water.

6.3 EFFECTS OF SOIL ACIDITY

6.3.1 Calcifuge and calcicole plants

Natural vegetation, farm crops and pasture species are all influenced by soil pH. The main exchangeable cations in soils are Ca^{2+}, Mg^{2+}, K^+, Na^+ and NH_4^+, of which Ca^{2+} is much the most abundant in all but the most acidic or most alkaline soils. In strongly acidic soils Ca^{2+} and other basic cations are replaced mainly by Al^{3+}, and plant sensitivity to pH is essentially a response to the relative abundance of exchangeable Al^{3+} and Ca^{2+}. Calcifuge plants require little Ca and are tolerant of Al; calcicoles are Ca-demanding and intolerant of Al. Among the cultivated crops, cocksfoot grass, rye, oats and potato are all calcifuges; lucerne, barley and sugar beet are calcicoles; and wheat, rape and maize are intermediate (mildly sensitive to acidic conditions). In general, tropical crop species are more tolerant of acidity than those that have evolved in the carbonate-rich soils typical of mid- and high-latitudes; consequently, many of them grow best at soil pH values of 5.0–5.8, whereas most temperate crops grow best at pH 6.5–7.0.

6.3.2 Metal uptake by plants

Strongly acidic soil conditions, as in acid sulphate soils, lead to rapid weathering of alumino-silicate minerals and the release of Si, Al, Fe, Ca, K and trace elements such as B, Fe and Cr (Sammut *et al.*, 1996). The effects of pH on concentrations of these and other elements in soil solution are shown in

Figure 6.2. Many of these elements are taken up by plants, often affecting their growth and the health of animals eating the plants. Also, if the ions are transferred to streams or other surface waters, they can poison fish or result in other major ecological changes. If there is limited leaching and outflow to streams, as in flat, low-lying, undrained areas where the water table is permanently close to the ground

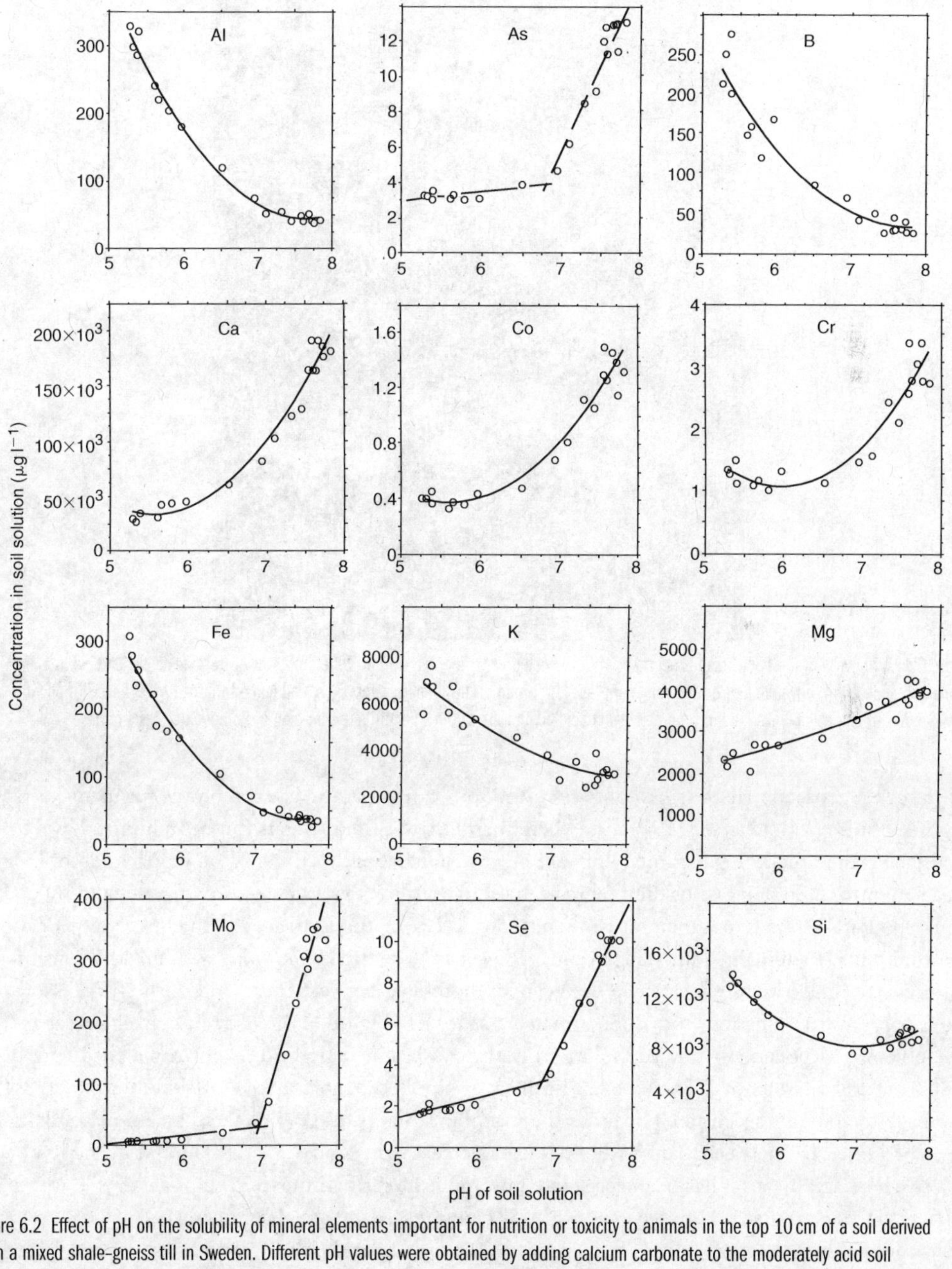

Figure 6.2 Effect of pH on the solubility of mineral elements important for nutrition or toxicity to animals in the top 10 cm of a soil derived from a mixed shale-gneiss till in Sweden. Different pH values were obtained by adding calcium carbonate to the moderately acid soil
Source: after Tyler and Olsson (2001).

Plate 6.1 Aerial view of the Park Grass Experiment at Rothamsted Experimental Station. Different fertilizer treatments, which have remained constant since 1854, have created a wide range of grass and broadleaved plant assemblages, and some treatments (nil and N applications as ammonium sulphate) have resulted in progressive soil acidification (reproduced with the permission of Rothamsted Research)

surface (e.g., coastal marshes), protons and metal ions can remain in the soil profile for many years or even centuries (White *et al.*, 1997). Elsewhere their loss to surface waters can be minimized by artificial containment (e.g., maintaining a high water table level), neutralization (application of lime to the soil or to drain outfalls) or dilution (e.g., limiting outflow to periods of high rainfall and runoff).

In the Park Grass Experiment at Rothamsted (Plate 6.1), the fertilizer treatment of many of the pasture plots has remained unchanged since 1854. Under natural conditions with no additions of fertilizer or lime (the nil plot), there has been a progressive decrease in the pH of the surface soil (0–23 cm) over this period from approximately 5.2 to 4.1 (Figure 6.3). As a result, exchangeable Al (soluble in 1M potassium chloride solution) in the soil has increased from approximately 10 mg kg^{-1} to 80 mg kg^{-1}, and since 1986 there has been an unexpectedly rapid uptake of aluminium by the herbage to amounts seven to eight times the maximum that would be tolerated by animals (Blake *et al.*, 1994). The increasing soil acidity has also increased the amounts of Pb and Ni in the hay. On plots where acidification has been accelerated by application of ammonium sulphate as a source of N, the pH is now around 3.5, and even more Al is being released from clay minerals and taken up in the herbage. However, use of sodium nitrate as a source of N over the same period has lowered the soil pH much less, from 5.7 to 5.5.

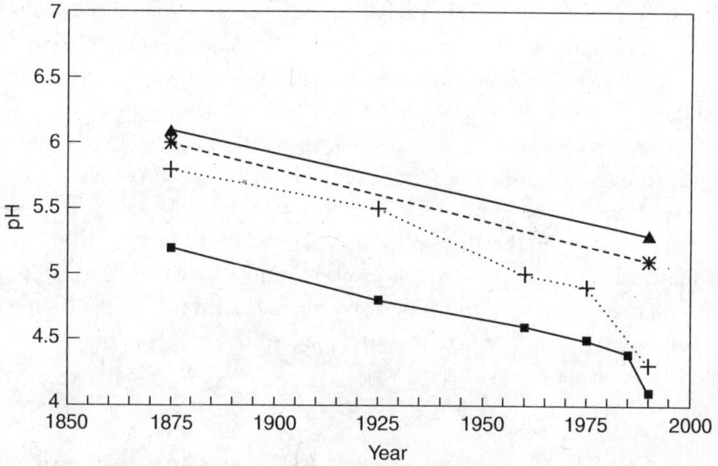

Figure 6.3 Decrease in pH with time in successive horizons of the soil from the unlimed, unmanured plot of the Park Grass Experiment, Rothamsted
Source: Blake *et al.* (1994).

6.3.3 Effects on the species composition of grassland and woodland

Acidification of the Park Grass plots has decreased the number of plant species, often to a startling extent. On plots where the natural acidification has been controlled by repeated liming to maintain a pH of 7, there are about 50 grass and broadleaved species, the assemblage varying according to the nutrients applied. Where pH has decreased under the natural effects of rain and dry deposition, the species list has slowly reduced to 30–40 species, the exact number varying from year to year, but where acidification has been accelerated by use of ammonium sulphate fertilizer there are now only 1–3 species of acid-tolerant grasses (Goulding *et al.*, 1998a). On a subsection of the ammonium sulphate-treated plot the number of species has recovered to about 15 by regular applications of lime since 1903. Applying N as sodium nitrate initially decreased species numbers from about 50 to 35 by eliminating legumes, and since 1880 there has been a further slow decline to about 20 species, because of increasing acidifying inputs from the atmosphere.

Natural acidification under woodland is often more rapid than under grass because the greater canopy slows the airflow and thus scavenges larger amounts of acidic compounds from the atmosphere. For example, compared with the decrease of 1.0 pH unit over about 150 years on the unfertilized and unlimed plot of the Park Grass Experiment, the topsoil (0–23 cm) of the Geescroft Wilderness Experiment at Rothamsted, which was an unlimed arable field until it was set-aside in 1885 and has since become a mature stand of oak (*Quercus robur*) trees, decreased in pH from 6.2 in 1883 to 3.8 in 1990 (Blake *et al.*, 1999). As a result Al^{3+} now occupies approximately 70 per cent of the exchange complex in the topsoil, and beneath the oak trees there is a very sparse vegetation understorey.

The topsoil of both the Park Grass Experiment and Geescroft Wilderness is a silty clay loam with 20–25 per cent clay. Soils with less clay to buffer the acidity, such as sands and sandy loams, acidify even more quickly under woodland and other land uses.

Acidity also influences the soil microbial biomass and animal population. For example, the decreased microbial biomass in very acidic soils is insufficient to feed the population of earthworms typical of neutral or alkaline soils. As a result, leaf litter tends to accumulate on the soil surface because it is not incorporated. For example, on the most acidic subplots of the Park Grass experiment, a peaty mat of unincorporated grass residues is accumulating, and beneath the oak trees of Geescroft Wilderness there is a 2–3 cm thick litter layer of unincorporated oak leaves. Larger animals such as moles, which feed on earthworms, are also absent in acidic soils. However, the absence of molehills is advantageous for recreational turf so, as most grass species can tolerate low pH, the soil of playing fields and golf courses should be kept fairly acidic.

Nitrogenous acidic inputs from the atmosphere can also affect the botanical composition of some semi-natural habitats, which are adapted to low levels of nitrogen supply. In parts of upland Britain, the recent increase in deposition of N from the atmosphere has led to changes from moss- or heather-dominated to grass-dominated communities, and to reductions in the species diversity of species-rich unfertilized meadows (Woods, 1993).

6.3.4 Effects of acidity on nutrient supply, crop growth and plant diseases

In acidic soils the turnover of organic matter and recycling of the nutrients it contains, especially N and P, are slowed because microbial activity is decreased. For example, the nitrifying bacteria that oxidize ammonium to nitrate (the form in which N is taken up through plant roots) are most active between pH 6 and 8. In acidic soils phosphorus is also rendered unavailable by strong adsorption on iron and aluminium oxides, though this problem is decreased by the chelating effects (formation of soluble Fe and Al complexes) of organic decomposition products, such as citrate, oxalate, tartrate, malonate and galacturonate. The availability of P to plants is greatest at pH 6.5–7.5. Acidity also leads to weak development of the nodules containing N-fixing bacteria (*Rhizobium* spp.) on the roots of legumes, and to increased solubility of several metals (Al, Mn, Fe, Zn, Cd, Pb and Ni) to concentrations that are toxic to plants. However, small increases in Al availability can stimulate the growth of some crops, such as tea, maize, clover, lucerne and ryegrass.

Evidence from field experiments has indicated optimum soil pH values for the growth of crops (Table 6.1), and also the values below which growth and yield are restricted, even if adequate nutrients are supplied (Table 6.2). For example, uneven liming many decades ago gave rise to a soil pH gradient along a narrow strip (Plate 6.2), which is planted to winter wheat each year; the wheat grows well and matures annually at pH > 5.5, but does not survive after germinating where pH < 4.5.

Acidity also encourages certain nematode infections, such as cereal cyst nematode (Johnston, 1997), and a range of fungal diseases, including club-root in brassicas (*Plasmodiophora brassicae*), fusarium-wilt (blight) in tomatoes (*Fusarium oxysporum*), damping-off in legumes, sugar beet and cucumber (*Pythium*) and take-all in cereals, which is caused by the root-infecting fungus *Guaemannomyces graminis* var. *tritici*. However, some pathogenic micro-organisms survive better in alkaline or neutral than acid soils, and liming may encourage them; for example, scab in potatoes (caused by the actinomycete *Streptomyces scabies*) often develops after heavy lime applications.

Overliming of soils to pH values >7.5 can also lead to poor plant growth because P, K and Mg become less available and various trace elements are immobilized, especially B, Co, Cu and Zn in sandy soils and Mn in organic soils. The trace element deficiencies can often be treated by applying salts such as sodium borate, cobalt chloride, copper sulphate or manganese sulphate to the soil, but it is important to calculate the application rates carefully, as even small excesses of these micronutrients

Table 6.1 Optimum soil pH values for arable crops and pasture

Cropping	Mineral soils	Peaty soils
Continuous arable	6.5	5.8
Grass with occasional barley crops	6.2	5.5
Grass with occasional wheat or oat crops	6.0	5.3
Continuous grass or grass-clover swards	6.0	5.3
Brassicas	6.5[a]	5.8
Other vegetables	6.5	5.8
Fruit, hops, vines	6.0-6.5	5.3-5.8

[a]7.0 if club-root is a problem.
Source: after Ministry of Agriculture, Fisheries and Food (2000).

Table 6.2 Soil pH values below which crop growth is adversely affected

Soil pH value	Crop
4.7	Fescue grasses, ryegrass
4.9	Potato, rye, blackberry
5.0	Apple
5.1	Tomato, strawberry
5.3	Oats, timothy grass, pear
5.4	Linseed, turnips, cabbage, parsnip, rhubarb, swede
5.5	Hops, maize, wheat, cucumber, raspberry
5.6	Oilseed rape, white clover, cauliflower, plum, red currant
5.7	Brussels sprouts, carrot, onion
5.8	Leek
5.9	Barley, sugar beet, peas, asparagus, red beet
6.0	Beans, black currant
6.1	Lettuce
6.2	Celery

Source: after Ministry of Agriculture, Fisheries and Food (1973).

can be toxic to plants. The optimum pH for supply of Mn to cereal, sugar beet and vegetable crops is 6.2 in mineral soils and 5.8 in peat soils, though the tolerance of wheat to Mn and its growth rate also depends on the Mg:Mn ratio in shoot tissue (Goss *et al.*, 1992).

Lime-induced chlorosis leading to lack of chlorophyll and inefficient photosynthesis in many calcifuge plants, including some fruit trees, is thought to result from changes in the metabolism of iron within the plant, rather than a lack of iron or the inability of the plant to take up iron. In soils limed to pH values >7.5, P is immobilized by precipitation of calcium phosphate and increased sorption on clay and organic matter. For every unit increase in pH, bicarbonate extractable P is decreased by 3–7 $mg\,kg^{-1}$ (Curtin and Syers, 2001).

Plate 6.2 The 'acid strip' at Rothamsted Experimental Station. Wheat is sown each year on soil with a pH gradient, and germinates but does not survive where pH < 4.5 (photo J.A. Catt)

Although neutral or weakly alkaline soils (pH 5.5–8.5) can usually provide sufficient calcium and magnesium for lime-demanding plants, those grown on more strongly alkaline soils may exhibit calcium or magnesium deficiencies. This is because the exchange complex becomes dominated by sodium and the calcium and magnesium are concentrated in the weakly soluble carbonates.

6.3.5 Effects on human and animal health

Human and animal health may be affected by soil acidity, resulting in increases in the Al and heavy metal contents of grass and arable crops or of ground- or surface waters used as sources of drinking water. The problem of Al in drinking water is sometimes made worse by the addition of aluminium sulphate to water supplies to coagulate the fine organic and clay particles that cause discoloration, or by the use of aluminium storage and cooking vessels. Nausea, skin and kidney complaints, arthritis, osteomalacia (bone softening) and neurological disorders such as Alzheimer's disease have all been attributed to excess Al in the human diet, though the links are often uncertain. The maximum concentration of Al in drinking water is set by the EU at 200 $\mu g\ l^{-1}$, but it has been estimated that in the UK over 2 million people regularly drink water with Al contents greater than this.

6.3.6 Subsoil acidification

Acidification initially affects the surface layers of soil, but if it continues unchecked it extends to subsoil horizons. In acid sulphate soils, where the sulphide-containing parent material is exposed to oxidation by intensive drainage operations, strongly acidic conditions (pH < 3.5) can extend to the depth of the lowered water table within five years (Bronswijk *et al.*, 1995). On Geescroft Wilderness and in the most acidic plots of the Park Grass Experiment at Rothamsted the decrease in pH that has occurred naturally or through use of ammonium sulphate fertilizer over the last 120–150 years can be detected at >69 cm depth (Figure 6.2). Over geological time, natural acidification can affect the subsoil to many metres depth. For example, gravelly terraces of the River Thames in southern England, which were rich in chalk fragments when they were originally deposited more than 500,000 years ago, have now lost their entire chalk contents to depths of 15 m or more.

6.3.7 Modelling the effects of soil acidification

Because of the complex chemical and physical processes leading to soil acidification and the influence of various soil management factors, such as drainage, the environmental consequences of acidification are best predicted using simulation models. Bronswijk *et al.* (1995) developed one such model, the Simulation Model for Acid Sulphate Soils (SMASS), to predict the rates of acidification in mangrove swamp (acid sulphate) soils and of loss of acid by leaching under various conditions of artificial drainage.

6.4 SOIL TREATMENT FOR ACIDITY

6.4.1 Lime requirement

Ground limestone or chalk ($CaCO_3$), burnt lime (CaO), slaked (or hydrated) lime ($Ca(OH)_2$), magnesian (dolomitic) limestone (dolomite is $MgCO_3{\cdot}CaCO_3$), marl (calcareous clay), basic blast furnace slag rich in easily hydrolysed calcium silicates, shell sand, calcified seaweed, lime sludge from sugar beet factories and lime-flocculated sewage sludge have all been used to decrease soil acidity. Farmyard manure, crop residues and other organic materials can also increase soil pH (Whalen *et al.*, 2000), though less effectively than calcium carbonate or lime. Organic additions are also useful for immobilizing Al, Fe, Mn and other metal ions in acid soils as insoluble organo-metallic complexes, thus minimizing uptake by plants or leaching losses to streams.

When a limestone such as chalk is added to wet soil, it slowly dissolves according to:

$$CaCO_3 + 2H_2O \quad Ca(OH)_2 + H_2CO_3 \tag{6.5}$$

and the base $Ca(OH)_2$ (or slaked lime if the neutralizing agent is added in this form) reacts with CO_2 in the soil air to form soluble calcium bicarbonate:

$$Ca(OH)_2 + 2CO_2 \quad Ca(HCO_3)_2 \tag{6.6}$$

Burnt lime (CaO) reacts rapidly with water also to form $Ca(OH)_2$. The resulting pH is indicated by:

$$2pH = K - \log(P_{CO_2}) - \log(Ca) \tag{6.7}$$

where P_{CO_2} is the partial pressure of CO_2 (in bars), (Ca) is the activity (M) of Ca^{2+} ions in the soil solution and $K = 9.6$ for pure calcium carbonate. For impure limestones the value of the constant K is >9.6 (usually around 10.4). In soils containing free calcium carbonate (Ca activity 0.001 M) the pH at low partial pressures of CO_2 may exceed 7.8, but where the soil air contains >1 per cent CO_2 ($P_{CO_2} > 0.01$ bar), as under some pastures, the pH in the presence of free carbonate is <7.7.

The amount of ground limestone or other liming material required to raise the soil's pH to a desired value over a given depth (usually 15–20 cm) is known as the soil's lime requirement. It is influenced by the soil's pH buffering capacity as well as its existing pH and by the neutralizing value (NV) of the liming material. The maximum theoretical NV for pure calcium carbonate is 56 (100 kg of $CaCO_3$ has the same acid-neutralizing power as 56 kg of CaO). Ground chalk and magnesian limestone have NVs of 50–55, most other limestones are in the range 45–50, sugar beet lime sludge is usually about 50, basic slag is in the range 35–45 and shell sand is often <40. The smaller its NV value the more of a liming material is required to reach a target pH.

The effectiveness of liming materials also depends on their particle size distribution. Coarse particles react more slowly with the soil water than small, and can remain unaltered in the soil for many months or even years. For ground limestone or chalk the statutory minimum quality in UK is that 100 per cent of material sold must be <5 mm, 95 per cent must be <3.35 mm and at least 40 per cent must be <0.15 mm. Shell sand and calcified seaweed must be entirely <6.3 mm.

A soil's lime requirement therefore depends on the following factors:

- the present and target pH values;
- the neutralizing value (NV) and particle size distribution of the liming material;
- the depth of soil to be treated;
- the buffering capacity of the soil;
- how thoroughly the soil and liming material are mixed.

Because the pH scale is logarithmic, the amounts of lime required to increase soil pH by one unit increase approximately ten times for each unit decrease in the present value; for example ten times as much lime is required to raise the pH from 5 to 6 as from 6 to 7. In soil containing little or no free $CaCO_3$, the buffering capacity is usually estimated by its cation exchange capacity (CEC). The greater the CEC, the greater the amount of lime required to meet a target pH. However, soils with a high CEC acidify less rapidly and therefore require less frequent liming than those of low CEC. CEC increases as the amounts of clay and organic matter in the soil increase.

Repeated use of dolomitic limestone to counteract acidity can result in an excess of Mg, which may lead to an imbalance between exchangeable K^+ and Mg^{2+}, with decreased availability of K. For some crops (e.g., potatoes) this can result in decreased yield because of K deficiency.

6.4.2 Estimating lime requirement

The lime requirement for a particular soil type can be estimated from field experiments in which the effects on soil pH of different application rates of a certain liming material are measured at various times after application. This method has been used in Britain for an upland reseeded grassland soil (Dampney, 1985) and for the lowland soil (Batcombe series) under permanent grassland and arable use at Rothamsted Experimental Station (Goulding *et al.*, 1989). Using early data from the Rothamsted liming experiments, Bolton (1977) proposed a simple model to predict lime loss as a function of temperature, soil pH, acidifying inputs from fertilizers and the atmosphere and the amount of water draining through the soil profile. This has recently been updated and computerized to form the

RothLime Model, which is available at the following website: http://www.Rothamsted.bbsrc.ac.uk (accessed 25 February 2004). RothLime uses knowledge of responses to lime applications obtained from various field experiments to predict lime requirement for a range of soil types.

However, field experiments are a slow and expensive way of accumulating the necessary data, and the results cannot be transferred easily or reliably to other soil and crop types or to sites with other rainfall values or where other types of fertilizer are used. More often, using material of a stated NV and statutory size distribution, the lime requirement is calculated from laboratory measurements of buffering capacity, which are multiplied by standard factors that take into account the initial soil pH, a target pH (usually 6.5, the optimum for growth of most crops in temperate soils), the depth of soil to be treated and the incomplete mixing resulting from standard soil cultivations. The Ministry of Agriculture, Fisheries and Food (MAFF) (1981) gave a simple laboratory method for measuring buffering capacity.

Lime applied to the soil surface and not mixed with the subsoil by deep cultivation has a very slow effect in neutralizing subsoil acidity. Even heavy dressings may take many years to increase subsoil pH, especially in clay-rich soils. In the first few years after application, lime spread on the soil surface and not incorporated affects only the uppermost 2–5 cm of soil. The persistence of subsoil acidity even after incorporation of lime to 15–20 cm is most likely to cause problems with growth and development of deeper-rooting crops such as sugar beet and fruit trees. Noble *et al.* (1995) described two Ca-saturated organic (humic and fulvic) products derived from coal, which are more mobile in soil and can therefore increase the pH and exchangeable Ca and decrease exchangeable Al to greater depths than surface-applied lime. However, they are not so readily available as lime, and the Ca they contain may displace exchangeable Mg and K, which are then lost by leaching, leading to Mg and K deficiencies in some crops.

6.4.3 Applying lime

In agriculture, lime is usually applied mechanically at a uniform rate over a field, but pH often varies widely within fields. To avoid general over- or underliming, it is therefore necessary to sample the soil of a field adequately. A representative soil sample is usually obtained by bulking at least 25 cores taken in a regular pattern (e.g., a square grid of points) across the field. However, this approach can still lead to localized over- or underliming, for example where there are patches of acidic non-calcareous sandy or peaty soils within a neutral or alkaline calcareous clay or chalky soil. In such circumstances uneven applications of lime applied with computer-controlled machinery using a template based on a detailed (large-scale) map of soil pH (i.e., precision farming) are likely to be cheaper and more effective than a blanket uniform application.

Because lime is required less frequently than fertilizers or other soil and crop treatments, the need for it is often neglected, even by the best farmers and gardeners. In the UK, this problem was exacerbated by withdrawal of the government-funded lime subsidy in 1976. As a result, crop nutrient deficiencies and diseases encouraged by soil acidity have become more common on soils that lack a natural supply of calcium or magnesium carbonate. Rather than wait for these problems to arise, it is cheaper in the long run to take regular pH measurements every 3–5 years and apply lime frequently in small dressings to maintain the optimum pH range. If the soil is allowed to become strongly acidic, the lime requirement is very large and, even after it is applied, the problems associated with acidity are rectified slowly and unevenly.

6.5 TREATMENTS TO ACIDIFY NON-ACID SOILS

Occasionally non-acid soils need to be acidified for the cultivation of acid-loving (calcifuge) plants. This is usually accomplished by applications of elemental sulphur, which produces sulphuric acid by bacterial oxidation:

$$6S + 9O_2 + 6H_2O \quad 6H_2SO_4 \tag{6.8}$$

However, in weakly buffered (e.g., sandy) soils, very acidic conditions can be achieved quite rapidly by excessive applications of sulphur. A safer (though more expensive) acidifier is potassium aluminium sulphate (potash alum), which produces acidity by the reaction:

$$2KAl(SO_4)_2 + 6H_2O \quad 2Al(OH)_3 + 3H_2SO_4 + K_2SO_4 \tag{6.9}$$

With alum the pH is limited to approximately 4.0 by the low solubility of $Al(OH)_3$.

Summary

Soils in humid regions are acidified naturally at a fairly slow rate because of carbon dioxide and acidic organic decomposition products dissolved in percolating water. However, since the industrial revolution, soil acidification has been greatly accelerated by acidic inputs from the atmosphere (NH_3, HCl and oxides of S and N), which has arisen mainly from the burning of fossil fuels. In soils rich in calcium carbonate and clay, acidification is much slower than in sandy, non-calcareous soils. Acidification is a problem because it limits the growth of most crops and eventually mobilizes aluminium, lead and other metals. Grass is more tolerant of acidic conditions, but in very acidic soils it takes up more of these metals than grazing animals can tolerate. Acidic soils also have limited faunal and microbial populations, so that the nutrients in plant litter are released less efficiently than in neutral or mildly alkaline soils. Some crop pests and plant diseases are more prevalent on acidic soils. In agricultural soils these effects can be minimized by applications of ground limestone or other materials that neutralize the acidity, but application rates must be calculated carefully to prevent development of problems typical of alkaline soils, such as deficiencies of certain micronutrients.

FURTHER READING

Archer, J., 1985, *Crop Nutrition and Fertiliser Use*, Ipswich: Farming Press Ltd.

Goulding, K.W.T. and Annis, B., 1998, 'Lime, liming and the management of soil acidity', *The Fertiliser Society Proceedings*, 410, 1–36.

Kennedy, I.R., 1992, *Acid Soil and Acid Rain*, 2nd edn, New York: John Wiley.

7 Modification of soil structure

The plough is one of the most ancient and most valuable of man's inventions; but long before he existed the land was regularly ploughed ... by earthworms.

Charles Darwin

Introduction

Soils contain mineral particles of various sizes: stones (>2 mm), sand (0.06–2 mm), silt (0.002–0.06 mm) and clay (<0.002 mm). In most soils, other than those very rich in sand and stones, the particles are bound together more or less strongly to create large structural units (peds or aggregates) of characteristic shapes, which are separated from one another by fissures and often contain voids or pores larger than most of the individual particles. Clay and organic matter play major roles in glueing coarser mineral particles together and, as Darwin recognized, the activities of soil organisms such as earthworms are important in creating structural units. The shapes of peds, aggregates, fissures and pores, and their stability, are important for soil properties, essential for crop growth and many soil processes of environmental importance. The natural structural organization of soil can be modified for better or worse, deliberately or unintentionally, by a range of treatments common in agriculture and civil engineering. This chapter outlines the nature of soil structure, the main factors influencing it and management techniques that may improve it. Where the original soil is lost by civil engineering projects, such as mining, quarrying or construction of landfill sites, the regeneration of a new soil with structural properties rendering it suitable for agriculture, forestry or recreational purposes involves special techniques, which are outlined at the end of the chapter.

7.1 THE NATURE OF SOIL STRUCTURE

The structural organization of soil is an important factor in determining a favourable environment for plant growth in terms of temperature, oxygen, moisture and nutrient supply, rapid and uniform germination of seeds and ease of root penetration (Curmi *et al.*, 1996). It is essential for infiltration of water rather than runoff over the land surface (Bresson and Boiffin, 1990), which has the attendant risk of surface water pollution by agro-chemicals. It also influences soil biological activity (Sims, 1990) and emissions of 'greenhouse gases' such as N_2O (Ball *et al.*, 1997).

Development of structure is perhaps the main feature that distinguishes soils from unaltered parent materials, such as sediments deposited by water, wind or ice. In these sediments the particles of various sizes are closely packed, separated mainly by packing voids that are usually no larger than the particles themselves. However, in most soils the particles have been semi-permanently rearranged to form larger peds or aggregates that are frequently separated by voids (interpedal or interaggregate voids) much larger than the individual soil particles. Peds are natural, relatively permanent structural units formed throughout the soil profile, whereas aggregates are less permanent units formed near the soil surface by cultivation or seasonal weather effects, such as freeze–thaw cycles.

Various natural soil development processes are initially responsible for the rearrangement of particles to form peds. They result from physical forces associated with the growth of plants and the movement of animals living in the soil, the activities of micro-organisms, expansion when water in pores freezes and shrink–swell resulting from changes in the interlayer water content of alumino-silicate clay minerals. Boundaries between peds become semi-permanent planes of weakness, along which the peds usually separate with each drying or freezing event.

7.1.1 Interparticle attractive and repulsive forces in soils

The strength and stability of peds and aggregates depends upon the balance between forces of attraction and repulsion between clay particles and between clay and organic matter. The attractive forces between clay particles arise mainly from electrostatic interactions resulting from charges on the surfaces and edges of layer silicate minerals. Charges on the extensive flat cleavage surfaces of layer silicates result from ionic substitutions within their molecular structures (e.g., Al^{3+} for Si^{4+}, Mg^{2+} or Fe^{2+} for Al^{3+}). Of the layer silicate clay minerals, vermiculite and smectite carry the largest surface charges (80–200 $cmol_c\,kg^{-1}$); clay mica (illite) and chlorite have smaller values (10–40 $cmol_c\,kg^{-1}$) and kaolinite very small values (2–15 $cmol_c\,kg^{-1}$). In most layer silicate minerals the charges on the broken edges are smaller than those on the cleavage surfaces, because the area involved is much less. However, in kaolinite, where there is little or no charge generated by ionic substitution, the edge charges account for most of the total charge. They are pH-dependent, having a maximum value of -2 per 0.4 nm^2 at pH >9, -1 per 0.4 nm^2 at pH 7–9, 0 at pH 6–7 and $+1$ per 0.4 nm^2 at pH 5.

Other soil minerals, including oxides such as quartz (SiO_2) and haematite (α-Fe_2O_3) and the hydrated oxides goethite (α-FeOOH) and gibbsite ($Al(OH)_3$), also have pH-dependent charges, positive in acid conditions and negative in alkaline conditions, originating from OH^{2+}, OH^- and O^- ions on broken edges. The effects of surface charge on quartz grains are usually small because, although this mineral is abundant in most soils, it mainly occurs as large equidimensional grains with a rather small total surface area. However, iron and aluminium oxides and hydrated oxides can play important roles in stabilizing soil structural units; they have large surface areas because they occur as very fine particles often coating ped faces and the surfaces of larger particles. Consequently soils rich in these minerals, such as the strongly weathered Oxisols of the humid tropics, have very stable peds.

Soil organic components also have surface charges contributing to electrostatic forces of attraction with other organic particles or with mineral surfaces. The charge is usually equal to or greater than that of the most strongly charged layer silicate clay minerals (vermiculite and smectite), with values of 200–300 $cmol_c\,kg^{-1}$, compared with 100–200 $cmol_c\,kg^{-1}$ on vermiculite and 80–150 $cmol_c\,kg^{-1}$ on smectite. In most soils 90 per cent or more of the charge on the organic matter is pH-dependent. In electron micrographs of clay–humic complexes, the organic matter often forms filamentous films on the extensive cleavage surfaces of finer particles ($<0.2\,\mu m$) of layer silicates such as smectite, vermiculite and illite (Laird, 2001). Root exudates consisting of mucilages (high molecular weights) and soluble exudates of small molecular weight (e.g., sugars and other carbohydrates) play important roles in forming and maintaining bonds between soil particles (Swift, 1991; Watt *et al.*, 1993; Traoré *et al.*, 2000). In soils with weak structure, anaerobic conditions stimulate the roots of some plants (e.g., maize) to produce exudates, possibly as a mechanism to improve the root environment (Boeuf-Tremblay *et al.*, 1993).

Opposing the attractive forces that stabilize peds, repulsive forces between mineral particles can destabilize the soil structure. The negative charge on the surfaces of layer silicate minerals attracts positively charged cations (counter ions) from the soil solution and repels negatively charged anions. Some of the cations are electrostatically attached to the mineral surface (forming an inner sphere complex) and lose most or all of their water of hydration (water molecules surrounding the cation). Others remain hydrated and form a diffuse outer sphere complex, in which the concentration of cations decreases exponentially away from the mineral surface. This double-layer surrounding clay particles accounts for many of the physico-chemical properties of soils (Bolt, 1982). In particular, where the diffuse outer sphere complexes of adjacent mineral particles overlap they produce an osmotic swelling pressure by inflow of water molecules, which acts as a repulsive force opposing the electrostatic attraction between the particles. The strength of the swelling pressure depends on the concentration and composition of cations in the outer sphere complex.

Uptake of interlayer water by smectitic clay minerals creates another repulsive force tending to weaken soil structure, and the extent to which this can occur is also influenced by the concentration and composition of cations in the soil solution. Small monovalent cations, such as Na^+, create greater interlayer uptake of water and greater osmotic swelling pressure than larger bivalent ions, such as Ca^{2+} and Mg^{2+}. When the repulsive forces resulting from osmotic swelling exceed the electrostatic forces of attraction between particles, cohesion between particles is lost and the peds can disperse to release individual particles into the soil water. When this dispersion is complete, the soil is in a deflocculated (structureless) condition. This can happen if Na^+ ions dominate the soil solution, as in saline soils (Section 3.3). In theory it can also occur by repeated wetting and drying cycles in soils with low concentrations of cations (Collis-George, 2001), though in most soils the balance of divalent ions to Na^+ is maintained by weathering of minerals. The most effective treatment for structural degradation associated with a dominance of Na^+ ions is application of a mineral such as gypsum ($CaSO_4 \cdot 2H_2O$), which dissolves slowly to release Ca^{2+} ions and so reduces the osmotic swelling pressure and increases the electrostatic attraction between particles.

7.1.2 Field description of natural soil structure

Soil structure is defined in terms of (a) the size, shape, degree of development and spatial arrangement of peds and aggregates and (b) the abundance, size, shape and extent of connectivity of interpedal or interaggregate voids (usually fissures) between peds and of the generally smaller intrapedal or intraaggregate voids (mainly subspherical pores) within peds.

Some peds and aggregates are large and readily described in the field; these constitute the soil's macrostructure. Others are much smaller, and can only be studied with a hand lens, in thin section using a polarizing light microscope or even by electron microscope; these define the soil's microstructure. The size boundary between macropeds/aggregates and micropeds/aggregates is usually drawn at approximately 1 mm. Very often the units are compound in that larger peds or aggregates break naturally into smaller ones when gently crushed in the field. Also, under the microscope, micropeds or small macropeds can often be resolved into numerous smaller units separated by fissures.

Peds can usually be assigned to one of the following shape types:

- *platy*: shorter in the vertical direction than the two horizontal directions and separated by essentially horizontal planar fissures;
- *lenticular*: platy peds that are thicker in the middle than at the edge;
- *prismatic*: greater in the vertical direction than the two horizontal directions and separated by vertical planar fissures often arranged as a polygon. The upper and lower surfaces of peds are flat. These are common in clay-rich subsoil horizons;
- *columnar*: prismatic peds with rounded caps. These usually occur in the B horizons of aridic soils rich in exchangeable sodium; they are rare in temperate humid countries such as the UK;
- *blocky*: polyhedral peds that are roughly equidimensional, and the faces either flat so that they meet at sharp angles (angular blocky) or rounded so that the angles between them are less sharp (subangular blocky). The faces are casts of those of surrounding peds;
- *spheroidal*: rounded and roughly equidimensional peds with no accommodation to surrounding peds. Granular peds have few intrapedal voids. Crumbs are very porous and usually <5 mm across.

Ped grade or degree of development depends on the ratio of cohesion within peds to adhesion between them. It is assessed in the field from the proportion of soil forming peds and by the frequency and distinctness of their surfaces. Soil is apedal if it shows no observable structural units and no orderly arrangement of planes of weakness. Precise definitions of soils with various grades of structure from weakly to strongly developed are given by Hodgson (1974).

It is important to assess soil structure in freshly exposed pits, because ped grade increases as the soil dries, especially in soils rich in clay or iron oxides. Structure is more strongly developed in unsaturated horizons above the permanent water table than in any saturated horizons below. This is because above the water table there is more faunal activity and root growth, more shrink–swell activity in clay components and more illuviation of clay, iron oxides and organic matter, with deposition of these materials on ped faces. Faunal activities important for the development of peds include the formation of faeces by a wide range of soil animal groups (e.g., production of earthworm casts) and the movement of nematode worms through the thin water films on ped surfaces.

In the laboratory, ped strength and degree of development is usually assessed by determining the percentage of aggregates in certain size ranges that remain stable on wet-sieving. The amounts of aggregates in each size range are first determined on air-dried soil sieved under standard conditions, and the exercise is then repeated with the sieves held under water. The difference in size distribution is attributed to slaking and dispersion of larger peds to produce smaller aggregates or individual particles. The stability of aggregates in this type of test has often been correlated with total soil organic matter or microbial biomass.

7.1.3 Artificial aggregates

In cultivated surface (Ap) horizons, the natural peds have often been partly or wholly replaced by artificial aggregates created by various processes of cultivation or by wetting/drying and freeze/thaw processes. Aggregates are less persistent than natural peds, and usually do not survive many cycles of wetting/drying, though two to three freeze–thaw cycles can increase their stability (Lehrsch, 1998).

In all but the driest soils, ploughing usually creates large aggregates (clods). These are later broken down by further (secondary) cultivations or by changes in temperature or moisture content into either smaller aggregates (sometimes termed fragments, or crumbs if they are spheroidal, porous and <5 mm across) or into the pre-existing natural peds. The size and stability of fragments and crumbs are important for germination and early growth of seedlings. Their stability depends mainly on clay and organic matter contents, but usually they have to be recreated afresh for each new crop. In temperate regions producing no more than one crop per year, this means an annual cycle of ploughing and seedbed preparation, but in warmer regions there may be two or more soil cultivation cycles per year, and this can lead to progressive loss of structural stability by increased oxidation of organic matter.

The natural structure of subsurface horizons can also be modified artificially by subsoiling (Section 4.4.7). Although this is mainly used to increase hydraulic conductivity and improve the efficiency of field drains, in forestry it can also improve the development of tree root systems where subsurface horizons are compacted or cemented.

7.1.4 Soil voids

The other fundamental aspect of soil structure is the abundance, shape, size distribution and connectivity of voids. A distinction is usually drawn between pores within peds/aggregates (intrapedal or intra-aggregate pores) and those between peds/aggregates (interpedal or interaggregate pores); the two types are also known as textural and structural voids, respectively (Fiès and Stengel, 1981). Textural pores are more strongly affected by the shrinking and swelling of clays than are structural pores.

Total porosity (Section 4.1) cannot be estimated in the field, mainly because voids <60 μm across cannot be seen by the unaided eye. However, the size and abundance of larger (>60 μm) structural voids can be estimated in the field either by comparison with charts, such as those of Hodgson (1974: figure 10), or from the assessment of peds and aggregates. As the voids >60 μm contribute most of the air-filled porosity, measurements of air permeability can also be used to estimate the size distribution of pores >60 μm (Iversen *et al.*, 2001). Fiès and Braund (1998) discussed the effects of varying sand, silt and clay percentages on textural (=packing) pores, and once these have been assessed the total volume of structural voids can be obtained by subtracting textural void space from total porosity (Nimmo, 1997).

The abundance, size distribution and shape of structural voids are strongly modified by mechanical soil processes such as cultivation and compaction, but these anthropogenic activities have little or no effect on textural porosity. The porosity and distribution of macropores in undisturbed resin-impregnated blocks can also be determined by the X-ray photon mapping technique of Brandsma *et al.* (1999b).

7.1.5 Soil microstructure

Microscopic peds, aggregates, pores and fissures 2–60 μm across can be seen in thin sections of soils viewed with a polarizing microscope, and their abundance within certain size ranges can be estimated from the amount of water retained in an undisturbed soil sample at a range of suctions (Section 4.1.1).

Some microscopic peds and voids seen in thin section are similar in form and origin to the larger macroscopic structural features, but others are different and are described using special soil micromorphological terms (Stoops, 2003). A major limitation of individual thin sections is that they provide only a two-dimensional view of the soil. To some extent this can be extended into three dimensions using parallel serial sections cut at closely spaced intervals (1–2 mm), but linkage of small features across the gaps between successive sections is often uncertain. Techniques for estimating pore shape, pore connectivity and other measures of soil microstructure from thin sections were described by Horgan (1998). Scanning electron microscopy can provide a three-dimensional impression of ultramicroscopic structural features within very small volumes of soil. At this scale the structural features often again show some similarity in shape to larger features.

7.1.6 Other methods of assessing soil structure

A unified approach to describing peds and voids over the wide range of sizes possible in soil, and to modelling their effects on physical, biological and chemical processes, is needed. In future it may be provided by the use of fractal geometry (Crawford *et al.*, 1997), neural networks (Koekkoek and Booltink, 1999) and network models (Vogel, 2000). However, all these approaches are in early stages of development.

The three-dimensional visualization of soil structure, which is essential to relate structure to important soil functions, such as water and nutrient movement, gaseous diffusion and mobility of soil organisms, is most likely to be achieved by non-invasive techniques that can produce computer-assisted tomography (CAT) images of undisturbed soil. Various approaches have been proposed, including the use of X-rays (Crestana *et al.*, 1986), gamma rays (Aylmore, 1993), electrical resistance, ultrasound and detection of single gamma-ray photons emitted by a radioactive tracer (Perret *et al.*, 2000). Initially the resolution by such methods was 1–2 mm, but it is now $<10\,\mu m$ for X-rays. As X-ray attenuation is related to water content, X-ray CAT scanning can be used to plot the distribution of water-filled pores, and water movement can be traced by scanning the same soil core before and after a period of water infiltration (Mooney, 2002). An interesting result of recent work on soil macropore connectivity by CAT scanning of undisturbed soil cores is that a large proportion (~80 per cent) of the void network in uncultivated subsoil horizons is often linked to a few independent pathways (Perret *et al.*, 1999). Similar results have been obtained from serial thin-sectioning of cores, in which dyes have been used to trace solute movement.

7.2 ZERO, MINIMUM AND CONSERVATION TILLAGE

There are several crop cultivation treatments that involve planting seeds directly into the previous crop residues. Currently, these systems are known as crop residue management systems, but they are also referred to as no-tillage, minimum cultivation, conservation tillage or direct drilling. The range of terminology is partly a reflection of the diversity of systems in use. In essence, the residues from the previous crop are left on the soil surface, to simulate the protective effects of vegetation. Fundamentally, this is a form of mulching. Then the next series of crops are planted and grow into the residue, and the new crops eventually provide the vegetative protective cover. These techniques have become popular over recent decades, especially in North America (Uri, 1998). Zero tillage involves planting seed by pushing it a few centimetres directly into soil that has not previously been disturbed at all by ploughing. In minimum tillage the seed is dropped into a narrow, shallow (~5 cm deep) fissure produced by drawing a thin blade (tine), chisel or coulter through the uppermost soil layer. Machinery for these methods is described by Davies *et al.* (1982: chapter 16). With either system the developing seedlings consequently grow through the stubble and unincorporated residues of the previous crop, unless these have previously been burnt or removed. In conservation tillage the crop residues are left as a mulch on the soil surface to limit evaporation and runoff, thereby conserving soil water. The seedlings also have to compete with weeds, unless these are killed with a suitable selective herbicide.

7.2.1 Advantages and disadvantages of zero tillage

The main advantages of direct drilling and minimum (shallow tine) cultivation compared with conventional tillage (ploughing and seedbed preparation) are:

- it is usually cheaper, as less labour and power inputs (Table 7.1) are required (Sijtsma *et al.*, 1998);
- it is quicker, so that larger areas of land can be sown at the optimum time, sometimes within a few hours of harvesting the previous crop, to ensure early crop establishment and maximize the length of the growth period (Christian and Ball, 1994);
- the natural structure of the surface soil is retained almost intact, so that percolation by preferential flow through the macropores created by old root and faunal (e.g., earthworm) channels is better than where the continuity of these features is destroyed by ploughing (Petersen *et al.*, 2001);
- crop residues left on the soil surface increase earthworm activity (Mele and Carter, 1999), in turn increasing infiltration rates and thus decreasing runoff and erosion (Edwards *et al.*, 1988). The residues also prevent the formation of surface crusts and insulate surface soil from temperature extremes at times when there is no crop canopy. Consequently, sandy and other low grade soils can be used more effectively and with less erosion risk (Quinton *et al.*, 2001), even on steeper slopes. Earthworms help incorporate crop residues, and in time their casts generate a well-structured surface layer;
- the soil is less likely to become compacted and anaerobic, because there are fewer passes with heavy machinery and each pass deforms the soil less (Figure 7.1). Conventionally tilled soils often go through an initial phase of increased surface bulk density (D_b) on conversion to zero tillage but, after a few years, the negative effects of natural compaction on crop growth are overcome by the positive effects of improved subsoil structure, increased infiltration and improved root growth (Merrill *et al.*, 1996);
- more soil water is retained, especially during germination and early crop growth, because the mulches of crop residues on the surface decrease evaporation (Logan *et al.*, 1991). In dry regions, such as the Great Plains of the USA, this has allowed increased cropping intensity, and the increased inputs of C to the soil may decrease atmospheric CO_2 (Peterson *et al.*, 1998). With minimum tillage, heavy rolling after sowing can also help minimize evaporation losses;

Table 7.1 Power requirements of cultivation implements used on previously undisturbed land

Cultivation method	Typical power requirement (MJ ha^{-1})
Mouldboard plough	166
Shallow plough	112
Direct drilling	53
Chisel plough	93
Heavy disc harrow	93
Shallow rotary cultivation	298

Source: Davies et *al.* (1982).

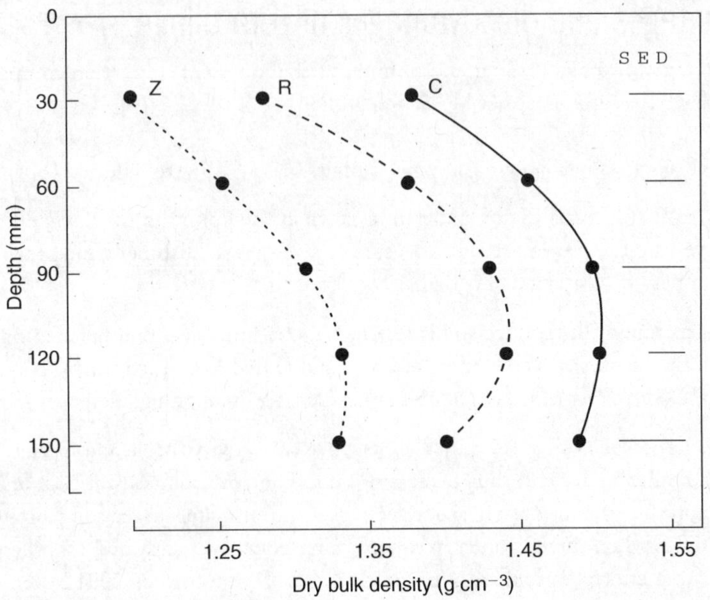

Figure 7.1 Variation in bulk density with depth in grassland soil under zero traffic (Z), reduced ground pressure (R) and conventional traffic (C). Bars to the right are standard errors of the differences between measurements
Source: Ball *et al*. (1997).

- initially there is less oxidation of soil organic matter because there is decreased contact with the atmosphere, and consequently there is less degradation of soil structure, less leaching of nitrate released by mineralization (Power and Peterson, 1998) and less leaching of metal ions released by soil acidification. After several years the organic matter may become concentrated in a thin (often <2.5 cm) surface layer, because it is not incorporated more deeply by cultivation, and the rate of mineralization may then increase, so that in the long term less fertilizer N may be required (Ismail *et al*., 1994);
- some plant diseases (e.g., nematode infections and the 'take-all' fungus in cereals) are more restricted, possibly because they are not spread as widely as they are by soil cultivation.

The overall effects are usually an improvement in soil quality and a decrease in pollution of surface waters by soil erosion. Also, in many situations, zero, minimum and conservation tillage techniques have increased rather than decreased crop yields, often because of better germination and crop establishment (Christian and Bacon, 1990). Even where yields are less than with conventional tillage, the difference is usually small, and the savings in machinery, labour and fuel costs increase profit margins compared with conventional tillage.

When these systems were extensively developed, particularly in the 1970s and 1980s, many considered them to be the panacea to many soil management problems, especially soil conservation (Section 2.4). However, there are various negative aspects of zero or minimum tillage:

- the decomposition (mineralization) of crop residues is slower on the soil surface, often approximately 60 per cent of that of buried residues (Douglas *et al*., 1980), so that the turnover of nutrients is slower; residues incorporated into the soil are exposed to a larger microbial population and mineralize quicker. Decomposition of surface residues is most rapid in climates where warm and preferably moist conditions prevail for most of the year;

- initial root development can be delayed because soil close to the surface (within about 25 cm depth) may be more compact than with frequent ploughing (Schjønning and Rasmussen, 2000);
- surface residues prevent the soil from warming so rapidly in spring, so that germination and early growth are slower (Kaspar *et al.*, 1990);
- if rain falls soon after application of N fertilizer as a spring top dressing, the greater continuity of vertical macropores can transfer nitrate to horizons beyond the reach of the roots of a developing crop (Goss *et al.*, 1993). This by-pass flow may also lead to greater loss of P to field drains, which may increase the eutrophication of surface water bodies (Addiscott and Thomas, 2000);
- there is greater loss of N by denitrification (Rice and Smith, 1982) or volatilization of urea-based fertilizers (Dick *et al.*, 1991) under zero tillage;
- the accumulation of crop residues over several years as a result of slow mineralization in a cool climate can impede crop establishment and provide a favourable environment for slugs and other pests (Christian and Miller, 1986);
- slow microbial decomposition of surface residues in cool, wet conditions also produces phytotoxins (Elliott *et al.*, 1978), which may limit the germination and early growth of subsequent crops, especially if the same species is planted repeatedly (Martin *et al.*, 1990). Wheat seems to be particularly susceptible to this problem. Production of toxins such as fatty acids has been attributed to deleterious rhizobacteria, including certain types of *Pseudomonas*, and soil inoculation with these bacteria might be used to control grass weeds such as downy brome (Elliott and Stott, 1997);
- one of the aims of ploughing is to bury weeds, to present a clean weed-free surface for subsequent cultivation. The lack of tillage can allow weed infestation and this is particularly a problem with grasses such as blackgrass (*Alopecurus myosuroides*) and sterile brome (*Bromus commutatus*), and volunteers (self-sown plants of previous crops) (Christian and Ball, 1994). Control often involves increased use of expensive herbicides and so these systems are not necessarily compatible with organic farming;
- plant diseases, such as eyespot foot rot (*Cercosporella herpotrichoides*) and *Septoria* leaf spot on wheat, overwinter on surface residues and are consequently propagated more effectively by conservation tillage (Cook *et al.*, 1978);
- these systems can produce compaction, especially on weakly structured silty soils with low soil organic contents. It seems that a crucial factor in success is the amount and activity of soil fauna, especially earthworms. Essentially, a rich soil faunal population can effectively till the soil on behalf of the farmer. That is one of the reasons why crop residue management systems have been particularly adopted on organic-rich soils in both North and South America.

As a result, the relative values of conservation and conventional tillage are still debated (Cannell and Hawes, 1994). It is likely that no single conservation tillage practice can be used universally, and different methods should be developed for different climates and soil types. For example, in cool wet climates, such as Scandinavia, zero and minimum tillage are best suited to clay-rich soils (Rasmussen, 1999), whereas the important benefits of moisture conservation are realized most in coarser soils of hot and relatively dry regions. On clay soils in Finland, Aura (1999) reported that minimal tillage increased the N uptake and yield of spring cereals mainly if the early summer was dry, probably because it decreased losses of water by evaporation compared with ploughing. Conservation tillage may also be better suited to certain crops; for example, in field experiments comparing direct drilling and mouldboard ploughing on clay soils in England, Christian and Bacon (1990) noted yield increases after direct drilling with oilseed rape but not with cereals. There are many gradations of cultivation between full 'no-till' and traditional ploughing. The actual selection means balancing the differing advantages and disadvantages of the two systems. Often, adopting a minimum cultivation system, with occasional full ploughing, acts as an acceptable balance.

In most countries a minimum cover of 30 per cent crop residues is usually recommended in conservation tillage to prevent runoff and erosion on gentle or moderate slopes (Section 2.4). The percentage can be measured and modelled by image analysis (Moran and McBratney, 1992) or predicted from the mass and type of residue using the equation developed by Gregory (1982).

7.3 CONVENTIONAL TILLAGE

7.3.1 Mouldboard ploughing

The most widespread method of soil cultivation practised throughout the world is mouldboard ploughing. The steel share or blade of the plough is shaped so that when it is drawn through the soil behind a tractor it penetrates to a fairly constant depth and overturns the surface layer to one side, thus leaving an initially empty V-shaped furrow. The next pass of the plough is displaced a short distance (20–30 cm) to the other side, so that the first furrow is filled with overturned soil from the second furrow. This process is continued until the soil of the whole field has been overturned and disturbed to a constant depth, usually 20–25 cm with modern equipment.

Powerful tractors can pull ploughs with multiple shares, so that four to eight furrows may be drawn with a single pass. Weeds and residues of the previous crop are almost completely buried by this procedure and usually decompose more rapidly than if left on the soil surface. Methods of ploughing with different types of machinery are described by Davies *et al.* (1982: chapter 9).

Repeated annual or more frequent ploughing leads to a homogenized uppermost soil layer (the Ap horizon), which has a uniform thickness and a sharp boundary over less organic horizons beneath. At the base of the Ap horizon a thin layer of smeared and compacted soil (the plough sole or plough pan) is often formed (Francis *et al.*, 1987). This may limit root penetration or periodically create anaerobic conditions within the Ap horizon, which can prevent germination (Richard and Guérif, 1988).

As power for drawing ploughs through soil has increased, the depth of disturbed soil (thickness of Ap horizons) has also increased. Long-term field experiments in Sweden have shown that in clayey soils ploughing to depths of 22 cm and 28 cm increases crop yields compared with shallower (15 cm) ploughing (Håkansson *et al.*, 1998). This is mainly because of deeper loosening and better aeration in the root zone and better control of weeds. However, in coarser soils with structurally less stable Ap horizons, there was no yield benefit from deeper ploughing. Also the increased energy costs of deep ploughing can outweigh the economic benefit of increased yields. Ploughing to depths >25 cm usually leads to a decrease in the organic matter content of the Ap horizon because of dilution with subsoil material containing very little organic matter. This decreases the chemical fertility of the soil and weakens the soil structure, so that water or wind erosion may be increased and compaction may occur more easily.

7.3.2 Secondary cultivations

In some dry, coarse and weakly structured topsoils with little clay and organic matter to bind particles, the soil disturbed by ploughing breaks into small natural peds or aggregates (fragments) on overturning. However, most soils do not break so easily, especially if they are ploughed when wet (moisture content close to *FC*) or after compaction by heavy machinery. Clods are then created, which usually become harder and more persistent on drying. Further natural breakage usually occurs only by frost-shattering (expansion of soil water on freezing) or shrinkage during drying.

The structural voids between unbroken or even naturally broken clods are usually much too large to retain water, so that seeds planted between them would fail to germinate through lack of water. A suitable seedbed (or tilth) of finer aggregates (ideally crumbs 0.5–5.0 mm across) must therefore be created by further (secondary) breakdown of clods using harrows, disc cultivators or other machinery (see Davies *et al.*, 1982: chapters 10 and 11). The property determining the ease with which soil crumbles under these applied stresses is known as its friability. It changes seasonally, and is linked to moisture and organic matter contents (Watts and Dexter, 1998). The moisture content at maximum friability is approximately equivalent to the soil's plastic limit. In field experiments with soils containing >15 per cent clay, Berntsen and Berre (2002) confirmed that ease of fragmentation depends on moisture rather than clay content.

If soil clods have become very dry and hard, considerable energy must be expended to make them smaller, perhaps by multiple passes with a power harrow. Damage to equipment can result in these circumstances, and the main effect of repeated passes is surface abrasion of the clods with production of dust rather than further fracturing of aggregates. The dust is an environmental hazard because it may be blown or washed away and pollute surface waters with particulate nutrients, particularly phosphorus. In contrast, if the soil is clay-rich and too wet, the clods absorb energy by deforming plastically rather than breaking. In cool regions, secondary cultivation is most easily achieved when the soil is drying from a frozen condition, as freezing decreases interaggregate bond strengths yet preserves the integrity of aggregates (Aluko and Koolen, 2000).

Although semi-empirical models have been proposed for predicting the structural conditions of soil following various cultivation systems (Roger-Estrade *et al.*, 2000a), in practice the choice of machinery and number of passes for secondary cultivation are usually decided by trial and error or past experience with soils of similar texture and moisture content. Seedbed cloddiness or structural quality can be assessed by sieving or image analysis to determine the size distribution of fragments, or by estimating visually the percentage of compacted clods >2 cm across (Roger-Estrade *et al.*, 2000b). Visual estimates of large clods are less accurate than sieving or image analysis, but can provide a rapid method for the farmer needing to decide quickly whether implement adjustments are required.

Secondary cultivation treatments crudely sort aggregates, with the coarsest at the surface (Aubertot *et al.*, 1999), so that visual or other assessments of surface tilth suitability for sowing usually err on the side of caution. Sorting of this type occurs mainly in dry soils and results from small aggregates falling through the gaps between larger aggregates when they are disturbed. It is known as kinetic filtering, kinetic sieving or interparticle percolation. It also brings larger stones to the surface, which can have the effect of increasing infiltration rates and decreasing evaporation and erosion of finer soil particles (Oostwoud Wijdenes and Poesen, 1999).

Seeds sown into a bed of coarse aggregates do not germinate as quickly or reliably as those in soil composed of finer fragments or crumbs. This is partly because they have less contact with the soil and do not obtain sufficient moisture, but also because many sink too deeply into the Ap horizon and pathways to the soil surface become long and tortuous. Even in coarser soils, which break down easily to produce a fine tilth, contact between seeds and aggregates can be too limited in dry conditions for adequate germination. In these circumstances, germination can be improved by recompacting the soil using a heavy roller after seeding. The effects of seedbed structure (seed–soil contact), water content and temperature were incorporated into models of seed germination by Bruckler (1983) and Dürr *et al.* (2001). There are critical temperatures, soil water contents and shoot pathway lengths, and also critical oxygen contents (Richard and Guérif, 1988), below which germination is impossible.

7.3.3 Effects of cultivation on soil organic matter, structure and nutrients

An important effect of ploughing and secondary cultivations is to incorporate organic manures and crop residues (Staricka *et al.*, 1991). Most implements result in a clustered arrangement of residues rather than a homogeneous distribution, and chopping crop residues after harvest and before ploughing often improves the distribution. It also accelerates the decomposition of residues and thus the recycling of nutrients. Natural breakdown of aggregates by frost, wetting and drying cycles or secondary cultivations exposes organic matter that would otherwise be protected within clods or undisturbed soil (Balesdent *et al.*, 2000). This also accelerates release of nutrients by microbial activity. In field and laboratory experiments, mechanical energy inputs typical of those used in cultivation increased soil microbial activity, as indicated by respiration (CO_2 emission) rates, for periods ranging from a few hours to over a week (Watts *et al.*, 2000).

Incorporation of fresh organic materials by ploughing can also improve the abundance and strength of small aggregates (Watts *et al.*, 2001). As it also increases microbial respiration, this structural improvement seems to depend at least partly on microbial activity, and increases with temperature.

The increased release of nutrients from soil organic matter by tillage may improve crop growth. However, it can also present environmental problems if the mineralization occurs when crops have not yet germinated or are growing very slowly and therefore require little or no nutrients, as in the cold winters of mid- and high-latitude regions. In such circumstances, zero or minimum tillage techniques can initially decrease losses of nitrate compared with ploughing, because organic matter is stored in the soil rather than mineralized. However, ploughing up soils that have been under zero or minimum tillage for several years can lead to large increases in winter losses of nitrate, as the additional, fresh and therefore less stable organic matter is rapidly mineralized (Catt *et al.*, 2000).

7.3.4 Soil crusting

Soil crusting or surface sealing is a common problem of seedbeds created in weakly structured sandy and silty soils with little clay and organic matter. If they are unprotected by vegetation, the surface aggregates of such soils can disperse when saturated by heavy rain or irrigation, both of which may have sufficient kinetic energy to break weak bonds between soil particles. The finer clay and organic particles are then moved a short distance (0.1–1.0 mm) below the surface, leaving a thin layer of bare sand or silt particles above. On drying, the two layers form a hard stratified crust containing only a few, horizontally elongated and poorly interconnected pores (Usón and Poch, 2000). This can restrict the emergence of seedling shoots (Whalley *et al.*, 1999). Often the shoots can only emerge through cracks in the crust (Velde, 1999) or when it has been resoftened by further rain. Dicotyledonous species are often more strongly affected than monocotyledons, because their shoots have a rounded hook-like shape, whereas monocotyledon shoots are usually pointed (Goyal *et al.*, 1982). Surface crusts also affect seedling emergence by limiting gaseous exchange with the atmosphere, and by decreasing surface evaporation so that the seedbed remains saturated and anaerobic (Rapp *et al.*, 2000). Because they are almost completely impermeable, with infiltration rates $<5\ \mathrm{mm\ hour^{-1}}$ (Shainberg *et al.*, 1997), crusts also greatly increase runoff, erosion and loss of nutrients and pesticides to surface waters.

Crust formation is a function of aggregate stability and can therefore be limited by increasing soil organic matter content. It can also be prevented by mist or other gentle irrigation and by protecting the soil surface under plastic, glass or a mulch of crop residues. Many of the loess-derived soils in Britain, which are prone to crusting but otherwise very productive, have given rise to important local glasshouse industries (e.g., on the West Sussex coastal plain).

7.4 PLAGGEN SOILS

In some areas, anthropogenic additions of organic soil materials have been so large that arable or horticultural crops can be grown with their roots entirely in a man-made layer overlying or mixed with the original natural soil. Such soils are generally known by the Dutch term 'plaggen soils', though other names such as 'man-made soils' are also used. In the USDA Soil Taxonomy (Soil Survey Staff, 1998) they are grouped as Plaggents. Usually a minimum thickness of 40–50 cm of new material is specified for these soil classes. The maximum recorded thickness is 140 cm (Pape, 1970).

In the Low Countries and north Germany, plaggen soils were formed by the long-continued practice of using turves, forest litter or heather sods as winter bedding for farm animals, and spreading the resulting manure on sandy arable land. Similar procedures have been used in parts of Scotland since Mediaeval times to deepen and improve shallow soils on hard crystalline bedrock (Glentworth, 1944). Very often strips 1–3 m wide, known as lazy beds, were constructed of imported peat and other organic materials laid between low retaining walls built of boulders. In parts of western England, Ireland and northwest France, other bulky soil amendments, such as calcareous beach sand, crushed shells, seaweed or organic industrial refuse were used, either alone or mixed with FYM (Conry, 1974).

Plaggen soils have also been formed by disposal of occupation residues (night soil, kitchen middens) around ancient settlements, such as early Indian sites in the USA and Roman towns or deserted Mediaeval villages in Europe. A black soil material known as 'Dark Earth', widespread on almost all Roman settlements in Europe, is 50–100 cm thick, rich in P from animal residues and contains numerous anthropogenic materials, including brick, tile, mortar, pottery, oyster shell and bone (Dekker and de Weerd, 1973). The more extensive black, fertile 'Terra Preta' soils of Amazon terraces and the 'Lou' soils of southern parts of the Chinese loess area (Zhu *et al.*, 1983) are probably also plaggen soils.

Although the main intention in building plaggen soils was to improve their physical condition and nutrient content for crop growth (Conry and MacNaeidhe, 1999), some are quite acidic and require liming and others are deficient in trace elements, such as Cu where the amendments were essentially organic, or Mn and B where calcareous materials were incorporated. Improvements in soil physical properties included increasing the available water capacity (AWC) by addition of loamy organic materials to sandy soils, improving the tilth of clay-rich soils and raising the soil surface above a shallow groundwater table. However, some plaggen soils are water-repellent, and may have variable moisture contents because of irregular infiltration (Dekker and Ritsema, 1997).

7.5 MARLING

Another long-standing, though now little used, procedure for improving soil structure, AWC and resistance to water erosion is the spreading of large amounts of clay (marl) onto weakly structured sandy or gravelly soils. Many such soils are naturally acidic and, as the marl was often calcareous, it usually had the additional advantage of raising soil pH. Markham (1636) recorded that marling was widely practised in the Weald of Kent (southeast England) since at least the early fourteenth century. It was also recorded by Pliny as widespread in Britain during the Roman occupation. In Norfolk (eastern England) the history of marling dates from at least AD 1252 (Prince, 1964), and there is also documentary evidence for marling in the Mediaeval period and up to the early nineteenth century on sandy and peaty soils in Dorset, Hampshire and many other parts of England (Prince, 1962, 1979; Darby, 1976: 116). Marling can also prevent wind erosion of sandy or peaty soils (Middleton, 1949). However, large amounts (500–1000 t ha^{-1}) may be necessary to stabilize the soil surface, because a minimum clay content of 10 per cent is required.

7.6 SOIL COMPACTION

Compaction is becoming a serious problem in agriculture (Soane and Van Ouwerkerk, 1995), mainly because of the increasing weight of farm machinery and the decrease in aggregate strength resulting from the slow loss of organic matter under a regime of repeated soil cultivation for arable crops. It occurs when the external stress or force applied at the soil surface exceeds the strength of bonds at points of contact between individual particles or soil aggregates, so that compression and shearing lead to a decrease in the size of voids and thus to an increase in bulk density (D_b). As the applied stress increases, D_b reaches a maximum because the soil particles cannot be packed more closely without being fractured. Degree of compaction is usually assessed as observed D_b divided by this maximum D_b measured in laboratory tests.

The extent of soil compaction can also be assessed semi-quantitatively using a soil penetrometer, which measures the force needed to push a spring-loaded rod a certain depth into the soil. More precise values can be obtained with gamma ray-neutron gauges, which measure D_b and soil water content simultaneously. They are useful for rapidly assessing the extent of existing compaction to about 30 cm depth and the soil's susceptibility to further compaction (Cassaro *et al.*, 2000).

The strength of very loose soils and their resistance to further compaction can be increased by mild compaction, for example by repeated passes with light machinery or treading by animals. Strength is also affected by the size, shape and degree of development of peds/aggregates, and by particle size distribution, particle shape, clay mineralogy, pedogenic cementation, water content and abundance and type of organic matter. Most of these factors change slowly with time or not at all, but water content often varies seasonally and may have an overriding influence on soil strength. Consequently, timing of field operations in relation to soil moisture content is important if excessive compaction is to be avoided (Ball *et al.*, 1997). Depending largely on their moisture content, soils can behave as solids, plastics or liquids. These geotechnical properties are important soil properties and are usually considered in engineering projects. We can determine whether soil is in a solid, plastic or liquid state in the laboratory and these states are separated by geotechnically determined limits, known as the Atterberg limits. Soils are particularly prone to compaction when they are in a plastic state, especially around field capacity (i.e., the soil moisture state when free gravity water has drained away, usually two or three days after being in a saturated state). Therefore, we should try to avoid compacting the soil when it is too moist and this helps diminish erosion problems (Section 2.4).

7.6.1 Factors affecting soil strength and compressibility

Surface soils of low D_b ($<1.3\,g\,cm^{-3}$) and with many large interaggregate voids are easily compressed by deformation and fracturing of aggregates. However, coarser soils are less compressible than finer (especially clay-rich) soils, because platy clay minerals have a smaller coefficient of friction than equidimensional sand and silt grains, and easily slide over one another so that the soil deforms plastically. This effect becomes noticeable in soils with >15 per cent clay, in which deformation can occur by orientation of the platy particles along short shear surfaces (Lupini *et al.*, 1981). However, it is often counteracted by the increased bond strengths in aggregates containing greater percentages of clay. The potential length of shear surfaces increases with clay content, and is virtually unlimited in soils with >40 per cent clay. Clay mineralogy is also important as cohesive forces between particles of illite, smectite and vermiculite are greater than between kaolinite particles. The presence of hydrated Fe and Al oxides, carbonates, silica or sulphates (e.g., gypsum) and other soluble salts deposited from percolating solutions can greatly increase the strength of subsurface horizons.

Pores that are completely filled with water and contain no air decrease soil strength because the water pressure acts in all directions, tending to push the particles apart and thus decreasing the number and strength of interparticle and interaggregate bonds. Increases in water content decrease the friction between layer silicate clay particles in particular, and this increases the plasticity of clay-containing soils, so that they deform more easily. Compaction that initially closes macropores in the surface soil decreases infiltration rates, so that after rain the soil becomes wetter than it would be without the initial compaction. Increased pore water pressures then render the soil easier to compact when external stress is renewed.

In weakly to moderately organic mineral soils, compressibility tends to decrease with increasing total organic matter content, because of the large number of bonds between organic compounds and clay minerals. Certain organic compounds, such as condensed lignins, fatty acids and aliphatic polymers, increase cohesion more effectively than others (Hempfling *et al.*, 1990), and compounds such as polysaccharides have less effect than more strongly humified components on bond strength in waterlogged conditions. In strongly organic soils with little mineral material, strength depends mainly on the size distribution of the organic constituents and the extent of humification. Strongly humified peaty soils are very easily compressed, but often exhibit some elastic recovery, which is rare in mineral soils (Tobias *et al.*, 2001).

7.6.2 Effects of compaction

Apart from decreasing infiltration rate, compaction restricts root growth, decreases nutrient uptake rates and crop yields, and increases erosion, leading to transfer of nutrients and pesticides to surface waters (Soane and Van Ouwerkerk, 1994; Ball *et al.*, 1997). With lack of oxygen, the biological activity of the soil is also decreased, resulting in slower mineralization of organic matter and reduced availability of plant nutrients.

The effect of pressure applied at the soil surface, for example by a tractor tyre, decreases downwards in the soil. The downward attenuation rate is influenced by the same range of factors as those determining surface strength, namely bulk density (D_b), structure, particle size distribution, organic matter content and pore water pressure. It is transmitted deeper in less dense, finer (clayey and silty), wetter, less organic and more weakly structured soils (Burger *et al.*, 1988). Because of their smaller contact area with the soil, rubber-tyred tractors with large lugs usually cause more compaction than metal- or rubber-tracked tractors. Low-aspect-ratio tyres with small lugs, as used on 'low-ground-pressure vehicles' (LGPV), can cause less compaction than even rubber tracks (Figure 7.1) because of more uniform distribution of weight and lateral stresses (Servadio *et al.*, 2001).

The worst compaction by farm implements ($D_b > 1.8\,g\,cm^{-3}$) is usually achieved by heavy tractors, the largest of which now weigh in excess of 20 t, and wheeled sugar beet harvesters, some of which are over 30 t when fully loaded, and are on the land in autumn when the soil can be very wet. Measured depths of compacted soil under these conditions are usually 15–25 cm (Gysi *et al.*, 2000), but can reach 60 cm (Tobias *et al.*, 2001). In forestry, compaction of topsoil and subsoil to 60 cm often occurs by the use of heavy timber harvesting equipment, but can be lessened if a thick mat of thin branches (brash) is first laid on the soil surface (Hutchings *et al.*, 2002).

Many of the problems associated with compaction can persist for over five years (Håkansson, 1985). Usually they can be rectified to about 25 cm depth by normal cultivation and soil loosening techniques (Gysi, 2001), but at greater depths the effect can persist for several years (Arvidsson, 2001), unless the soil is loosened by subsoiling (Section 4.4.7).

Deeper horizons in unconsolidated soil materials may also be compacted naturally by the growth of subsoil ice lenses and laminae. This occurs at the present time by ground freezing in arctic regions, but also persists as a relict feature (or fragipan) in loamy soils of many temperate regions that were subject to periglacial conditions during cold stages of the Pleistocene period before 10,000 years ago (FitzPatrick, 1956).

Compacted subsoil is often distributed very unevenly over fields, and subsoil loosening is not necessarily required over the entire area. Patches of compacted or clay-rich subsoil within rooting depth (<1.5 m) can often be located by shallow geophysical methods, such as surveys of electrical conductivity using a multi-electrode array such as the Offset-Wenner system (Barker, 1981). Such surveys are very rapid and the cost is often much less than that of unnecessarily subsoiling areas of uncompacted soil. The survey results can also be used for other aspects of precision agriculture, such as improved targeting of K fertilizer to areas of sandy subsoil.

Compaction by agricultural machinery is especially a problem in headlands, where tractors and other machines turn. Not only does this produce compaction, but the turning wheels can smear the soil. This can produce a tightly compacted soil with a platy structure and very low infiltration rates. Incorporation of grassland into headlands can act as a type of 'cushion', decreasing compaction. Moreover, such grassed areas can be integrated with nature conservation plans which encourage wildlife in the countryside.

7.6.3 Poaching

Treading of wet soil by large hooved animals, such as cattle and horses, is another common cause of soil compaction. It is often known as poaching, from the pocketed appearance (French: *poche*) where hooves have sunk into the soil surface. It occurs mainly in pasture where animals congregate, such as at feeding or watering areas, at the entrance to shelters or in field gateways. As with other causes of compaction, initial treading often decreases soil porosity, so that infiltration rates decrease and the increase in pore water pressure then makes the soil easier to compress with further treading. Susceptibility to poaching increases with increasing soil clay, organic matter and moisture contents, and decreases with bulk density and the strength and continuity of the grass cover. It occurs most

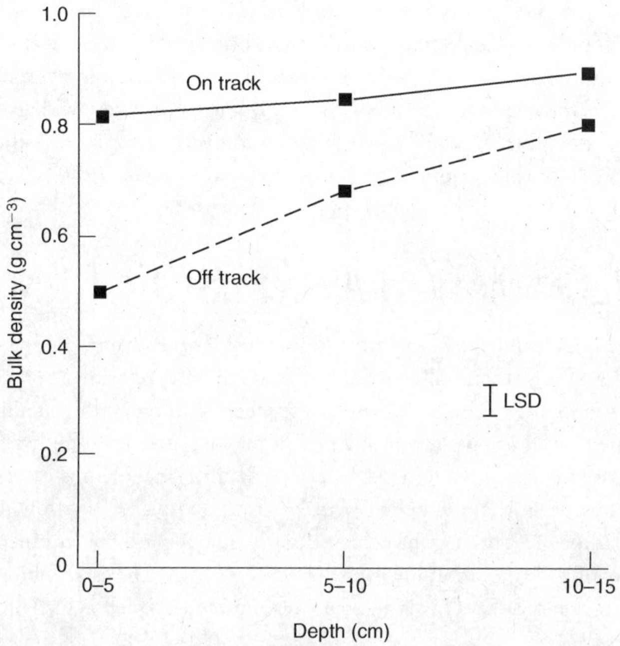

Figure 7.2 Effect of human trampling for four years on soil bulk density at three depths beneath the St James Walkway, New Zealand
Source: Stewart and Cameron (1992).

commonly in winter when sward growth is weakest and soil moisture content likely to be large. Scholefield and Hall (1985) devised a method for assessing susceptibility to poaching by measuring soil strength with a purpose-built penetrometer during repeated treading and concurrent application of water.

Localized surface soil compaction can also result from repeated treading by human walkers, for example on popular footpaths (Figure 7.2). Loss of natural vegetation from footpaths often reflects the poor growth of grasses and other plants in the compacted soil as much as abrasion of the vegetation by repeated treading. Accelerated erosion resulting from compaction also leads to poor growth through loss of nutrients in runoff or transported organic matter and clay.

7.6.4 Increasing the soil's resistance to compaction

Various methods of increasing a soil's resistance to compaction have been proposed. Strengthening of aggregates by application of soil conditioners (Section 2.4) often helps to a considerable extent, but only for a limited period. For soil under sports fields, Groenevelt and Grunthal (1998) have suggested incorporating large amounts (up to 40%) of rubber tyres shredded to crumbs of about 8 mm size and with the steel reinforcing belts removed.

However, for arable fields the most effective strategies involve zero tillage methods, increasing the organic matter content by repeated applications of organic manures or the introduction of perennial vegetation such as a pasture ley into the crop cycle, and avoidance of trafficking when the soil is still close to field capacity. The value of a pasture ley depends on its duration; short-term leys (<3 years) improve aggregation in dry conditions, but the bonding between soil particles depends on compounds such as polysaccharides that do not persist in waterlogged conditions. Longer leys are usually required to produce humic substances that contribute to bonds which resist waterlogging. Where trafficking in wet conditions, such as sugar beet harvesting in early winter, cannot be avoided on weakly aggregated soil, the use of low-ground-pressure vehicles with very wide tyres is probably the only option. Use of these vehicles can bring structural improvements even to soil which has previously been compacted by conventional machinery (Douglas *et al.*, 1998).

7.7 SOIL RECLAMATION AND RESTORATION

Landscaping after various civil engineering operations, such as quarrying, mining, road-building or disposal of waste in landfill sites, usually involves reclaiming and restoring a soil cover for agricultural, forestry or amenity use. Some of these operations, especially mining and waste disposal, are also likely to cause a wide range of well-known environmental problems, such as groundwater or atmospheric pollution, but these are not considered in detail here. The initial problem is creation of a stable cover with suitable structure and nutrient-supplying capacity for good plant growth with minimal runoff and erosion. This part of the work can be achieved fairly quickly, usually within 1–2 years, and is often termed reclamation. Restoration involves the further stage of re-establishing a typical soil macro- and microfauna and flora. This is an ecological problem, which is likely to take much longer, probably several decades.

Many national and regional authority regulations simply specify that land is restored to its pre-mining use or vegetative cover and level of agricultural productivity. However, some types of semi-natural vegetative cover may be difficult to achieve, and other original aspects may not necessarily be desirable. Some post-restoration uses may be ecologically better or more useful to the local community than the original. It is important to decide at the outset which post-restoration use is intended, because this will determine the exact procedures adopted.

7.7.1 Stockpiling soil

At many quarry, landfill and road-building sites, the original soil cover is initially removed and stockpiled awaiting the reclamation phase. Stockpiles should be designed so as to separate carefully and avoid mixing of the more organic, and therefore more nutrient and biotically rich and better structured, topsoil (usually O, Ah or Ap horizons), from subsurface (B) and deeper subsoil (C) horizons (Dunker and Jansen, 1987). Something like the original soil profile can then be reconstructed over the site by first respreading subsoil, then subsurface soil and finally topsoil. However, where the B horizon is acidic, it may be advantageous for agriculture if it is mixed it with less weathered, calcareous C horizon material (Dancer and Jansen, 1981).

Because of insufficient oxygen, the microbial biomass of topsoil rapidly decreases in size after even quite shallow burial (Harris and Birch, 1989). Stockpiled soil quickly becomes anaerobic below depths of only 0.3 m for clay to 2.0 m for sand (McRae, 1989). Consequently, stockpiles of topsoil should be thin and respread as soon as possible, or they should be avoided completely by moving fresh topsoil directly from areas being excavated to parts of the same site that are being reclaimed (Figure 7.3). Long-term stockpiling of topsoil also leads to loss of structure, especially if the soil is coarse-textured, and structural recovery after respreading may take as long as ten years.

An important effect of decreased microbial population is the limited ability of topsoil to release N and P by mineralization of organic matter. This effect can be partially mitigated by incorporation of

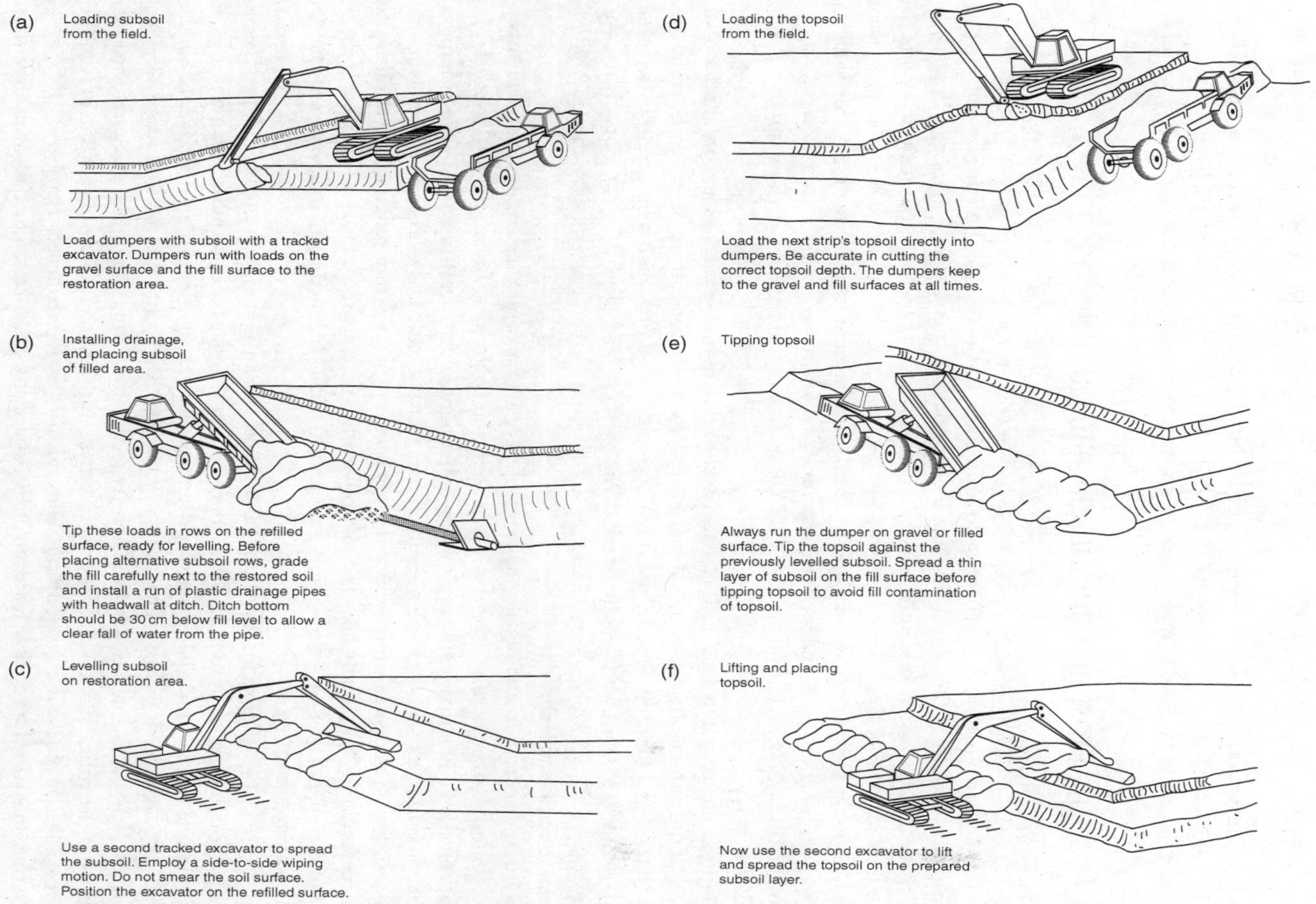

Figure 7.3 Method for progressive reclamation of gravel pit using topsoil transferred from fresh land to areas where gravel has already been removed
Source: Bransden (1991).

manure or inorganic fertilizer during or immediately after respreading. Microbial populations can then recover in as short a period as 1–2 years. However, the loss of *Rhizobium* strains suitable for symbiotic nitrogen fixation by legumes cannot be rectified chemically. Instead, inoculation with the specific strains appropriate to the chosen legume species is almost always essential. Mycorrhizal inoculants are also useful for improving P uptake by trees or arable crops sown on a newly reclaimed site. Live inoculants are usually incorporated into sugar- or clay-based seed coatings immediately before sowing.

To avoid damage to the topsoil beneath storage mounds, it is usually stripped from areas where subsurface and subsoil materials are to be stockpiled. Also, any mounds that are likely to remain for more than a few months should be seeded with grass to minimize erosion and growth of weeds.

7.7.2 Avoiding erosion, compaction and poor drainage in land reclamation

To avoid runoff and erosion, it is important to limit the final slope angles on any restored surface. This may mean reconstructing something like the original contours of the site, though in many hilly or mountainous regions it is often possible to improve the potential for agriculture, sports fields or other post-restoration land uses by creating flat land where none existed before. However, rather than an absolutely level surface, Jansen and Melsted (1988) recommended creating gentle slopes of 1–1.5° in humid regions intended for arable farming. This minimizes erosion and prevents the development of local waterlogged patches, as it allows lateral drainage by interflow through the new soil cover.

Attempts to create exactly level land can also lead to differential settlement and an uneven final thickness of the redistributed soil cover. Uneven settlement often results in depressions, which may remain waterlogged for long periods, and often become enlarged by erosion because they serve as conduits for movement of water to the water table (Barnhisel and Hower, 1997). For this reason, it is useful to delay redistribution of subsoil and topsoil for up to several years to allow the landfill or minespoil to settle and stabilize.

Removal, transportation and replacement of soil by heavy earth-moving equipment frequently leads to compaction, especially if the work is done in wet conditions. Bransden (1991) suggested that wet soil should be dried before removal by cropping it with ryegrass up to the date of removal. Compaction during lifting of soil often results from smearing caused by wheel slip or the blades of bulldozers or earthscrapers. During transportation, it results mainly from running over undisturbed soils or material that has already been laid. During soil replacement, compacted layers are often created in the upper layers of graded minespoil/landfill or just below the interface between the topsoil and subsurface layers.

The increased D_b of compacted layers results in runoff and erosion of the restored soil, poor aeration and mechanical impedance to root development (King, 1988), which can be more important in determining the yields of early post-restoration arable crops than nutrient supply (McSweeney and Jansen, 1984). Root development is increasingly impaired as the width of fissures decreases to $<500\ \mu m$ (Dexter, 1986) and as D_b increases from as little as 1.20 g cm^{-3}. Finer soil materials are more easily compacted to the critical point where root development is almost completely impeded than those richer in sand (Barnhisel and Hower, 1997). The critical D_b is approximately 1.60 g cm^{-3} for clay- and silt-rich soils and 1.75–1.85 g cm^{-3} for sandy loams and sandy clay loams. The heavy earth-moving equipment often used for redistributing soil (scrapers, large bulldozers and end-dump trucks) can easily compact soil materials above these critical values. Excavators, draglines, small dump trucks and conveyor belt-spreader systems cause less damage (Ramsay, 1986; McSweeney *et al.*, 1987). However, soil should not be moved by any equipment when its moisture content exceeds the plastic limit; this condition is indicated if the soil remains coherent and does not crumble when it is hand-rolled on a flat surface into a thread 3 mm in diameter.

To minimize compaction during replacement of soil, planned haul routes for each soil layer should be strictly adhered to, so that compaction is limited to repeatedly trafficked 'tram-lines'. Along the compacted haul routes, the physical condition and hydraulic conductivity of the soil are then improved by subsoiling in dry conditions before the next layer is spread. However, the effects of ripping may last for only a few years (Larney and Fortune, 1986), especially if any subsequent trafficking is in wet conditions. Barnhisel and Hower (1997) suggested that injection of organic matter into the voids created by subsoiling can help stabilize the new structure and delay the need for repeated operations. For replacement of topsoil, transportation by dumptruck and spreading by excavator are usually recommended (Ramsay, 1986). Moffat and Bending (2000) demonstrated experimentally the benefits of these techniques in reclamation for woodland.

Poor drainage resulting from compaction can also be rectified by installation of a pipe and mole field drainage system. Its effectiveness may need to be increased by subsoiling. However, performance may soon be impaired by siltation resulting from disaggregation of particles in weakly structured topsoil. Because of this problem, a long-term pasture ley may be a better initial land use than frequent soil cultivation for arable crops.

Landfill sites often require a capping of slowly permeable clay or synthetic material to prevent water entering the fill and to control gases such as methane generated within it. The capping should be thick enough (>1.5 m) to allow installation of vents for control and possible flaring of gas and to allow agricultural operations such as underdrainage and subsoiling to take place without puncturing the cap.

7.7.3 Improving soil fertility in land reclamation

Landfill, minespoil and even subsoil materials available on site are usually deficient in plant nutrients, especially N and P. Some are also too acid for production of crops, including grass. It is therefore important to cover these materials with fertile topsoil; up to about 25 cm, the thicker the layer the stronger and quicker is the establishment of the initial plant cover (Table 7.2). Applications of lime and fertilizers, or (on well-drained sites) of organic manures or sewage sludge, at rates indicated by soil analyses are also important for successful restoration to arable agriculture, woodland or amenity grassland. However, nitrification of ammonium to nitrate is initially much slower in restored soils because of depleted bacterial populations (Hons and Hossner, 1980). Also the availability of fertilizer P is frequently limited by enhanced fixation. To overcome this problem, Powell *et al.* (1980) recommended a mixed application of sulphur-coated and normal triple superphosphate to meet both the initial and the longer-term P requirements of crops.

The acidity of shale-rich coal mine spoil is often increased after restoration because of oxidation of pyrite (FeS_2). Very acid conditions ($pH < 3.5$) can severely limit the growth of most plants and the availability of nutrients applied in fertilizers. On reclamation the initial pH may be quite favourable, but as oxidation proceeds after reclamation the pH can decrease fairly rapidly. Pyrite oxidation rates are five to ten times greater in summer than in winter, because temperature influences the activity of

Table 7.2 Effect of thickness of clay loam soil cover on yields of grass–clover mixture (t ha^{-1}) grown on colliery spoil at Mainsforth, northern England

	Soil thickness (cm)			Least significant difference ($P < 0.05$)
	0	10	25	
First season				
Harvest 1	0.49	1.64	3.12	0.75
Harvest 2	4.16	5.96	7.75	1.79
Second season				
Harvest 1	1.50	3.21	3.35	1.30
Harvest 2	0.90	2.32	2.20	0.67

Source: Data from Gildon and Rimmer (1993).

Thiobacillus ferrooxidans (Backes *et al*., 1993). Total potential acidity includes any unoxidized pyrite, and this should be determined before a lime requirement is calculated (Section 6.4). Methods for determining pyrite content were assessed by Dacey and Colbourn (1979).

7.7.4 Pasture establishment in land reclamation

The most common initial land use on reclaimed soils is pasture, often for rearing of sheep or other small animals. However, in the early stages livestock should be removed when the soil is wet, so as to prevent poaching. Also, overgrazing should be avoided to prevent erosion. Usually seed mixtures of various grass species with two or more different legumes are broadcast into a seedbed prepared by harrowing or discing. Factors that need to be considered in the choice of grass and legume species include soil pH, nutrient availability, initial growth rate, life span, drought resistance, frost tolerance and potential for grazing. To ensure erosion control, grasses that establish rapidly, such as perennial ryegrass (*Lolium perenne*) should be a major component of the mixture. However, many such species are short-lived, and they should be accompanied by long-lived species, which may not establish so quickly. Grasses such as red fescue (*Festuca rubra*) can tolerate moderately acid conditions (pH 4.0–5.5), but perennial ryegrass and Timothy grass (*Phleum pratense*) prefer pH 5.5–8.5. In regions with harsh winters or very dry summers, it is important to establish a strong sward in seasons before frost or drought threaten its survival. Alternatively, the broadcast seed can be covered by a protective mulch of straw, wood chips, shredded bark or compost.

Winter cereals are the most suitable alternatives to grass as an initial crop. However, spring cereals should be avoided, because in spring the restored soil is likely to be too wet for cultivation and sowing without structural damage. Similarly, damage to wet soil with late harvesting in the autumn and early winter precludes root crops, such as sugar beet and potatoes.

7.7.5 The restoration stage

Following re-establishment of a soil profile with a structure, microbial population, nutrient content and drainage status able to support growth of grass, arable crops or trees, further work on reclaimed

Table 7.3 Fertilizer rates for different sports turfs

Sport	N (g m^{-2} yr^{-1})	P_2O_5 (g m^{-2} yr^{-1})	K_2O (g m^{-2} yr^{-1})
Golf putting greens and bowling greens			
Soil root zone	11-20	2	8-15
Sand root zone	25	5	20
Cricket pitches and lawn tennis courts	8-12	4	4-10
Coarse turf areas			
Grass clippings removed			
Soil root zone	16-20	8-10	8-10
Sand root zone	25	8-10	8-10
Grass clippings returned			
Soil root zone	8-10	2-5	2-5

Source: from Lawson (1989).

sites is often essential to develop a naturally functioning ecosystem. This final restoration stage may take many years, especially if a woodland community with slow-growing trees is desired. As the soil microbial community is responsive to and develops parallel with the plant community, stages in the development of the whole ecological community are most easily monitored by periodically measuring the size of the total microbial biomass and assessing which of its components are most active (Section 8.1.4). At this stage, introduction of microbial symbionts, such as *Rhizobium* and mycorrhizae, by inoculation of the soil may encourage particular plant species. It may also be necessary to introduce soil mesofaunal elements, such as earthworms.

7.7.6 Soil restoration for sports fields

Many new sports fields are laid on disturbed ground ('brownfield sites'), such as disused quarries, landfill sites, minespoil or old urban or industrial areas. Often they need specially constructed soils to ensure adequate winter drainage and strong growth of grass, and these can be produced as cheaply on disturbed sites as elsewhere. Even where they are sited on undisturbed agricultural land, the natural soil profile almost always requires extensive modification or complete reconstruction to make it suitable for use in very wet conditions (Adams, 1986).

Adequate drainage is usually achieved by a closely spaced shallow pipe system designed to give drainage rates greatly exceeding those required for arable use, and connected to the surface by a dense rectangular network of narrow slits, 20–50 mm wide and 100–350 mm deep, filled with gravel overlain by sand. Such narrow slits help preserve a completely flat surface, because gravel can collapse down wider slits and create shallow surface furrows. This drainage system is installed in a thick layer of homogeneous artificial rooting medium composed mainly of fine–medium sand (100–500 μm) with <5 per cent clay and about 5 per cent organic matter (Baker, 1989), which is well aerated, has a high hydraulic conductivity and large AWC. It is separated from a bed of fine gravel beneath by a blinding layer, which prevents it collapsing into voids between the gravel clasts. Recommended fertilizer rates for turf used for different sports are given in Table 7.3.

Summary

Maintenance of a strong three-dimensional pattern of structural units with large (drainable) interpedal and intrapedal pores is essential for plant growth and for limiting some environmentally damaging processes in soil. Peds are created by various natural processes, such as growth of plant roots and the activities of soil organisms. However, the peds in the surface soil layer are easily damaged by trafficking with heavy machinery, treading by large animals or cultivation in wet conditions. This can lead to compaction, loss of larger (air-filled) pores and increased bulk density, which lead to runoff and erosion, restricted root growth and decreased crop yields. A new assemblage of topsoil aggregates suitable for germination and growth of crops can be created by careful primary and secondary soil cultivations when the soil moisture content is within a certain range, but in clay-poor soils the newly created aggregates are weak and easily destroyed by heavy rain, leading to the problem of surface crusting. Alternatively, zero or minimal tillage techniques allow topsoil peds to redevelop naturally and can help improve other soil properties important for crop growth. However, they may also lead to problems such as increased weeds, plant diseases and loss of N as a nutrient.

Additions of large amounts of organic matter (as in plaggen soils) and clay (marling) can greatly improve soil structural conditions, though these older techniques are effective only when practised frequently over long periods. In many countries, they have been ignored for several centuries, but the increasing problems of weak soil structure encountered in modern intensive arable farming suggest they could well be reinstated as part of modern soil management programmes for sustainable agriculture. Reclamation of sites from which the soil cover has been removed for engineering purposes involves careful stockpiling of the original topsoil and deeper horizons, the respreading of each in such a way as to avoid compaction, poor drainage and erosion, and improving the chemical fertility of the reinstated topsoil. Initially grass, trees or winter cereal crops are established to help reclaim the soil as a functioning ecosystem, but full restoration for any purpose usually takes many years.

FURTHER READING

Allen, H.P., 1981, *Direct Drilling and Reduced Cultivations*, Ipswich: Farming Press Ltd.

Bolt, G.H. (ed.), 1982, *Soil Chemistry B: Physico-chemical Models*, 2nd edn, Amsterdam: Elsevier.

Bradshaw, A.D. and Chadwick, M.J., 1980, *The Restoration of Land*, Oxford: Blackwell.

Carter, M.R. and Stewart, B.A. (eds), 1996, *Structure and Organic Matter Storage in Agricultural Soils*, Boca Raton FL: CRC Press.

Davies, D.B., Eagle, D.J. and Finney, J.B., 1982, *Soil Management*, Ipswich: Farming Press Ltd.

Emerson, W.W., Bond, R.D. and Dexter, A.R. (eds), 1978, *Modification of Soil Structure*, Brisbane: Wiley Interscience.

8 Soil organic matter and its conservation

Whoever could make two ears of corn or two blades of grass grow upon a spot of ground where only one grew before, would deserve better of mankind, and do more essential service to his country, than the whole race of politicians put together.

Jonathan Swift

Introduction

The importance of soil organic matter has been stated repeatedly in previous chapters. Among other things, it helps strengthen and stabilize soil structure, minimize runoff and erosion, degrade and absorb some pollutants, and by microbial breakdown provides a source of nutrients for new plant growth. Soil organic matter includes numerous living and dead components, which are involved in a wide range of processes. The roles of many components are poorly understood at present, and even determining their amounts is either difficult or impossible. This chapter initially summarizes present knowledge of the various living and dead components, how some of them can be measured and their importance for soil chemical, biological and physical properties. Compared with the inorganic (mineral) components of soil, the organic components are much more reactive, their decomposition eventually leading to production of simple inorganic compounds, such as water, carbon dioxide and nitrate, a process known as mineralization. Rates of mineralization vary with composition of the different components and with environmental conditions including soil management. Later in the chapter methods of maintaining or increasing soil organic matter content and dealing with large quantities of crop residues such as cereal straw are considered. In very wet areas of the land surface (mires), organic matter accumulates naturally because mineralization rates are exceeded by growth of fresh plant material and accumulation of partially decomposed residues. The resulting peat deposits present special problems in agricultural and environmental management, which are discussed in the final part of the chapter.

8.1 THE NATURE OF SOIL ORGANIC MATTER

8.1.1 The main components

Soil organic matter (SOM) includes all the natural material derived from plants and animals incorporated into the soil or lying on its surface, which is either living or in various stages of decomposition after death, but excludes above-ground parts of living plants. The living components of SOM include plant roots (the phytomass), soil-dwelling animals (the faunal biomass) and micro-organisms <200 μm (the microbial biomass). Micro-organisms are by far the most abundant in both numbers and total weight, though they account for only 1–4 per cent of the total organic C in most soils (Jenkinson and Ladd, 1981). Their abundance increases with clay content and is greater in forest and grassland soils than those under arable cultivation.

Killham (1994) estimated that a typical temperate grassland soil contains approximately 10 t ha^{-1} microbial biomass, of which 5 t are fungi, 2 t bacteria, 2 t actinomycetes and 0.5 t protozoa. Soil micro-organisms are very diverse and able to perform a multitude of different functions, though <5 per cent of the types likely to be present in soil have been named and classified (Kennedy and Gewin, 1997). As they are so small, most soils contain enormous numbers of them; for example, most bacteria are <1.0 μm across, and soils usually contain several hundred million per gram (Pelczar *et al.*, 1993).

Fungi are larger than bacteria and occur as threads or hyphae 2–10 μm across, which are segmented or branching, often forming filamentous aggregates known as mycelia. Although most soil fungi are non-pathogenic, some are phytopathogens, either at the clinical level (causing visual symptoms of disease) or subclinical level (disrupting tissue functions and causing localized cell death). *Rhizoctonia solani*, *Pythium irregulare* and *Fusarium oxysporum* are examples of clinical fungal pathogens. Others, such as the mycorrhizae, have beneficial roles in plant growth, increasing the uptake of poorly mobile nutrients (e.g., P) or of water in drought conditions, improving soil structure by channelling C into the soil, and protecting plants from root diseases by interacting with the fungal and other plant pathogens (Sylvia and Chellemi, 2001). For example, Datnoff *et al.* (1995) observed that *Fusarium* rot in tomato can be controlled by inoculation with the mycorrhizae *Trichoderma harzianum* and *Glomus intraradices*.

Bacterial cells are prokaryotic (lack a nucleus), and most are attached to soil particles and surrounded with mucilage for protection from desiccation. However, they are generally intolerant of acid conditions; the total bacterial population decreases sharply at pH values <4. In contrast, many fungi, especially those involved in the decomposition of the lignin in woody tissue, prefer weakly acid conditions (pH 3–5). Actinomycetes are a type of bacteria that often form filaments thinner and shorter (<15 μm long) than those of fungi. They prefer weakly alkaline soils and, as a defence against competitors, some (e.g., *Streptomyces*) can produce antibiotics, which have been isolated for use in medicine. Another group of bacteria, the Cyanobacteria, Cyanophyceae or 'blue-green algae', can utilize atmospheric CO_2 in photosynthesis, and some can fix atmospheric N_2.

The bacteria are classified according to shape (e.g., spherical cocci, rod-shaped bacilli and curved threads or spirilla), their ability/inability to retain dyes (Gram-positive/Gram-negative, respectively), mode of nutrition, oxygen requirement and reactions they can perform. Like fungi, heterotrophic bacteria require organic molecules for energy and growth, whereas autotrophic bacteria obtain energy either from sunlight or from the oxidation of inorganic compounds (chemoautotrophic bacteria). Aerobic bacteria require oxygen for respiration; facultative anaerobic bacteria normally require oxygen, but can adapt to anaerobic conditions by using inorganic compounds (e.g., nitrate) as a source of oxygen; obligate anaerobic bacteria can grow only in the absence of oxygen.

Soils also contain algae, which are green, unicellular, eukaryotic (i.e., with a nucleus) organisms, occurring singly or in filaments or colonies. Like the Cyanobacteria, they mainly inhabit the soil surface because they need energy from sunlight for photosynthesis.

Soil protozoa are in the 5–20 μm size range, and comprise three groups distinguished by their mode of locomotion: the flagellates (with whip-like flagellae), the amoebae (with finger-like projections of protoplasm) and the ciliates (with a coat of short hairs). They live on soil bacteria and generally inhabit moisture films around soil particles and aggregates, but can survive periods of drought by forming resistant cysts.

The soil faunal biomass is divided on size into mesofaunal components (0.2–10 mm) and macrofaunal components (>10 mm). The mesofauna include the thread-like nematode worms, the small predatory annelid (enchytraeid) worms and a wide range of smaller arthropods, such as the arachnids (mites), myriapods (centipedes and millipedes), termites, ants, springtails, rotifers and beetles. Like the protozoa, most of these inhabit water films and graze on soil micro-organisms, but the termites, which mainly inhabit the surface layers of tropical savannah soils, live on plant litter. The macrofaunal components include the larger annelid (lumbricid) worms, such as the European earthworm (*Lumbricus terrestris*), molluscs (slugs and snails), larger arthropods such as isopods (woodlice) and burrowing vertebrates such as moles and rabbits.

Like the termites, earthworms play an important role in the physical breakdown and incorporation of plant litter, thereby accelerating its decomposition and mineralization by fungi and bacteria. The organic matter they ingest is mixed with inorganic soil material in the gut and excreted as a cast on or near the soil surface or as a lining to their burrows. Earthworm casts are richer in plant nutrients such as P (Sharpley and Syers, 1976) than the surrounding soil, and contain aggregates which are more water-stable (Marinissen, 1994). Consequently, enhanced earthworm populations improve soil physical conditions (Scullion and Malik, 2000), and increase plant growth (Stockdill, 1982). Earthworms are most abundant in neutral soils (pH near 7) and virtually absent in acid conditions. Their abundance is often increased by 'organic' farming methods (Scullion *et al.*, 2002).

The non-living components of SOM consist of unincorporated plant and animal remains on the soil surface (the litter), visible fragments (>50 μm) of plant and animal remains contained within the soil matrix (macro-organic matter), organic particles <50 μm not soluble in water (humus), water-soluble organic compounds, mainly sugars, organic acids and amino acids, in the soil solution (dissolved organic matter), and inert fragments carbonized by burning (charcoal) or deep burial (coal and graphite). The humus is further divided into compounds with identifiable molecular structures, such as polysaccharides, proteins, amino acids, fats, waxes, lipids, lignins, tannins, sporopollenins and algaenans (Derenne and Largeau, 2001), and compounds with molecular structures that are not readily identifiable by standard wet chemical methods (humic substances).

Humic substances (Hayes and Clapp, 2001) are conventionally divided into the humic acid fraction, which is soluble in alkali but precipitates on subsequent acidification, the fulvic acid fraction, which remains soluble on acidification after alkali extraction, and the humin fraction (Rice, 2001), which is insoluble in alkali. Further fractions can be produced by other extractants, such as the ethanol-soluble hymatomelanic acid fraction, the electrolyte-precipitable brown humic acid fraction, and the non-electrolyte precipitable grey humic acid fraction (Beyer, 1996). All these fractions are variable mixtures of numerous compounds, some of which may be generated or modified by the alkali extraction procedures.

Modern analytical techniques, such as infrared spectroscopy, soft ionization mass spectrometry, pyrolysis gas chromatography/mass spectrometry, thermochemolysis with tetramethylammonium hydroxide coupled to gas chromatography/mass spectrometry, and solid-state ^{13}C nuclear resonance spectroscopy (Hatcher *et al.*, 2001), allow more specific identification of many humic substances and avoid the problems of incomplete extraction and synthesis of artefacts during alkali extraction. However, they have shown that there is considerable overlap in composition between the various fractions; for example, polysaccharides, paraffins, alkanes, fatty acids and phenols occur in both fulvic acid and humic acid fractions from various soils (Beyer, 1996). So the rather arbitrary division of humic substances based on these extraction techniques has limited value in understanding the origin and behaviour of SOM.

Another subdivision of humus is into the forms associated with soils containing different amounts of nutrients, especially N, P and Ca, and with different types of vegetation. In the alkaline, neutral or weakly acid soil conditions beneath deciduous forest or grassland, rapidly decomposing mull humus is distributed over the top 30–50 cm of the profile by strong faunal activity, especially that of earthworms. In more acidic, nutrient-poor soil conditions under coniferous woodland or heathland, organic matter is decomposed more slowly by fungal mycelia rather than earthworms, and is termed mor. Mor humus can usually be separated into three thin layers, totalling 5–20 cm in thickness: a layer of fresh surface litter (L horizon) overlies partly decomposed and comminuted litter (F or fermentation horizon), and this in turn overlies a slightly more strongly decomposed (H or humified) layer, which is scarcely mixed with the mineral soil beneath. In some soils, the humus shows features of both mull and mor and is known as moder.

In the last decade or so, much information on the nature of SOM and rates and processes of decomposition in different soil environments has been obtained by separating it into size fractions (Christensen, 1992). Fractions >20 μm usually consist of little-altered plant material, whereas the SOM in finer fractions is more decomposed (Amelung *et al.*, 1998). Where the activity of certain components of the soil biomass is retarded, for example by strongly acidic, alkaline or saline conditions, much of the SOM remains in coarser particles, often rich in lignin. SOM can also be physically fractionated according to density by flotation in liquids of various specific gravities. Certain size-density fractions occurring in the aggregates or as discrete particles seem to result from recent organic inputs, and may be useful indicators of changes in SOM composition important for its role in soil behaviour.

8.1.2 Chemical composition of soil organic matter

The range of important roles played by SOM in soil properties and processes (Section 8.2) can be attributed to its diverse chemical nature, which has still to be fully resolved, though the modern methods of study listed above have provided much semi-quantitative information on the chemical composition and structure, particularly of humic substances (Beyer, 1996; Swift, 1996). The summary of this work by Baldock and Nelson (2000) includes the following main conclusions:

- multiple carboxyl, hydroxyl, carbonyl and alkyl substitutions in aromatic rings are common;
- alkyl C chains generally of shorter lengths (C1–C20) and often substituted with O-containing functional groups are also common;
- the aromatic and alkyl groups are linked in random sequences mainly by C–C bonds and ethers to form a 'backbone structure' of humic molecules;
- protein and carbohydrate polymers are bound to the 'backbone';
- the humic molecules exist as random coils (Schulten and Schnitzer, 1993), which are tighter and more strongly linked in the centre;
- molecular weights range from 10^3 to $>10^6$.

8.1.3 Determining total organic matter content

It is usually impossible to determine the percentage of SOM in a soil sample accurately. A common approximation is the loss of weight on ignition for 30 minutes at 850°C of a sample previously dried to constant weight at 105°C to remove moisture. However, this often overestimates SOM, because some soil minerals retain water at 105°C but lose it at higher temperatures and others lose volatile elements by oxidation (e.g., sulphur from pyrite) or by decomposition (e.g., carbon dioxide from carbonates). Many of these losses are avoided by slower ignition at a lower temperature, such as 375°C for 16 hours (Ball, 1964).

An alternative is to remove carbonates by acidification and then determine the C content of the SOM by wet oxidation under reflux using boiling 0.4 M potassium dichromate solution acidified with sulphuric and phosphoric acids. The amount of unused dichromate is then measured by titration against standard ferrous ammonium sulphate solution. Organic C can also be measured by dry combustion and use of an infrared detector to measure the amount of CO_2 generated (Nelson and Sommers, 1996). However, these methods ignore the H, O, N and other elements present in SOM, so organic C content is usually multiplied by an arbitrary factor to obtain SOM. A factor of 1.72 is widely used but, in some soils, a factor nearer 2.0 would be more appropriate. For comparisons between and within soil profiles, it is therefore preferable simply to report measured organic C percentages.

8.1.4 Determining soil biomass and its components

The most convenient method for measuring the total size of the living soil biomass (essentially the microbial biomass) is the fumigation-extraction procedure (Vance *et al.*, 1987), in which living cells are killed by fumigation with ethanol-free chloroform at 25°C and the C they contain is extracted with 0.5 M K_2SO_4 solution, then measured by oxidation with acidified potassium dichromate, as outlined in Section 8.1.3. Wu *et al.* (1990) suggested a more rapid automated procedure, in which the C extracted by K_2SO_4 is oxidized to CO_2 with sodium persulphate under ultraviolet radiation and the CO_2 measured by infrared absorption. Biomass can also be determined by measuring the amount of adenosine 5′ triphosphate (ATP) in a soil sample (Tate and Jenkinson, 1982). ATP is a useful measure of biomass because it occurs in approximately the same proportion in all living cells, but decomposes rapidly after death.

Both the fumigation-extraction and ATP methods measure total biomass and so do not indicate the relative abundance of individual components, such as bacteria or fungi. Previously these could only be estimated by counting individual types under a high power microscope, using morphology, presence/absence of chlorophyll and staining techniques to assist identification. More rapid methods for determining the population structure of the active soil microbial biomass depend upon the measurement of chemical markers, such as fatty acid methyl esters (FAMEs) or phospholipid fatty acids (PFLAs) derived from living micro-organisms. PFLA analysis is probably best for determining amounts of fungal and bacterial components, which have specific PFLA signatures.

Components of the living population can also be determined by the rate of utilization of various substrates added to a soil. For example, the Biolog™ method (Garland and Mills, 1991) is based upon the extent of reduction of tetrazolium dye, which develops a purple colour on receiving electrons donated by oxidation of nutrients through bacterial respiration. More than 500 bacterial species can be identified by this method (Garland, 1996).

Known inputs of gaseous substrates, such as CH_4 and $^{15}N_2$, and measurements of end-products can also be used to determine the amounts of functional microbial components involved in various soil processes, such as methane oxidation, N-fixation or the mineralization of N in SOM. ^{15}N has been used to measure gross N mineralization (total release of NH_4 by microbial activity) by the isotopic pool dilution method (Murphy *et al.*, 1999).

8.2 THE IMPORTANCE OF SOIL ORGANIC MATTER

SOM usually influences soil chemical, biological and physical properties to a greater extent than would be expected from its abundance relative to the inorganic (mineral) components.

8.2.1 Effects of soil organic matter on soil chemical properties

SOM contributes most of the cation exchange capacity (CEC) of A horizons, even of mineral soils with small amounts of SOM. It is especially important in sandy soils and those with clay fractions dominated by minerals with low CEC values, such as kaolinite. Measured CEC values of SOM range from 60 to 300 $cmol_c\ kg^{-1}$ (Leinweber *et al.*, 1993) and, in soils of neutral pH, each weight percentage of SOM contributes up to 3 $cmol_c\ kg^{-1}$ (McBride, 1994). SOM also buffers pH. In many soils its buffering capacity is about an order of magnitude greater than that of clay (Curtin *et al.*, 1996). Additions of organic matter usually increase the pH of acid soils (Pocknee and Sumner, 1997), but can decrease that of alkaline soils, especially under waterlogged conditions, and decrease soil sodicity (Nelson and Oades, 1998).

SOM can also sorb metals, especially polyvalent cations, by formation of complexes with carboxylic, phenolic, amine, carbonyl and other groups. This process of chelation is important in reducing the concentrations of soluble toxic metals such as Al, Cd and Pb in the soil solution (Anderson, 1995), in releasing nutrient cations such as K^+ from silicate minerals and in increasing the availability of P (Cajuste *et al.*, 1996). Downward movement of Fe, Al, Mn and other metals to subsoil horizons, as in podzols, also depends upon the mobility of organo-metallic complexes.

8.2.2 Biological effects of soil organic matter

SOM strongly influences all biological activity in soils. It provides a substrate from which soil organisms obtain metabolic energy, and the macronutrients it contains (mainly N, P and S) are converted by soil micro-organisms into inorganic forms, which are then recycled into the microbial biomass, faunal biomass and phytomass.

Even in soils generously dressed with inorganic fertilizers, the SOM provides most of these macronutrients. For example, apart from the sulphate- and sulphide-rich soils of desert and intertidal regions, almost all the S taken up as a nutrient from other soils is recycled from SOM (Nguyen and Goh, 1994). Also, much of the nitrate produced by mineralization of SOM probably comes from microbial oxidation of amino sugars, and in field experiments Mulvaney *et al.* (2001) found a strong correlation between the yield of unfertilized plots and their soil amino sugar-N contents. Some organic components in soils of the humid tropics are so rapidly mineralized that they can be regarded as equivalent to inorganic N and P fertilizer, and Palm (1997) proposed a decision tree based on the C/N ratio and lignin and polyphenol contents of crop residues for calculating the N- and P-supplying capacities of tropical arable soils. Some SOM components, such as humic acids, tannins and melanins, are also thought to stimulate or inhibit plant enzyme activity (Gianfreda and Bollag, 1996).

8.2.3 Physical effects of soil organic matter

In terms of soil physical properties, the small amounts of SOM play important roles in binding clay particles by the formation of clay–humus complexes. For example, degradation of SOM in an oxidizing environment increases carboxylic and other organic acid groups on the edges of humic particles; these develop a negative charge, causing them to attract hydrated cations attached to the basal (001) cleavage faces of layer silicate minerals (Oades, 1984). These and other complexes assist in the formation of stable microaggregates (<60 μm) and in binding microaggregates together to form macroaggregates (Golchin *et al.*, 1998). In turn the microaggregates are important in determining the plant AWC of soils. Consequently organic C is used in most pedofunctions for estimating AWC from easily measured soil properties (Da Silva and Kay, 1997). In addition, SOM itself can absorb and retain up to 20 times its own mass of water. The increase in soil aggregate strength and stability with increasing SOM content is especially important for reducing soil erosion (Section 2.4), increasing resistance to compaction, improving infiltration and hydraulic conductivity and providing a more favourable environment for micro-organisms and root development.

Polysaccharides, mucigels released by growing roots and rhizosphere micro-organisms, and fungal hyphae are important in stabilizing aggregates (Tisdall and Oades, 1982). However, their effect decreases after only a few weeks or months, which explains why aggregate stability is often affected by short-term soil management changes, even though they do not change total SOM contents (Haynes and Swift, 1990). Longer-term stability of aggregates has been attributed to less reactive humic substances (Haynes *et al.*, 1991). Maintenance of optimal structural stability therefore depends upon frequent inputs of fresh organic matter, as in 'organically' farmed soils (Shepherd *et al.*, 2002).

Even small amounts of SOM impart a dark brown, dark grey or black colour to surface horizons. This increases soil temperature compared with paler surfaces, and in the cool spring seasons of temperate regions encourages earlier germination and growth of crops.

8.3 FACTORS DETERMINING SOIL ORGANIC MATTER CONTENT

8.3.1 Effects of climate, parent material and soil structure

Soils contain very variable amounts of SOM. Usually the topsoil contains more than subsoil horizons because the decomposition products of surface litter are incorporated to a limited depth by faunal activity and other soil disturbance processes. However, in some podzolic soils soluble organic decomposition products are leached downwards in acidic water and reprecipitated to form very organic subsoil (Bh) horizons. In very acidic and seasonally or permanently waterlogged or frozen soils, the restricted microbial and faunal populations achieve very little decomposition and incorporation, so that litter mainly accumulates as a peaty layer on the soil surface. In contrast, in hot desert soils (Aridisols), inputs of litter are small and mineralization is rapid, so that the amounts of SOM are very small – usually <0.5 per cent C in the surface layer. At the other end of the scale, peat soils (Histosols) are usually defined as containing a minimum of 20 per cent SOM (12 per cent organic C), though this threshold is increased to 30 per cent SOM (18 per cent organic C) if the remaining (mineral) material is mainly clay (Avery, 1980; Soil Survey Staff, 1998). However, many peat soils, especially in cold and temperate regions, contain much more organic matter than these minimum values.

SOM content is determined by the balance between inputs of plant, animal and microbial residues and losses by leaching and evaporation of the soluble and volatile products of decomposition. Losses may also occur by soil erosion. SOM inputs include those derived from the soil parent material (e.g., incorporated into sediments deposited in anaerobic conditions), autochthonous components accumulated by plant and animal growth *in situ* and allochthonous components, such as those derived from other soils by lateral transportation. Where much of the above-ground biomass is regularly removed by harvesting, as in arable farming, the main autochthonous inputs are from the annual stubble (often incorporated by ploughing), roots and root exudates. With cereal crops, root exudates form about 10 per cent of the inputs (Barber and Martin, 1976) and consist of fairly reactive compounds that are fully decomposed within a year. Roots and stubble decompose more slowly.

In general, cold, anaerobic (waterlogged) or acidic soil conditions lead to accumulation of SOM even though annual inputs are often small, because decomposition rates are slow, especially in winter. In contrast, in warmer, well-aerated or neutral to weakly alkaline soils, inputs are often much greater, but there is little or no accumulation because mineralization is rapid throughout the year. Consequently, it is difficult to conserve or accumulate SOM in the tropics, especially under arable cultivation (Powlson *et al.*, 2001a), though slow accumulation even in sandy tropical soils has been achieved by introducing grass leys or legume crops into tropical arable systems (Wu *et al.*, 1998).

Under natural conditions, there is often a positive correlation between SOM content and rainfall, and at constant rainfall there is a negative correlation with mean annual temperature (Post *et al.*, 1982). Ladd *et al.* (1985) showed that the rate of mineralization of ^{14}C-labelled plant residues doubled with an increase in mean annual temperature of 8–9°C.

Clay content also influences SOM contents, as some humic components are protected from decomposition by bonding to clay minerals. In allophane-containing soils derived from volcanic ash (Andosols) and acidic soils rich in exchangeable Al, SOM is often protected from mineralization by the formation of stable Al-humus complexes (Boudot *et al.*, 1988). Even colloidal and soluble intermediate decomposition products of SOM may be protected from further chemical or microbial attack by bonding to the surfaces of minerals such as iron and aluminium oxides and hydrated oxides (Kaiser and Guggenberger, 2000). SOM is also protected from decomposition where it is enclosed within peds possessing pores that are too small for access by decomposer micro-organisms; according to Kilbertus (1980), this means pores <2 μm. Small micropores are also likely to be water-filled and therefore restrict the access of oxygen necessary for decomposition.

8.3.2 Effects of topography

There are major and minor influences of topography on SOM content. The strongest results from elevation. The lower mean annual temperatures at greater heights result in decreasing production of both natural vegetation and crops, so that SOM inputs are less than at lower elevations. However, this effect is counterbalanced to some extent by the slower mineralization of SOM at lower temperatures. The main result of increasing elevation is therefore a decrease in SOM turnover rate, resulting in reduced nutrient availability. Minor influences of topography on SOM content result from the downslope movement of water, nutrients, clay and SOM itself, all of which tend to increase SOM in footslope and valley floor positions (Burke *et al.*, 1995).

8.3.3 Effects of management

Soil management in both agriculture (Paustian *et al.*, 1997) and forestry (Johnson, 1992) often overrides natural factors in determining SOM content. Soil disturbance resulting from deforestation or ploughing up long established grassland usually leads to a decrease in SOM content because of enhanced decomposition resulting from increased soil aeration and water content. Under arable agriculture, there is also less input of organic matter to the soil than under grass or woodland. Conversely, reafforestation or establishment of semi-permanent pasture after arable cultivation usually lead to an increase in SOM content. Zero or minimal cultivation often also increases SOM content, especially in the uppermost soil layers. Frequent use of fertilizers also increases SOM content, because the increased crop yields result in greater inputs of roots and crop residues to the soil.

For any constant combination of climatic, parent material, topographic and soil management factors, SOM levels gradually approach an equilibrium value. However, rates of change in total SOM content following a change in any of the environmental factors are very variable. For example, after a change from forest to arable agriculture, a new steady state in SOM content may take as long as 100 years in temperate regions (Dick *et al.*, 1998) or 2000 years in the Arctic (Liski *et al.*, 1998). Such slow changes in total SOM content following a change in management are difficult to detect from measurements of total organic C repeated at intervals of a few years or even decades. However, they are often quite easily detected in measurements of the total microbial biomass or of its individual components, as these are responsible for the processes by which the bulk SOM is changing (Powlson *et al.*, 2001b).

Johnston (1991) discussed other rates of change based on analyses of archived soil samples taken from long-term field experiments with recorded changes in management at Rothamsted and other sites in southeast England (e.g., after ploughing up old grassland, sowing arable land to grass or repeated applications of fertilizer or manure to arable land). The importance of soil clay content on changes in SOM content under arable cultivation were clearly demonstrated by two similar experiments on the continuous cultivation of cereals, one on the silty clay loam soil at Rothamsted and the other on sandy loam at Woburn Experimental Station, about 30 km away. At Rothamsted the topsoil (0–23 cm) organic C content has remained stable since 1850 at approximately 0.9 per cent where barley has been grown annually without fertilizer or manure. With NPK fertilizer, it has stabilized at a slightly greater value (1.1 per cent) because of the inputs from stubble and roots. With annual applications of 35 t FYM ha^{-1} it increased rapidly from 1.0 per cent to 3.0 per cent in the first 50 years, and over the last century has increased more slowly to reach what is now an almost steady value of about 3.4 per cent. In contrast, on the sandy loam soil at Woburn, organic C declined under all treatments (nil, NPK fertilizer and FYM as part of a four-course rotation) from 1.5 per cent when the experiment began in 1876 to approximately 0.8 per cent when it was terminated in the 1960s. A similar decline in SOM occurred in a long-term continuous cereal experiment on sandy loam soil at Askov in Denmark (Christensen and Johnston, 1997). Evidence of this type is useful for predicting sequestration of C in soils as a strategy for decreasing CO_2 in the atmosphere to combat global warming (Swift, 2001).

8.4 RATES OF ORGANIC MATTER DECOMPOSITION

8.4.1 Influence of chemical composition

The various non-biomass SOM components decompose at different rates. For example, sugars and hydrolysable compounds, such as polysaccharides and proteins, are often degraded within hours or days; lignins and other non-hydrolysable macromolecules persist for centuries or longer; and charcoal produced by forest or grassland fires or by burning of crop residues often survives unaltered for 10^4 years or longer. C–C and C–O–C bonds in organic molecules are more difficult to break than those between C and other atoms, and the resistance of SOM components to decomposition is often related to the abundance of these bonds. The SOM protected within aggregates often contains larger amounts of the more decomposed components, such as aliphatic hydrocarbons, carboxylic anions and aromatic compounds, than does 'free' SOM in the bulk soil (Sohi *et al.*, 2001).

8.4.2 Modelling organic matter decomposition

As SOM decomposition is an important soil process, considerable effort has been put into developing conceptual models that predict rates of decomposition in various soil environments and management situations. Some of these models (e.g., Paustian, 1994) are organism-based, that is, they divide soil environments according to the different abundances and activities of decomposer organisms. However, most are substrate compartment-based, that is, they recognize a range of SOM pools or compartments, which differ in their relative decomposabilities; the most commonly used models of this type are the RothC (Coleman and Jenkinson, 1996) and CENTURY models (Parton, 1996).

Models such as RothC, which attempt to predict changes in total SOM as indicated by per cent organic C, operate on a long timescale (decades or longer), usually with a monthly time-step. However, to predict nitrate supply from SOM mineralization, which is influenced by rapid changes in soil temperature and moisture content, a shorter time-step of days or even hours is required. Also it is necessary to take into account important nitrate sinks, such as plant uptake and leaching, by linking the changes in SOM to soil–plant–water models.

Compartment-based mineralization models usually assume that each SOM pool decomposes according to first-order kinetics (i.e., its rate of decomposition decreases exponentially over time), though some assume second-order (double exponential) kinetics (Whitmore, 1996). However, rates for individual pools do not equate to the rate for the soil as a whole, because decomposition of some SOM compartments leads to transfer of the residual components into other pools, some less reactive and some more reactive. As most of the decomposition reactions involve microbial activity, with C, N and other elements being temporarily immobilized in microbial cells, the microbial biomass is an important SOM pool, with a reactivity determined by microbial mortality rates. At least two other pools are usually also recognized, including those for easily decomposable and resistant plant inputs, and sometimes one for physically protected occluded (intrapedal) but otherwise easily decomposable SOM.

The size of the microbial biomass compartment is easily measured (Section 8.1.4) and its reactivity can be determined by simple experiments (Ladd *et al.*, 1995). However, the size and reactivity of other model compartments are not so easily determinable. One approach is to measure the amounts of fractions separated by physical methods; for example, the coarser particulate SOM separated by sieving and density fractionation probably corresponds with a more resistant pool (Christensen, 1996). Another approach is to measure the reactivity of more resistant pools from radiocarbon decay rates (Jenkinson and Coleman, 1994; Paul *et al.*, 1997).

The RothC model divides incoming plant residues into decomposable plant material (DPM) and resistant plant material (RPM), both of which decompose to form microbial biomass (BIO), humified organic matter (HUM) and CO_2. There is also a pool of inert organic matter (IOM). Each pool except IOM decomposes by first-order kinetics, the maximum rates depending on soil moisture content, temperature and plant cover. The sizes of the BIO, HUM and CO_2 pools are also affected by clay content.

The CENTURY model deals with N, P and S as well as C turnover. It divides incoming plant residues into resistant and structural components, both of which decompose to active, slow and passive organic matter pools. For C there is also a CO_2 pool. As in RothC, the maximum decomposition rate of each compartment is influenced by soil moisture and temperature, and clay content determines transfers to the passive pool. The C turnover rate in CENTURY is slightly greater than in RothC.

Although the RothC and CENTURY models were originally designed to simulate SOM turnover in different situations, arable land in southern England and prairie soils in North America, respectively, both have been successfully used to simulate turnover in a range of other environments. For example, Falloon and Smith (2002) showed that they work equally well in simulating measured changes in the SOM content of long-term arable experiments in Hungary and Sweden, and in regenerated woodland, grassland and ley–arable experiments in southern England.

8.5 ORGANIC MANURES

As SOM plays vital roles in supplying plant nutrients and improving the physical properties of soil, it is important to maintain the quantity and quality of SOM in soils used for agriculture. Repeated ploughing, seedbed cultivation and use of artificial inorganic fertilizers for production of arable crops can lead to a slow decline in SOM content. In this type of farming regime, especially where almost all the above-ground parts of the crop are removed at harvest, organic inputs to the soil are limited to the roots, stubble and root exudates of each successive crop, which may be insufficient to match the losses resulting from microbial mineralization of SOM during and after soil cultivation. SOM decline in these circumstances is more rapid in coarse-textured than in clay-rich soils, but both types require periodic additional organic inputs if soil fertility and the sustainability of the agroecosystem are to be maintained in the long term.

8.5.1 Types of organic manure

Various organic manures can be applied to the soil of arable land. They include farmyard manure (FYM), slurry, pig and poultry manures from intensive animal rearing enterprises, composts (plant residues humified in well-aerated heaps), raw or treated sewage sludge, waste from food factories (e.g., sugar beet sludge, dairy salt whey) or abbatoirs (e.g., dried blood), peat, seaweed, paper-mill sludge, sawdust and wood chippings. Some of these are more valuable than others, both in terms of nutrient content (Douglas *et al.*, 2003), as indicated by C/N ratio (Table 5.4), and their physical effects on soil structure. FYM, slurry, animal manures, composts, sewage sludge and waste products from food manufacture can provide significant amounts of N, P and K, and improve soil aggregate stability, whereas peat, seaweed, wood products, waste paper and other organic materials are generally nutrient-poor and have less effect on aggregate strength; their main value is in increasing the soil's water-holding capacity.

The nutrient content of the more nutrient-rich manures is also very variable and unpredictable (Ministry of Agriculture, Fisheries and Food, 1976; Agricultural Development and Advisory Service, 1982). On a fresh weight basis, the N contents of different samples of FYM, cattle slurry, pig slurry, poultry manure and pig manure can vary from <0.01 to >5 per cent, the P contents from 0.001 to >3 per cent, and the K contents from <0.001 to >3.5 per cent. Poultry manure is often richest in all three, especially N. Plant composts are less variable in composition, usually containing approximately 1 per cent N, 0.25 per cent P and 0.5 per cent K on a fresh weight basis. Animal manures, such as FYM, pig and poultry manures and slurries, also contain very variable amounts of micronutrients such as Cu, Mn, Mo and Zn, depending to some extent on the composition of animal feedstuffs. The quantities of several micronutrients supplied by an average application of FYM are equivalent to those removed in several arable crops in succession (Cooke, 1982), so annual FYM applications progressively increase micronutrient reserves.

Sewage sludge is sometimes available in raw form, but more often it has been treated by aerobic fermentation (composted sludge) or anaerobic fermentation (digested sludge) to decrease odours and pathogens. These processes result in dark-coloured liquids containing only 2–5 per cent solids; flocculation with lime or $FeSO_4$ and dewatering to produce sludge cake increases the solids content to 40–50 per cent. However, much soluble N is lost in the process, and this is usually the most valuable component for improving the growth of arable crops or forest plantations (Ferrier *et al.*, 1996).

Raw sewage sludge and some of the treated forms often contain heavy metals (Cd, Cr, Cu, Hg, Mo, Ni, Pb, Zn) and other elements (e.g., As, B, F, Se) in amounts that are toxic to plants, farm animals and humans (Section 5.3). Ideally large proportions of these elements should be removed before the sludge is spread on land, but special chemical pretreatments for this purpose are expensive. In most countries the present policy is therefore to analyse the sludges before application and also the soils receiving them, so that predetermined levels of these elements in the soil are not exceeded. However, opinions differ on acceptable levels of some elements, so the total amounts permitted in soil vary between countries. Uptake from the soil also depends on soil properties, especially pH and the redox potential Eh (Rowell, 1981), and the crop grown. The different toxicities of these elements and their different rates of removal from soil by leaching or other processes are further complicating factors.

FYM, slurry, compost and some animal manures are often available locally in areas of mixed farming. However, in areas dominated by arable crop monoculture, most of these products are usually not available on or close to the farm in sufficient quantities to treat more than a small proportion of the arable land, and transporting the large quantities required from distant sources is usually too expensive for the farmer.

8.5.2 Green manuring

One way of increasing the SOM content in arable-only farming is the practice of green manuring, or ploughing in a quick-growing green crop before it matures. Although many plants have been used, the most suitable green manure crops are legumes, such as clover, or grass–legume mixtures, which fix atmospheric N and thus quickly increase the total soil N content (Kumar and Goh, 2000). Legumes also take up P from sources in the soil that are not available to many non-leguminous plants (Singh *et al.*, 1992). They have considerable benefit for subsequent crops when incorporated as residues (Haynes, 1997), when grazed by animals (He *et al.*, 1995) or even after they have been harvested as a crop in their own right (Holford and Crocker, 1997). However, as with winter cover crops (Section 5.4.2), the release of N by mineralization of incorporated green manures may occur at times when subsequent crops need little for growth, and the main benefit of the green manure (increased soil N) may then be partially lost by leaching.

Green manuring has been practised in China for at least 2000 years. According to Jiao (1983), originally self-sown weeds were ploughed in, but purpose-sown legumes such as mung beans have been used for over 1500 years. In the loess-derived soils of central China, one season of green manuring, in which 9–29 $t\,ha^{-1}$ plant material are added to the soil, increases the yields of cereal grain by a cumulative total of 0.8–2.65 t ha^{-1} over the succeeding 2–3 years. Leguminous green manures are also useful in the nutrition of wetland rice (Singh *et al.*, 1991).

8.5.3 Grass-clover leys

The benefit of a single green manure crop can be extended by growing grass–clover mixtures for longer periods. Leys of this type may be grazed by animals or ungrazed, and left for several years before they are ploughed in. Johnston *et al.* (1994) showed that the amount of N released by mineralization of the incorporated residues increases with age of ungrazed leys up to three years, but not beyond. In Britain, leys are usually incorporated in July, and most of the accumulated N is mineralized within a few months, though beneficial effects on crop growth can continue for 2–3 years and, in very dry conditions, may be delayed even longer. The ley should contain at least 20 per cent clover (Davies *et al.*, 1996), and ungrazed swards usually release less N than grazed pastures. Provided the grass and clover are grown closely together, much of the N fixed by the clover is mineralized and taken up by the grass (McNeil and Wood, 1990), and the amount transferred is often greater under a grazing regime than where the grass is mown (Høgh-Jensen, 1996).

Grass–clover leys are an important component of 'organic' farming systems (Watson *et al.*, 2002), as they provide a means of increasing soil nutrients without the use of artificial fertilizers. However, the mineralization of the organic N in grass and legume residues occurs at unpredictable times and rates. As a result, there is often insufficient N available when non-leguminous crops, especially cereals, are growing rapidly, so that their yields and N contents are less than those of crops grown with artificial fertilizers (Berry *et al.*, 2002). Also much of the N released at other times may be lost by leaching. Crops with a smaller N requirement or prolonged period of N uptake, such as sugar beet, potatoes or maize, may therefore be better suited to 'organic' rotations than cereals. A further problem with 'organic' systems is that annual income for the period of the ley is less than when arable crops are produced. Grazed leys support the grazing animals and reduce the need for purchased fodder, but in an all-arable system ungrazed leys provide no income for the length of the ley.

8.5.4 Sapropel

A valuable source of organic matter in some all-arable systems is the organic mud ('sapropel') extracted from lake floors. Lakes are very common in areas affected by the Last (Weichselian) Glaciation; for example, in Lithuania (total area 65,200 km^2) there are 2830 lakes, each exceeding 1 ha in area. The sediment accumulations on their floors are often quite thick, sometimes exceeding 20 m, and are rich (15–90 per cent by weight) in both autochthonous and allochthonous organic matter. Some also contain calcium carbonate. As in the long-established procedure known as marling (Section 7.5), spreading of sapropel onto sandy soils can improve their physical quality, nutrient status, pH and resistance to erosion (Baksiene, 2002). Removal of mud is also beneficial to the lake ecosystems, helping to prevent eutrophication by release of nutrients from the sediment. Sapropels are now available commercially in Latvia and the Ukraine.

8.6 CROP RESIDUE DISPOSAL

In addition to the food, fuel or industrial products for which arable farm crops are grown, large amounts of less valuable residues such as straw are often produced. For many crops plant breeding has increased the harvest index, that is, the ratio of the valuable product (e.g., grain from cereals) to total above-ground biomass. For example, wheat straw is now usually <0.5 m long, whereas older varieties before the 1950s produced straw up to 1.5 m long. Despite the decrease in amounts of unwanted crop residues, disposal can present a major problem for the farmer.

In the past, cereal straw was used in the UK and other countries for several purposes, including bedding for animals on mixed farms, thatching (roofing for barns and cottages) and production of straw-board for use in building construction. As animal bedding, the straw was later returned to the soil as farmyard manure. However, the move in many regions from mixed to intensive arable farming led to increased production of straw, and at the same time eliminated its main on-farm use. Initially the residues were destroyed by in-field burning after harvest, and this had some further advantages, such as reducing weeds, pests and diseases. However, the nuisance and damage that residue burning often caused led to legislation prohibiting the practice in many countries, and to consideration of other methods of disposal. The most obvious is incorporation into the soil, but this introduced new problems.

8.6.1 Straw mulching versus incorporation

The main alternative methods of disposal are incorporating cereal straw and other residues into the soil by ploughing, leaving them as a surface mulch or composting them in heaps and then spreading them as part of a conservation tillage regime. In a dry climate, a surface mulch can decrease evaporation and thus increase crop yield by increasing water storage in the soil (Van Doren and Allmaras, 1978). However, there is considerable experimental evidence that in humid temperate regions, such as Britain, crop residues left on the soil surface or incompletely incorporated by shallow tining can delay crop establishment and decrease yield (Christian and Miller, 1986; Christian and Bacon, 1988, 1991; Christian *et al.*, 1999). For example, on a clay soil where winter wheat was direct-drilled through a mulch of chopped straw, Christian *et al.* (1999) reported a mean yield reduction over nine years of approximately 30 per cent compared with direct-drilling after straw burning. This could have resulted from lignin and polyphenols in the straw, which can decrease mineralization rates (Vanlauwe *et al.*, 1997), or from production of phytotoxins during decomposition of the straw by deleterious bacteria. These problems may be decreased by aerobic composting of straw (Elliott and Stott, 1997).

In contrast, experiments in which large amounts of cereal straw were deeply incorporated into various soil types in the UK did not show a long-term yield decrease, and this practice has the added advantages of recycling nutrients more rapidly and improving the structure of the subsurface soil. Jenkyn *et al.* (2001) recorded small decreases in wheat yield but only in the first year of straw incorporation, probably because the decomposition of straw initially makes a demand on the soil's N resources, so that less is available for crop growth. The early decomposition of straw is by soil fungi (Cheshire *et al.*, 1999), with bacteria probably involved in later stages. Straw from a typical cereal crop contains only about 35 kg N ha^{-1}, and its C/N ratio is much larger than that of fungal and bacterial cells (Table 5.4). Consequently, soil mineral N is taken up (immobilized) by the micro-organisms to balance the additional C from the straw (Jenkinson, 1985). In later years the experiments discussed by Jenkyn *et al.* (2001) showed no effect of repeated straw incorporation, even at very high rates. This was probably because N immobilized in the enlarged microbial biomass decomposing straw in the first year was mineralized in the second year, thereby releasing N that could support the decomposition of straw in the second and subsequent years. Increased turnover of organic matter may also account for the greater uptakes of N, P and K by wheat observed after straw incorporation in the same experiments. Similar results have been obtained in field experiments in Norway, where Børreson (1999) reported little or no yield reduction of wheat after incorporation of chopped straw and, in the USA, where Bird *et al.* (2001) showed that rice yields after straw incorporation could be maintained with smaller inputs of fertilizer N.

Early calculations by Hutchinson and Richards (1921) suggested that 1 t of wheat straw would immobilize approximately 7.5 kg of soil N so that, at least for the first year of straw incorporation, fertilizer N should be increased by this amount to maintain yield. However, later field experiments, including those mentioned above, have suggested that the amount of N immobilized per tonne of straw is, in practice, now usually less than this (2–6 kg). The difference has been explained in terms of the larger amounts of N in the straw and other residues (roots, root exudates and stubble) of modern wheat varieties, which are given larger fertilizer N dressings (Glendining *et al.*, 1996).

One of the possible problems resulting from leaving straw and other crop residues as a surface mulch, especially in temperate humid climates, is the increased incidence of fungal and other crop diseases. However, in experiments on winter wheat comparing straw incorporation with straw burning (Prew *et al.*, 1995) and removal (Jenkyn *et al.*, 2001), there were no increases in the pests and diseases monitored, though straw incorporation resulted in small decreases in eyespot (*Cercosporella herpotrichoides*) and take-all (*Gaeumannomyces graminis* var. *tritici*).

8.6.2 Crop residues as fuel

Another possible use of crop residues is as a fuel for generation of electrical power. However, for various reasons, they have rarely been exploited for this purpose. First, the heating value of crop residues is approximately 3×10^6 kcal t^{-1}, which is only approximately 30 per cent that of oil (Epstein *et al.*, 1978), and losses involved in the various processes of converting heat energy to electrical power decrease the efficiency of most fuel types to approximately 30 per cent (Alich and Inman, 1975). Second, the cost of transporting large amounts of residues to distant power stations can be a considerable proportion of the value of the power generated. Third, compared with oil, crop residues leave large amounts of ash on burning, and this presents a disposal problem at the power station.

Local use of crop residues for fuel is a more viable proposition. For example, in some rural parts of the USA and India, crop residues and other waste organic products have been composted anaerobically to generate methane (synthetic natural gas), which is then used on-farm or piped to the houses within a village (Klass, 1976). Other possible processes that might be used locally are anaerobic pyrolysis to produce gas and conversion of organic wastes to oil by hydrogasification and liquefaction (Steffgan, 1974).

However, the recently recognized problem of global warming as a result of increasing atmospheric inputs of anthropogenic CO_2 (Section 9.3.1) has generated increased interest in the use of crop residues, or even specially grown crops, for fuel. In this context biofuels are seen as beneficial replacements for fossil fuels, despite the often increased transportation and handling costs, because they are 'CO_2-neutral', that is, the CO_2 released to the atmosphere when they are burnt is approximately equalled by the CO_2 they absorb from the atmosphere during growth. Also, Smith and Smith (2000) calculated that, in terms of the amount of C used by trucks, the transportation costs of biofuels is in fact very small.

Apart from using cereal straw as a fuel, there is considerable interest in the production of rapidly growing woody fuel crops, such as willow, and perennial grasses, such as *Miscanthus* (elephant grass) and switchgrass (Garten and Wullschleger, 1999; Riche and Christian, 2001). Of these, *Miscanthus* has considerable promise. Apart from the rapid growth of above-ground biomass for use as fuel, there is also considerable accumulation of organic matter (sequestration of C) in the soil as roots and rhizomes, very small losses of N by leaching of nitrate and beneficial effects on wildlife, such as spiders, mammals and birds (Powlson *et al.*, 2001a). More conventional food crops are also being grown increasingly to produce automotive fuels. In Britain, the amount of oilseed rape grown annually to produce biodiesel fuel is currently about 30,000 t, and will probably increase in future. Also it is likely that biopetrol will soon be produced from wheat, potatoes and sugar beet.

8.7 PEAT SOILS AND THEIR MANAGEMENT

8.7.1 Definition and types of peat soils

Peat accumulates where the rate of decomposition of organic matter is slower than its production in wet or acidic conditions. Organic accumulations of this type are often termed mires. They can be divided into two main types. Fen accumulates where the groundwater is neutral or alkaline and rich in plant nutrients (eutrophic), encouraging the growth of reeds, sedges, grasses, shrubs and trees. Where trees such as willow (*Salix*), alder (*Alnus*) and birch (*Betula*) occur, the vegetation is termed fen-carr. In contrast, bog forms either where the groundwater is acidic and nutrient-poor (dystrophic), or where the thickness of organic matter is such that its surface has been raised locally above the influence of groundwater so that continued growth of vegetation depends entirely on nutrient-poor rain. The first type is often termed blanket bog, and the second raised bog. Fen peats often accumulate in flat, low-lying coastal areas, and usually have greater ash contents than bog peats. The vegetation of raised bogs consists mainly of mosses such as *Sphagnum*, acid-tolerant grasses (e.g., *Eriophorum*) and other plants with small mineral nutrient requirements (e.g., *Calluna vulgaris*).

In addition to subdivisions based on pH, nutrient content and botanical composition, peat soils are also classified according to the degree of decomposition (humification) of the plant remains. The extent of humification on a ten-point scale (Von Post, 1924) can be determined in the field from the distinctness of plant material, the proportion that can be extruded between the fingers and the colour of liquid squeezed from the peat or extracted with dilute potassium hydroxide solution. Humification assessed in this way correlates positively with bulk density, and negatively with saturated hydraulic conductivity and water-holding capacity. Decomposition of peat results mainly from microbial activity. However, this is often enhanced by enchytraeids (potworms), which are the main invertebrate group in many peat soils. They fragment plant material and, by grazing on the microbial communities, probably spur them into an active growth phase. Their combined influence results in the formation of soluble organic compounds, which are leached from the peat to form brown, strongly coloured ground- and surface waters. Enchytraeid abundance and microbial activity increase with temperature, and are at a maximum, leading to increased loss of dissolved organic carbon in late summer (Cole *et al.*, 2002).

8.7.2 Peat wastage and degradation

Large areas of fen peat soils throughout the world have been drained in recent centuries to form arable land or improved pasture. Because of their large water-retaining capacities and ease of cultivation, drained peat soils can be very productive, though large amounts of fertilizer are usually required, as peat is generally deficient in mineral nutrients (P and K) and, where it is strongly acidic, there is little N mineralization. However, following drainage, peat shrinks through loss of water, and when its surface layers become dry in periods of drought they are easily eroded by the wind or lost by burning. Repeated cultivation also accelerates oxidation. This combination of processes leads to a rapid decrease in the thickness of peat and lowering of the land surface, an effect known as peat wastage. In some parts of the Fenland of eastern England, which have been drained and intensively cultivated since the late seventeenth century, 4–5 m of fen peat have been lost in the last 150 years (Hutchinson, 1980), and the current rate of wastage is estimated as 0.5–3.0 cm yr^{-1} (Burton and Hodgson, 1987). As the upper earthy (humified) layers are lost, the underlying layers of originally raw peat become humified, and eventually underlying inorganic deposits may be exposed. In several areas, such as the Somerset Levels, peat loss has also been exacerbated by wholesale extraction for horticultural use.

Areas of undrained lowland peat with their natural wetland vegetation have consequently become increasingly rare wildlife habitats in many countries, and many of the remaining areas are now protected sites. They often support rare plant and animal species, are important as breeding areas for birds and contain valuable records of archaeological and past environmental change. Consequently there is often pressure to restore many peatland areas, though once the peat layer has been completely or almost entirely lost, this is virtually impossible in the short term.

Perhaps the most important factor in conserving lowland peat is restricting or eliminating peat extraction by using suitable peat substitutes in horticulture, such as composts, coir and wood and paper waste. In addition, where peat soils remain in arable cultivation, some practices should also be avoided. Excessive lowering of the water table, frequent deep cultivations and repeated subsoil dehydration by deep-rooting crops, such as sugar beet, can lead to structural degradation. Under heavy rain after a period of drought, this may result in formation of an organic surface crust, downward illuviation of dispersed humus to infill voids in an organic B (humilluvic) horizon and precipitation of iron from the soil solution. Together these effects can decrease infiltration and hydraulic conductivity. Once dry, the massive, poorly structured and cemented surface and subsurface horizons can also lose the ability to reabsorb moisture, and ploughing creates hard, angular fragments, which behave as stones in decreasing the soil's AWC (Caldwell and Richardson, 1975). In the Fenland of eastern England, peat soils degraded in these ways are described as 'drummy'.

Some lowland fen peats contain mineral layers rich in sulphides, especially pyrite (FeS_2), which were deposited in estuarine or marine conditions when sea level rose sufficiently to temporarily inundate the peat swamps. Drainage and exposure to the air has then resulted in oxidation of the sulphides, leading to the development of very acid conditions (Section 6.1.1). If the original estuarine or marine mud was calcareous, gypsum crystals may also be present, and iron released from pyrite is often reprecipitated by filamentous bacteria as a felted mass of ochre, which can block drain outfalls (Bloomfield and Zahari, 1982). Because of these problems, peat soils with pyritic clay layers within 2–3 m of the surface should not be cultivated.

Upland blanket bogs are also subject to wastage by gully erosion, which often starts at the sloping margins and extends onto flat areas where the peat is thicker, leaving extensive areas of bare unvegetated peat. Much of the gully flow is derived from water moving rapidly through the peat in subsurface pipes, which in the blanket bogs of the Pennine uplands in northern England can be $>$150 m long and up to 70 cm in diameter (Holden and Burt, 2002). This type of degradation has been attributed to a range of factors (Tallis *et al.*, 1997). On the Pennines various experimental measures to prevent it have proved unsuccessful; the most promising is building small check dams in the gullies and sowing quick-growing grasses on the redeposited peat that accumulates behind the dams (Burt and Labadz, 1990).

Many bogs at lower altitudes have been afforested with fast-growing conifers, such as sitka spruce and lodgepole pine. The drainage improvement required for adequate growth of trees and stability in high winds is often achieved by ditches or pipe systems and by planting the young saplings on ridges created by deep ploughing. Dressings of N, P and K are usually required, especially for sitka spruce. Once established, the plantations help to dry the peat and increase aeration by extracting water and intercepting rainfall (King *et al.*, 1986).

Summary

Of all the components of soil organic matter, the living microbial biomass is by far the most important because it is responsible for the decomposition of all other components, taking up the nutrients they contain and on death releasing them in mineralized forms suitable for uptake by higher plants. The microbial biomass therefore acts as the 'eye of a needle', though which nutrients must pass before becoming available for plant uptake. The size of the microbial biomass therefore determines the rate of natural nutrient cycling in soils and reflects general soil health. It is also influenced by soil management, and is an early indicator of the much slower changes affecting amounts of total organic matter, such as the decreases resulting from cultivation of previous woodland or pasture soils or the increases in arable soils resulting from regular inputs of organic manures, sewage sludge, composts, peat or organic lake deposits (sapropel). The organic matter content of arable soils can also be increased by green manuring, especially with leguminous crops (e.g., grass-clover leys).

Many unwanted crop residues, such as cereal straw, have high C/N ratios and decompose fairly slowly after incorporation into soil. They also immobilize nitrogen for a period, as the microbial biomass decomposing the residues needs to take already mineralized nitrogen from the soil to match the large amounts of carbon in the residues. As there are problems with the alternative disposal mechanism of leaving large amounts of straw and other crop residues as a mulch on the soil surface, it may be more appropriate to consider using unwanted residues as fuel for power generation. Compared with fossil fuels, which increase atmospheric carbon dioxide, crop residues or even specially grown fuel crops, such as willow or elephant grass, are 'CO_2-neutral', in that the carbon dioxide they release to the atmosphere on burning is roughly equivalent to the amount they remove from the atmosphere during growth.

Soils that are very rich in natural organic matter, such as those formed in peat, were previously thought to be very useful for arable crop production because they have large available water capacities and are easily cultivated. However, they usually need draining, and this leads to accelerated wastage by oxidation and wind erosion. It also results in loss of a rare wildlife habitat and frequently also to rapid soil acidification by oxidation of sulphides. For various reasons, peat soils are therefore best left in a natural, undrained and uncultivated condition.

FURTHER READING

Killham, K., 1994, *Soil Ecology*, Cambridge: CUP.

McBride, M.B., 1994, *Environmental Chemistry of Soils*, Oxford: OUP.

Stevenson, F.J., 1994, *Humus Chemistry. Genesis, Composition, Reactions*, 2nd edn, New York: J. Wiley.

Wilson, W.S. (ed.), 1991, *Advances in Soil Organic Matter Research: the Impact on Agriculture and the Environment*, Cambridge: Royal Society of Chemistry.

Wood, M., 1989, *Soil Biology*, Glasgow: Blackie.

9 Soils and climatic change

We know more about the movement of the celestial bodies than about the soil underfoot.

Leonardo da Vinci

Introduction

Potentially one of the most important issues confronting humanity in the twenty-first century is that known as global warming. This is a progressive rise in temperature at the Earth's surface over the last century or so, which, if it continues, could modify the Earth's climate and environment in ways that threaten aspects of civilization, such as provision of sufficient food and water. The temperature rise has occurred at approximately the same time as some trace gases in the atmosphere (mainly CO_2, CH_4 and N_2O) have increased through human activities, such as burning fossil fuels, destroying large areas of natural forest, draining wetlands and using fertilizers to increase crop yields. The trace gases are often termed 'greenhouse' gases because they help retain heat in the lower atmosphere, rather like a greenhouse warms up in the sun and loses little of its heat to the cold air outside. Their main effect is to absorb parts of the near-infrared (heat) radiation, which originates in solar radiation but is reflected upwards from the Earth's surface. The radiation absorbed by these trace gases is re-emitted in all directions, including downwards to the Earth's surface, so that some of it does not return to space; see Harvey (2000) for a detailed explanation of the physical processes involved.

Reactions in soils, especially those resulting from microbial activity, have a considerable influence on 'greenhouse' gases in the atmosphere. Approximately 20 per cent of the CO_2, 40 per cent of the CH_4 and 65 per cent of the N_2O released globally into the atmosphere are thought to result from soil processes, including aerobic and anaerobic decomposition of soil organic matter. Consequently, more careful management of soils and vegetation may offer opportunities for combating the increases in 'greenhouse' gases resulting from anthropogenic activities.

However, 'greenhouse' gases are only one of numerous factors affecting climate. A wide range of evidence shows that the Earth's climate changed naturally in the past at times before there could possibly have been any human influence, and the same natural factors are equally likely to cause future climatic changes. So an important question is: are the decreases in 'greenhouse' gases that can be achieved by better management of soils and vegetation going to make any difference to future climate?

At present the answer to this question is far from clear. This chapter first considers some important aspects of the climatic system and then examines the results of climatic models that try to predict changes over the next century or so. Section 9.4 discusses how soils are likely to be affected by global warming and how they might be managed to limit their own emissions of 'greenhouse' gases or absorb those entering the atmosphere from other sources. Finally, other factors known to have influenced climate in the past will be considered, so as to put into perspective the contribution soils may make.

9.1 THE NATURE OF CLIMATIC CHANGE

Climate varies spatially according to latitude, distance from oceans, presence/absence of mountains and other topographic factors. It is determined by four main interacting components: the atmosphere, hydrosphere (oceans and other large water bodies), biosphere (vegetation, animals, humans and soils on the land surface) and cryosphere (ice and snow). The activities of these components are driven by various forcing mechanisms, of which the most important is solar radiation. The factors that determine the climate at any locality on the Earth's surface include:

- the balance between incoming energy (solar radiation) and outgoing long-wave radiation;
- the heat stored in and moved by the oceans and other large water bodies and transferred from them to the atmosphere;
- the heat and moisture moved through the atmosphere by winds.

9.1.1 Radiation balance

The annual input of solar radiation to a horizontal surface in the tropics is 2.4–3.5 times as much as at the poles (Budyko *et al.*, 1962). However, there are large seasonal differences: during the summer, when the polar region has a 24-hour day, the daily input of solar energy is 1.4 times greater than at the equator, but near the winter solstice it is zero. The mean radiation inputs translate into mean daily maximum air (shade) temperatures at sea level ranging from <7°C north of about 40° latitude in the northern hemisphere to >27°C over most of the tropics (Ransom, 1963).

Much of the solar radiation received at the Earth's surface is reflected back into the atmosphere, and some of this into space. The proportion reflected (the albedo) depends on the type of surface: ocean surfaces reflect only 2–3 per cent if the solar elevation angle is large and the water calm; forests reflect 9–18 per cent, depending on tree type and foliage density; for grassland the value is approximately 25 per cent, and for desert areas with little or no vegetation it is around 30 per cent; for fresh snow it is as high as 90 per cent (Barry and Chorley, 1998).

9.1.2 Atmospheric circulation

The Earth's major wind circulation pattern results from the unequal heating of different parts of the atmosphere. Over the warmer tropical regions, the lower atmosphere expands; this decreases the air pressure at the Earth's surface and causes the surfaces of equal pressure to bulge upwards into the troposphere (the lowest 8–16 km of the atmosphere). Conversely, in polar regions the colder and denser air causes downward contraction of the surfaces of equal pressure. This leads to an overall pressure gradient from poles to equator, which probably initiates the motion of the atmosphere. However, because the Earth is rotating, the angular momentum or Coriolis force generates an almost circumpolar flow or vortex in the atmosphere, strongly modifying the effect of the pressure gradients. Over each hemisphere, the circumpolar vortex created by the Coriolis force is west to east around the Earth and, at any given latitude, is fairly constant within the troposphere from about 1 km to 10 km above the Earth's surface. However, the maximum wind speeds within it (usually 45–70 $m\,s^{-1}$) are concentrated in a narrow band around 30° latitude, called the jet stream, which coincides with the latitudinal zone where the meridional temperature gradient (parallel to lines of longitude) is greatest.

Anticyclones and depressions are probably generated in the lower troposphere below a height of approximately 2 km at points of imbalance forming waves and eddies in the circumpolar vortex where its path is deflected by mountains, or where the pressure gradient in the troposphere is slightly stronger or weaker. Anticyclones usually form and are steered in direction on the warm side of the circumpolar vortex; depressions usually occur on its cold side. Consequently, anticyclones form belts in subtropical latitudes (around 30°) on either side of the equator and depressions occur mainly at higher subpolar latitudes on the poleward sides of the circumpolar vortex. However, another belt of depressions occurs close to the equator between the subtropical anticyclones, where insolation and heating of the lower atmosphere are greatest, and at very high latitudes anticyclones result from the cold (dense) polar air.

Winds blow clockwise around anticyclones and anticlockwise around depressions in the northern hemisphere, but the reverse in the southern hemisphere. In the lower atmosphere, this gives rise to the easterly trade winds between the equator and the subtropical belts of anticyclones, and to predominantly westerly winds north and south of the subtropical anticyclonic belts, i.e., between latitudes 30° and 60° approximately (Martyn, 1992). The trade winds of the northern and southern hemispheres converge in the somewhat discontinuous equatorial low pressure trough termed the Intertropical Convergence Zone (ITCZ). As anticyclones and depressions can reach a few thousand kilometres across, cold and warm winds from very different sources and directions can be brought together in the lower atmosphere, creating fronts along which warm humid air rises, cools and forms clouds.

The circumpolar vortex varies between a regular (zonal) west–east flow pattern and a more distorted or meandering pattern of large-amplitude waves with some north–south (meridional) component. A strongly meandering vortex can wander from around 30° north or south of the equator almost to polar regions, and is responsible for most of the heat transfer in the atmosphere from the equator towards the poles (Barry *et al.*, 2002). However, the meandering pattern may become slow-moving or even stationary for long periods, giving rise to a 'blocking situation', which results in almost stationary anticyclones or depressions in the troposphere of middle latitudes. As a result, areas such as northwest Europe can experience prolonged spells of abnormally dry and warm summer conditions or dry and cold winter conditions, whereas other areas have prolonged wetter spells, both replacing the succession of eastward-moving depressions with their accompanying fronts and rain belts.

Changes in the extent of distortion of the circumpolar vortex and their effects on air movement in the troposphere account for much of the daily variation in weather and the seasonal, year-to-year and possibly longer-term variation in climate in mid-latitude regions. However, their causes are at present unknown. Regular season-to-season changes in day length on either side of the equator move the wind zones by approximately 10° of latitude, but there are larger, less regular and shorter-term changes superimposed on this, so that the seasonal progression of wind and temperature change in mid-latitudes is often overridden by brief periods of unseasonal weather.

9.1.3 The Northern Hemisphere Annular Mode

An important feature of atmospheric circulation that has resulted in much of the recent interannual and interdecadal climate variability of the northern hemisphere are the oscillations in mean atmospheric pressure at sea level between the zone of mid-latitude cyclones, such as the Icelandic depression (~65°N), and the zone of subtropical anticyclones, such as that often persisting over the Azores (~40°N). These oscillations form what has been called the North Atlantic Oscillation (NAO), though its influence is more extensive than just the Atlantic region, and the term Northern Hemisphere Annular Mode or NAM (Thompson and Wallace, 2001) is perhaps more appropriate.

When the NAM index is high (strong north–south pressure gradient), as in the last 40 years, there is a tendency for strong westerly winds in northern parts of this latitudinal zone, with mean winter temperatures approximately 5°C higher and greater precipitation over the USA, Canada and Europe than when the index is low. A low index, as in the middle of the twentieth century, results in weaker westerlies between the Icelandic depression and Azores anticyclone, with frequent blocking conditions leading to a predominance of cold winter anticyclones over Canada, the US Pacific Northwest, Scandinavia, Russia and the Arctic. Measurements of tree-ring density and thickness in the northern hemisphere over the last 300 years suggest cycles of 8 and 24 years in NAM variation, with periods of especially high values between AD 1741 and 1758 and in the last two decades of the twentieth century. The cause of these cycles is unknown.

9.1.4 Oceanic circulation

In addition to the transfer of heat from equator to polar regions by atmospheric circulation, considerable heat redistribution also occurs by oceanic circulation. The ocean has a large heat capacity and can store heat for long periods before it is radiated back into space or released to the atmosphere in water vapour (i.e., latent heat). Oceanic circulation is much slower than atmospheric circulation, and results from:

- external factors, including the Coriolis force, insolation variations, tides, the influx of fresh water from rivers and glacial meltwater and the frictional drag of winds; and
- internal factors, such as density differences resulting from salinity and temperature variations.

Many of these variables are interconnected, and the exact cause of a particular ocean current and the likelihood of it changing with time are often difficult to establish. Nevertheless, studies of the planktonic microfauna from dated layers in deep ocean sediment cores suggest past changes in surface water temperature and salinity, which must have resulted from changes in the movement of ocean water masses.

Tides, driven by the gravitational pull of the moon and sun, and winds are the main sources of turbulent energy, which causes mixing of the upper ocean (Munk and Wunsch, 1998). This produces a thermally mixed oceanic surface layer up to 400 m thick, which is warmed principally by solar radiation and overlies a thermally stratified layer (the thermocline) extending to about 1 km depth. Below this there is a deep layer of cold, dense water, within which water movement (thermohaline circulation) results from density variations caused mainly by differences in temperature and salinity.

Many ocean currents are driven by wind patterns; for example, the North and South Equatorial Currents are driven by the persistent trade winds. Where the currents are driven against eastern continental coasts, the water cannot escape by sinking because of its higher temperature and stable vertical stratification. It is consequently deflected polewards along the coast under the influence of surface winds, and forms a narrow rapid current. Examples are the Gulf Stream or North Atlantic Drift of the North Atlantic and the Kuroshio Current of the western North Pacific.

In contrast, currents driven against western continental coasts are deflected equatorward and then westwards away from the coast, allowing replacement by slow upwelling of cold deep ocean water. This occurs, for example, in the southeast Pacific, where the Peru Current is deflected northwards and then westwards, and upwelling water normally cools the western South American coast. However, every 2–11 years the upwelling ceases for a period of weeks or months on either side of mid-December, and during these periods there is a change in the mean atmospheric pressure gradient measured between Tahiti and Darwin. This so-called El Niño Event results in sea surface temperatures approximately 6°C above the usual (~24°C) for this season, and causes intense rain and windstorms that batter the western coasts of Peru and Mexico. At the same time, the western Pacific becomes drier than usual because here the ocean is cooler and there is less evaporation; this leads to droughts, dust storms and natural fires in Indonesia and Australia. The cold, deep ocean water brought to the surface by upwelling is nutrient-rich, so its periodic failure can also greatly reduce the fish harvest on the eastern Pacific coast. The 2- to 11-year cycle of oscillations in the Pacific circulation and associated changes in atmospheric pressure gradients are known as the 'El Niño Southern Oscillation' (ENSO).

The temperature and salinity differences causing deep ocean circulation mainly result from sinking of warm saline water to the deep ocean basins in exchange for upwelling of colder and less saline water to the ocean surface in other areas. Most is known about present and past deep ocean (thermohaline) circulation in the Atlantic, where surface water warmed in subtropical regions, principally the Caribbean but supplemented in mid-Atlantic by water flowing from the exit to the Mediterranean (Bower *et al.*, 2001), currently moves northeastwards as the Gulf Stream. The warm surface water is slightly more saline than usual because of evaporation in the warm Caribbean and Mediterranean seas, and also because of evaporation by colder, but strong, winds in the Greenland, Iceland and Norwegian (Nordic) Seas. In colder parts of the North Atlantic around 60°N the greater water density resulting from the salinity increase and cooling causes it to sink, forming 'North Atlantic Deep Water' (NADW). The loss of heat on sinking results in a temperature decrease from approximately 10°C to 2°C and warms the atmosphere of coastal regions of northwest Europe by approximately 10°C. The NADW then initiates a slower and more diffuse global deep water (>1500 m) current flowing southwards to the South Atlantic. Here it is augmented by further cold and dense deep water, which has subsided in the Ross and Weddell Seas. Under the influence of the Coriolis force, the deep water flows eastwards and then northwards into the Indian and Pacific Oceans, where the salinity is decreased by mixing, allowing the water to rise and form a shallower return flow to the Atlantic. The complete global thermohaline cycle, which has been likened to a conveyor belt (Figure 9.1), is completed in approximately 1500 years.

Micropalaeontological studies of sediment cores from mid-latitude regions of the North Atlantic have indicated that the northern limit of Gulf Stream penetration fluctuated repeatedly and often very rapidly over the past 120,000 years (Section 9.5.2). Also on several occasions the formation of NADW and the whole circulation system seems to have virtually ceased. Episodes of southward retreat by the Gulf Stream coincided with southward excursions of cold polar water (the polar front), and probably resulted from influx of cold, fresh meltwater from glaciers in North America and Scandinavia. These, in turn, led to large and rapid decreases in atmospheric temperature in northwest Europe. The three types of circulation mode identified in Atlantic sediments have been termed cold, warm and off (Rahmstorf, 2002). In the cold (or stadial) mode, NADW formed in the open North Atlantic south of Iceland; in the warm (or interstadial) mode it formed in the Nordic Seas north of Iceland; and in the off (or Heinrich) mode NADW formation ceased.

Changes in Atlantic sea surface temperatures resulting from changes in the strength of the oceanic conveyor belt may also have affected the climate of regions remote from northwest Europe. Higher sea temperatures probably enhance the summer monsoon rainfall in West Africa, so lower Atlantic surface temperatures could lead to drier conditions in the Sahel. Changes in the amount of deep water reaching the Indian and Pacific Oceans could also affect the strength of the Asian Monsoon (Schulz *et al.*, 1998), and the off mode may even affect Antarctica, which at present receives some

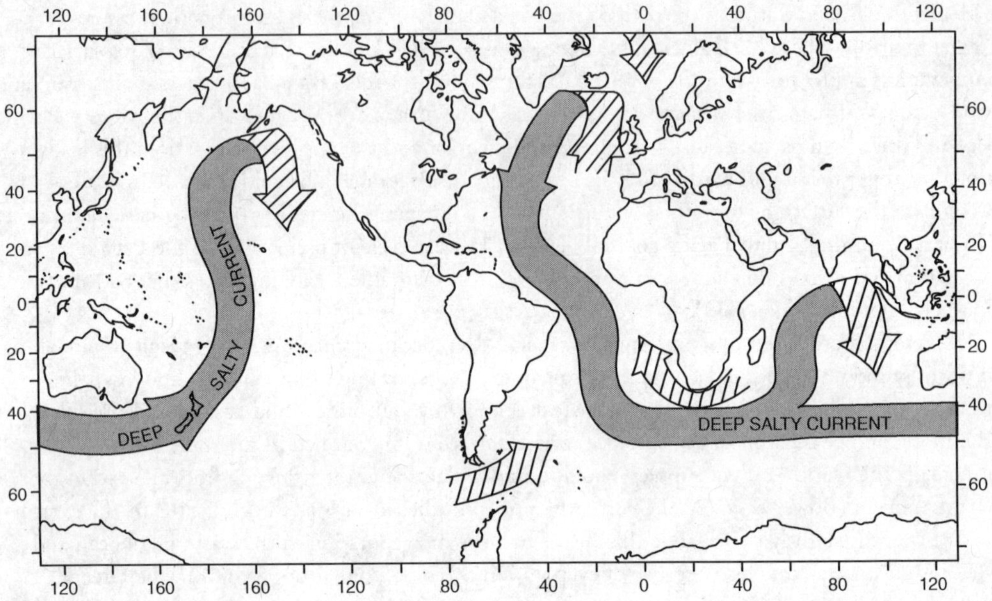

Figure 9.1 The global conveyor belt of oceanic thermohaline circulation
Source: Broecker and Denton (1990).

heat from the NADW carried southwards from the Atlantic to the Southern and Pacific Oceans. Part of this water resurfaces off the Antarctic coastline and, although it is then only a few degrees above freezing, it is warmer than the adjacent Antarctic continent and can melt patches of coastal sea ice. If it failed to appear because NADW formation ceased, the greater extent of sea ice around Antarctica would reflect more solar radiation and thus cool the region (Adams *et al.*, 1999).

9.1.5 Water vapour and clouds

Atmospheric water vapour, produced by evaporation from the oceans (84 per cent) and evapotranspiration from land surfaces (16 per cent), accounts for most of the heat transferred from the Earth's surface to the atmosphere. Water vapour is lighter than dry air and consequently rises, especially at higher temperatures when the proportion in the atmosphere increases. At the lower temperatures in higher parts of the atmosphere it then condenses to form clouds and the water is returned to the Earth's surface as precipitation. The condensation liberates the latent heat used to generate the water vapour, and most is lost to space as long-wave (infrared) radiation. Because warm air can hold more water as vapour than cold air, the water vapour content of the atmosphere in either hemisphere is greater in summer than in winter.

High clouds, such as cirrus, cirrostratus and cirrocumulus, reflect and absorb >30 per cent of the solar radiation and thereby have a cooling effect. However, an extensive cover at lower levels, as formed by stratus, nimbostratus and cumulus clouds, absorbs outgoing terrestrial radiation and thus has a warming effect in winter. Together with the 'greenhouse' gases, they contribute to the 'natural greenhouse effect', which currently maintains a global mean temperature near the Earth's surface of 14°C, compared with −58°C at the top of the troposphere (10 km above the surface).

9.2 MODELLING FUTURE CLIMATE

Models that simulate operation of the whole complex climatic system are built from databases of meteorological records for selected locations and time periods and parameters for climatic forcing mechanisms, feedbacks and energy transfers based upon physical laws controlling the atmosphere and oceans. The central core of such models are the coupled Atmosphere–Ocean General Circulation Models (AOGCMs), which trace horizontal and vertical movements of three-dimensional units of air and water and changes in their energy and composition according to equations of physical and dynamic relationships, such as conservation of mass and momentum. Calibration and adjustment are based on runs with actual meteorological data for recent decades. Once the simulations agree with observations for the past periods, the models are used for predicting future change according to scenarios defined by a limited number of modifications to the input parameters thought to influence future climatic forcing. AOGCMs require very large amounts of data to build, calibrate and run them, and it is often difficult to decide precise values for the numerous input parameters. Nevertheless, they are widely regarded as the best means of predicting the effects on increasing 'greenhouse' gases on future climate.

9.2.1 The Intergovernmental Panel on Climate Change (IPCC) predictions of twenty-first century climatic change

The most complex and sophisticated AOGCMs available are those developed by various groups within the United Nations Intergovernmental Panel on Climate Change (IPCC). With a 'business as usual' scenario (i.e., minor controls on industrial and agricultural emissions that would allow atmospheric CO_2 to increase to twice the concentration in the pre-industrial era by AD 2070), the first of the IPCC reports in 1990 predicted an increase of 3.5°C in global mean temperature by AD 2100. With adjustments to various input parameters, this value was decreased in 1992 to 2.5°C (range 1.5–4.5°C), though with greater increases in high latitudes of both hemispheres than elsewhere. These early IPCC results also suggested an increase in maximum temperatures, fewer very cold periods in winter, greater atmospheric turbulence (storminess) because of cooling of the stratosphere to restore the balance between incoming and outgoing radiation, and increased precipitation in mid-latitude storm tracks in winter and throughout the year in the ITCZ.

In 1995, the predicted increase in global mean temperature with a doubling of atmospheric CO_2 content by AD 2100 was decreased further by IPCC to a mean of 2°C (range 1–3.5°C), assuming emissions would increase according to a scenario of moderate world economic growth (2.3 per cent yr^{-1}) between AD 1900 and 2100. The largest increases (>3°C) were again predicted over polar regions because of decreased snow and sea ice covers, though increases over large continents, such as Africa, Australasia, South America and north Asia, were expected to exceed 2°C; increases less than the mean of 2°C were predicted mainly over the oceans.

Since 1995 AOGCMs have been expanded to include additional forcings and the effects of some 'greenhouse' gases other than CO_2. Most model runs have included 'idealized forcing', that is, a compound increase in atmospheric CO_2 of 1 per cent yr^{-1}, which includes the CO_2-equivalent effects of some other 'greenhouse' gases. Outputs from approximately 20 AOCGMs were compared to assess uncertainties in the climate responses to individual forcings. These were small, but agreement over mean annual temperature change was better than for precipitation change (IPCC, 2001: chapter 9). Agreement between the model outputs was greater for low than for high latitudes.

One recent IPCC prediction (IPCC, 2001) that has attracted considerable interest concerns the greater probability of extreme weather events, such as storms and prolonged droughts. This is partly because actual meteorological observations in recent decades seem to confirm some of the predicted weather extremes. The higher mean temperatures suggested by early model runs implied an increasing probability of extremely warm days and decreasing probability of extremely cold days. All later model predictions and some meteorological measurements in critical areas have confirmed these aspects. Temperature minima are increasing most in areas of high latitude and altitude where snow and ice are disappearing (Balling *et al.*, 1998), and temperature maxima are increasing most over large land areas, such as Australia. However, in the USA there is no discernible trend in maximum temperatures, and in China maxima have even declined (Easterling *et al.*, 2000). In the long British record, Jones *et al.* (1999) discerned a decrease in annual numbers of cold days, but no increase in the number of hot days over the last 150 years.

The recent modelling and meteorological observations have both indicated a general decrease in diurnal temperature range, because nocturnal minima increase faster than daytime maxima. Other consistent model predictions include an increasing frequency of deep low-pressure systems with strong winds in the northern hemisphere winter, leading to increases in annual rainfall and precipitation intensities, which have been observed in the USA, southern Canada and western Russia (Easterling *et al.*, 2000). Increased rainfall intensity could result from the atmosphere's increased ability to hold water vapour at higher temperatures and its greater instability because of the increased temperature gradient with height above the Earth's surface. Both have been confirmed by recent measurements. Greater precipitation intensity could increase runoff and erosion rates (Favis-Mortlock and Guerra, 1999) but, at present, there are no predictions of changes in the intensity of local thunderstorms, hailstorms or tornados, which are smaller than the grid scale at which AOGCMs currently operate. Other local weather patterns in areas less than approximately $10^4 km^2$, such as those influenced by topography, land use and the presence of inland water bodies, are also impossible to predict at present.

9.2.2 The Kyoto Protocol

The second report of IPCC (1996) led to the Kyoto Protocol of the UN Framework Convention of Climate Change held at Kyoto, Japan in 1997. The protocol called for large reductions in emissions of all 'greenhouse' gases by every country. This could be achieved by:

- limiting consumption of fossil fuels;
- decreasing emissions of artificial 'greenhouse' gases, such as hydrofluorocarbons (HFCs), perfluorocarbons (PFCs) and sulphur hexafluoride (SF_6); or
- increasing carbon sequestration through afforestation and changes in agricultural management that would decrease atmospheric CO_2.

To calculate a country's total emissions, all the 'greenhouse' gases are aggregated according to their CO_2-equivalent global warming potentials, which depend upon each gas's capacity to absorb long-wave (infrared) radiation and its lifetime in the atmosphere. Over 170 countries ratified the Kyoto Protocol and many industrialized countries signed 'Annex 1' to decrease 'greenhouse gas' emissions in the period 2008–2012 by at least 5 per cent below their 1990 levels.

9.3 EFFECTS OF GLOBAL WARMING ON SOILS AND MANAGEMENT TO DECREASE EMISSIONS OF 'GREENHOUSE' GASES

Evaluations of the impacts of global warming on soils and the possible influence of soils on 'greenhouse' gases are qualitative and rather speculative at present, because of uncertainties in the climatic predictions themselves and the complexity of possible interactions between soils, vegetation, land management and atmospheric composition. Since global warming was recognized as a major issue, there have been many attempts to evaluate some of the interactions by observation and experiment, but few clear patterns have emerged. The various changes expected in different regions must affect soils, natural vegetation and crops in different ways (Parry *et al.*, 1999).

9.3.1 Plant growth and carbon sequestration in soils

The current gross increase in atmospheric CO_2 (6.5 Gt C yr^{-1}) (1 Gt = 10^9 t) comes mainly from burning of fossil fuels and a further 1.6 Gt arises from deforestation (Smith, 1999). The CO_2 sink in terrestrial ecosystems (vegetation and soils) is thought to be approximately 2.0 Gt C yr^{-1}, and the oceans absorb a further 2.7 Gt C yr^{-1}, so the net increase in atmospheric CO_2 is approximately 3.4 Gt C yr^{-1}. Soils contain large amounts of C; Lal *et al.* (1995) estimated that the total amounts worldwide are 1550 Gt organic C and 1700 Gt inorganic C (mainly in $CaCO_3$), both of which are much greater than the pools in either the atmosphere (750 Gt C) or in all living organisms (550 Gt C). If it can be taken from the atmosphere, a small percentage increase in the soil organic C pool (0.1–0.2 per cent yr^{-1}) could therefore counteract the current increase in CO_2 content of the atmosphere (about 1.5 parts per million by volume per year).

As green plants remove CO_2 from the atmosphere as a raw material for production of organic matter (mainly carbohydrates) by photosynthesis, and much of the organic matter is incorporated into the soil, increasing plant growth should, in time, decrease atmospheric CO_2 and sequester much of the C in the plants themselves or in soils. However, much of the organic C in soils returns to the atmosphere through microbial oxidation (mineralization), so minimizing the return of CO_2 to the atmosphere by decreasing mineralization rates would also help decrease atmospheric CO_2. Soil management can affect both plant growth and mineralization of organic matter, and whether soils serve as a net source or sink of C depends strongly on their management (Lal, 1999).

With increases in soil temperature, microbial processes affecting SOM mineralization are likely to accelerate, especially where the soil remains moist but well aerated in summer (Leiros *et al.*, 1999). Worldwide this could result in decreased sequestration of C in soils and increases in atmospheric CO_2 and N_2O. Warmer soil conditions also depress microbial N-fixation, which may further decrease sequestration of C in vegetation and soils.

Land use history affects SOM content (Pulleman *et al.*, 2000), and some land use changes, such as conversion of arable to forest or pasture, can lead initially to greatly increased rates of SOM accumulation (Section 8.3.3). However, rates of increase usually decline rapidly, often reaching a new equilibrium value within a few decades (Mosier, 1998). For example, Neill *et al.* (1998) reported the initially high accumulation rate of 0.304 kg C $m^{-2} yr^{-1}$ after deforestation for pasture, but it soon declined to a much lower value. High rates of C accumulation in biomass or SOM really need to be maintained for decades or even longer if they are going to have any impact on rising atmospheric CO_2 content.

Decreasing SOM mineralization rates by zero or minimal tillage techniques (Section 7.2.1) or by fallowing increases C sequestration in the short term. However, other effects of minimal tillage, such as surface compaction, can have the opposite effect (Brevik *et al.*, 2002). Soil clay content, structural stability and drainage are also important in determining rates of SOM mineralization, in that bonding of organic compounds to clay minerals and protection within stable aggregates and waterlogged horizons can both confer some resistance to microbial decomposition. These soil characteristics should therefore be considered in developing strategies for increasing C sequestration, though their effects may be influenced by soil management to a greater extent in the tropics than mid-latitude regions (Rosenzweig and Hillel, 2000).

9.3.2 Carbon sequestration by afforestation

Past forest clearance, especially by burning, has increased the atmospheric concentrations of CO_2, so limiting or reversing these activities should reduce the rate at which atmospheric CO_2 is increasing. IPCC (2001) suggested that more extensive afforestation and improved forest management should be regarded as important strategies for mitigating global warming. As forest biomass has a much longer life than arable crops and pasture, expansion of forest areas or increasing tree production per unit area should sequester large amounts of C for the length of time necessary to influence rising atmospheric CO_2. However, net primary production in the early stages of afforestation is small and may not exceed loss of C to the atmosphere by SOM decomposition. As tree growth increases, the primary production of more mature forests can become an important CO_2 sink but, in later stages of forest maturity, the biomass decays faster than it is produced and forests can again become a source of CO_2 rather than a sink. Careful long-term management will therefore be required if forests are to maintain a significant role as a sink for atmospheric CO_2, but this will be expensive.

It was originally anticipated that the higher temperatures expected with global warming and the increasing atmospheric CO_2 should themselves accelerate the growth of plants, thereby partially compensating for increasing atmospheric CO_2 concentrations. In C3 plants (trees and mid- and high-latitude grasses, which produce compounds with three C atoms at an intermediate stage in photosynthesis), an increase in atmospheric CO_2 should increase the photosynthetic rate, because of the increased probability of CO_2 bonding with the rubisco enzyme (ribulose biphosphate carboxylase) that catalyses photosynthesis. This is often termed the 'CO_2 fertilization effect'. In C4 plants (tropical grasses producing intermediate compounds with four C atoms during photosynthesis), there is less likelihood of increasing photosynthetic rates in this way, because the CO_2 in contact with rubisco is maintained at a higher concentration by an internal transfer mechanism.

However, experiments on plants grown in glasshouses with elevated atmospheric CO_2 and in more natural conditions with CO_2 piped to the site at controlled rates (the 'free-air CO_2 enhancement' or FACE experiments) have not fully confirmed these expected results. Seedlings of many C3 plant species have shown initial increases in growth rates exceeding 30 per cent in CO_2 concentrations double present atmospheric values (Idso and Idso, 1994) but, over time, there is often a loss of the stimulatory effect of increased CO_2 on growth. This probably results from a decrease in the concentration of rubisco with increased CO_2 concentrations. Unexpectedly, short-term responses to elevated CO_2 have also been reported in some C4 plants (Poorter, 1993). Increasing atmospheric CO_2 can also cause premature senescence and leaf-fall in deciduous tree species (McConnaughay *et al.*, 1996) so that, although photosynthetic rates are greater in the spring and early summer, the annual period of photosynthesis is shortened and the end-of-season gain in C is no greater than with the present atmospheric CO_2 content. FACE experiments have shown that ozone (O_3), another atmospheric trace gas that has increased rapidly over the last century, also decreases forest productivity, even in the presence of enhanced CO_2, and also decreases C inputs to forest soils (Loya *et al.*, 2003).

Increased CO_2 may also lead to ecological changes in forests that could accelerate biomass C turnover rates, thereby creating a positive feedback mechanism causing further increases in atmospheric CO_2. For example, Phillips and Gentry (1994) noted an increase since 1960 in rates of tree mortality in many humid tropical forests. They attributed this to increasing atmospheric CO_2 stimulating the growth of parasitic lianas, and suggested that, with continuing global warming, decomposition of the dead trees would make the forests a net source of C rather than a sink. Increased decomposition over periods of up to a century or so could also result from the delay in migration of tree species adapted to higher temperatures while the existing poorly adapted species gradually decline in abundance (Bolker *et al.*, 1995).

Further evidence that natural vegetation does not necessarily grow more rapidly in a CO_2-enriched environment was provided by measurements in a 1-ha depression fed by a geothermal 'CO_2 spring' near Siena, Italy. At this site, Körner and Miglietta (1994) found no evidence for a difference in plant growth between the depression, where the atmospheric CO_2 concentration is maintained at 500–1000 ppmv, and surrounding areas, where it is close to the present atmospheric level (350–370 ppmv), even though the plant communities, soil characteristics and weather were the same.

The effect of increased atmospheric CO_2 on the size of stomatal openings in leaves could have a positive effect on C sequestration. During photosynthesis, leaf stomata open to permit CO_2 to enter but, simultaneously, water is lost by transpiration. Increasing atmospheric CO_2 concentrations should therefore lead to smaller stomatal openings for a given inflow of CO_2, and this will decrease transpiration rates (Field *et al.*, 1995). Consequently, with increased atmospheric CO_2 there should be less mineralization of SOM and more sequestration of C, because soils should on average become wetter and less aerobic. However, results from experimental plots of trees grown in elevated CO_2 environments have not yet confirmed any increase in C sequestration in soil. Much of the increased carbon uptake in forest trees is allocated to short-lived tissues such as leaves, which are rapidly mineralized in the soil despite the possible small increase in soil water (Schlesinger and Lichter, 2001).

Oren *et al.* (2001) also regarded the IPCC estimates of carbon sequestration by forest ecosystems as too optimistic. Their field experiments showed no increase in tree biomass with elevated CO_2 unless soil nutrients such as N are also increased. Consequently, as many mid- and high-latitude forest soils have very low fertility and N fertilizers are too expensive to use in this way, it is unlikely that afforestation will greatly decrease atmospheric CO_2. In any case, increasing the N contents of forest soils with fertilizers is unlikely to increase their C sequestration rates, because increased N accelerates the mineralization of labile SOM fractions with short (decadal or less) turnover times (Neff *et al.*, 2002).

As the largest temperature increases from global warming are expected in high northern latitudes, it is likely that expansion of evergreen boreal forests will be greater than that of forests at lower latitudes (Haeberli and Burn, 2002). However, extension of boreal forest into tundra areas could result in a positive feedback effect, decreasing the albedo (reflectivity) of the tundra during winter and thus causing further high-latitude warming. High-latitude warming would also cause increased mineralization and accelerated release of CO_2 from the large amounts of SOM stored in the peaty soils of higher latitudes.

Changes in total annual rainfall and its seasonal distribution, which are expected with global warming, are also likely to affect plant growth and SOM mineralization. Although these are predicted much less reliably than temperature change, especially at regional and local scales, one of the most likely changes is an increase in the frequency of drought, especially in continental interiors. This could lead to progressive disappearance of large areas of natural vegetation, leading to regional decreases in C sequestration and considerable input of CO_2 into the atmosphere because of accelerated SOM mineralization (Dixon *et al.*, 1994). So even careful management of extensive new forests could have little impact on the increases in atmospheric CO_2 that may well accompany global warming.

9.3.3 Carbon sequestration by arable crops

Agricultural crops may have a greater effect on removing CO_2 from the atmosphere than forests, because most of them are resown each year and are therefore more frequently in the immature (seedling) stage, in which growth is most strongly affected by increased atmospheric CO_2. Agricultural production is also likely to benefit from the expected temperature increases occurring mainly at night, in winter and in colder regions. Also any deleterious effects of global warming and increasing atmospheric CO_2 on growth are likely to be weaker than in forest ecosystems, mainly because farmers can adapt to environmental change by modifying dates of sowing and harvesting, planting species or varieties that are better suited to higher temperatures or different rainfall characteristics, increasing N and other nutrients so that crops can take advantage of the 'CO_2 fertilizer effect', and increasing or decreasing irrigation rates and field drainage.

However, with a moderate level of adaptations like these, Rosenzweig and Parry (1994) estimated that world cereal production should change little by 2060, when atmospheric CO_2 is likely to be double the present concentration and the mean world temperature could have increased by 2.5°C. Various FACE and glasshouse pot experiments on wheat indicated a mean increase in grain yield of approximately 28 per cent with a doubling of CO_2 (Downing *et al.*, 2000), but with a wide range of results (Figure 9.2).

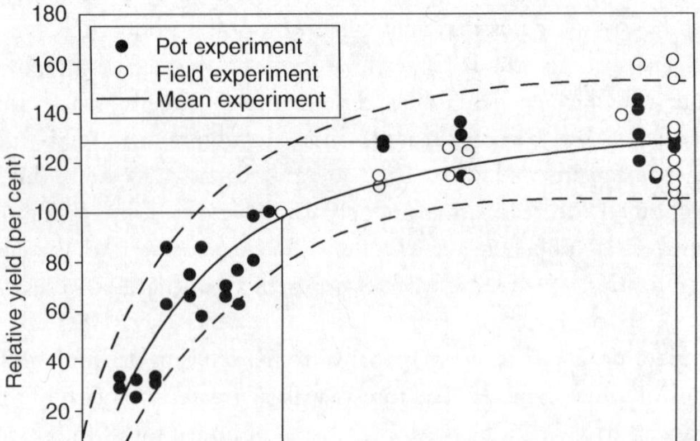

Figure 9.2 Effects of varying atmospheric CO_2 concentration on wheat grain yield in free air CO_2-enhancement (FACE) field experiments (open circles) and pot experiments in glasshouses (solid circles). The solid line shows the mean estimated effect; dashed lines show the 95 per cent confidence limits

Source: Downing *et al.* (2000) and Oleson and Bindi (2002).

With their greater resources and infrastructure for agricultural adaptation, crop yields in industrial countries will probably benefit more from higher temperatures and increased atmospheric CO_2 than those in developing countries, which could experience small decreases in crop production because of inadequate supplies of water and nutrients (Reilly and Schimmelpfennig, 1999). Oleson and Bindi (2002) predicted that in Europe the northward migration of climatic zones with global warming will expand the total area suitable for crop production and increase productivity, especially in the north. This would probably increase C sequestration over large areas. However, other regions may not be so fortunate. For example, southern Europe should experience water shortages, greater yield variability and a decrease in area suitable for many traditional crops, so that C sequestration would be less.

The mitigation strategy with perhaps the greatest potential is use of surplus agricultural land for production of rapidly growing biofuels such as willow and *Miscanthus* (Section 8.6.2). These are not only 'CO_2-neutral' (i.e., the CO_2 they produce on burning is matched by that sequestered during growth), but they also develop dense and extensive root systems in which large amounts of C are sequestrated for a long period. They provide no income for the farmer in the first few years after planting, but intercropping of biofuel, arable and horticultural crops using appropriate rotations is a compromise that may be profitable for the farmer as well as beneficial for sequestration of atmospheric CO_2.

Increased variability of future climate, with increasing incidence of extreme events, such as droughts in mid-latitude lands, may be more significant in determining crop yields than changes in mean climatic factors, such as mean annual temperature and rainfall or even monthly means over the growing season. Using a wheat growth simulation model, Porter and Semenov (1999) showed that grain yields in southern Spain should change little with a future climatic scenario based only on likely changes in mean weather factors. However, it could decrease by 30 per cent in a scenario incorporating a likely increase in variability of rainfall, on account of the greater probability of drought in the period of vegetative growth. This would make wheat an unreliable crop to grow in southern Spain. In contrast, the climatic variability expected in the UK should have little effect.

9.3.4 Meeting the European commitment to the Kyoto Protocol for carbon sequestration

Under the Kyoto Protocol, the European Union (EU) is committed to a decrease in CO_2 emissions by 2008–2012 of 8 per cent compared with the 1990 level. Using data from long-term experiments, in which changes in soil organic carbon (SOC) have been measured following recorded changes in land management, Smith *et al.* (2000b) suggested that much of the EU commitment under the Kyoto Protocol might be achieved by combining several management options for increasing either the above-ground plant biomass (e.g., woodland regeneration and production of fast-growing bioenergy crops) or the organic matter content of soils (e.g., straw incorporation and zero/minimal tillage).

Within the European Union, the member states rearranged this overall commitment to allow some to achieve CO_2 decreases greater than the 8 per cent mean value and others decreases less than this. The agreed UK commitment is for a 12.5 per cent decrease, and further calculations by Smith *et al.* (2000a, c), also based on long-term experiments, suggested that 31 per cent of the 12.5 per cent decrease (49 per cent of the mean EU commitment) might be achieved by various realistic combinations of policies. The main individual policies and their modelled effects are shown in Figure 9.3. The combinations of policies giving the largest C sequestration potentials were bioenergy crops + animal manures + no-till, woodland regeneration + animal manures + no-till and extensification + animal manures + no-till.

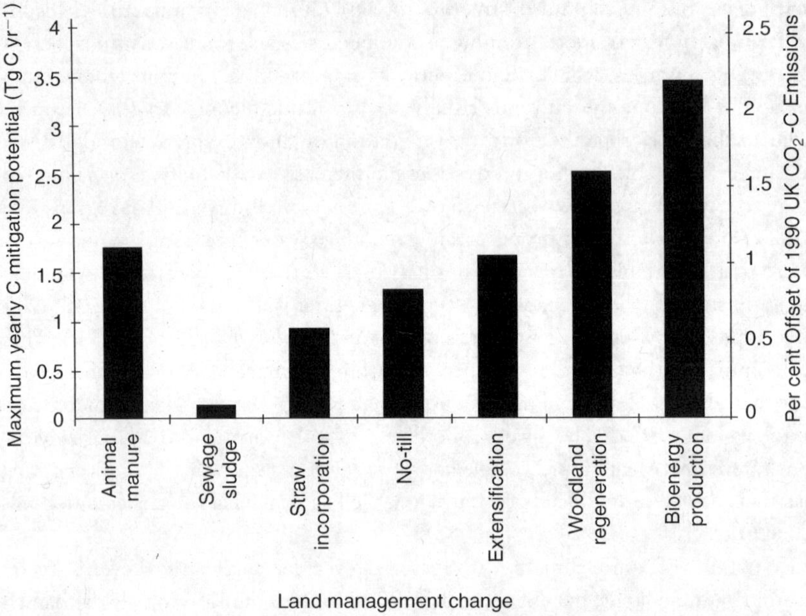

Figure 9.3 Estimates of the yearly carbon mitigation potential of various changes in individual land management practices as total amounts of carbon and as percentages of UK 1990 CO_2-C emissions. Animal manure was applied at 20 t ha^{-1}yr^{-1} to 45.3 per cent of arable land; sewage sludge was applied at 1 t ha^{-1} yr^{-1} to an extra 5.3 per cent or arable land compared with 1990 levels; straw was incorporated at a rate of 10 t ha^{-1} yr^{-1} to an extra 40.4 per cent of arable land compared with 1990 levels; no-till was applied to all suitable land (36.8 per cent of arable land); extensification was applied to 33 per cent of all arable land under the same management as 1990 (i.e., 28.6 per cent of arable land); woodland regeneration and bioenergy production were each applied to surplus (i.e., 10 per cent of arable land)
Source: Smith *et al.* (2000a).

Of these, the best combination was the first. However, as much surplus arable land is probably too far from power stations able to use bioenergy crops, a more realistic scenario (Combined Policy Opt) is to split the surplus arable land equally between bioenergy crops and woodland regeneration. This gave a C mitigation potential only slightly smaller than that of the less realistic combination of bioenergy crops + animal manures + no-till.

9.3.5 Other ways of sequestrating organic carbon

Because much of the C in plant material incorporated into soils is returned to the atmosphere as either CO_2 or CH_4, often within a few decades (Mosier, 1998), there may be more effective ways of sequestering the atmospheric C taken up by plants. For example, the net primary production of plant material in lakes (Dean and Gorham, 1998), peatlands (Turunen *et al.*, 2001) and coastal wetlands such as salt marshes (Howes *et al.*, 1985) is large, little of this C is mineralized and returned to the atmosphere because of the anaerobic conditions typical of these environments, and sedimentation is often rapid. As a result, the C is sequestered for long periods by burial and is not returned to the atmosphere anywhere near as quickly as that in aerobic arable, pasture and forest soils (Connor *et al.*, 2001). Rabenhorst (1995) calculated that wetlands contain 14.5 per cent of the C sequestered in the world's soils, even though they form only 4 per cent of the total land area of the Earth. Coastal salt marshes have an additional advantage in mitigating the effects of global warming in that they release much smaller amounts of CH_4 and N_2O than freshwater wetlands (De Laune *et al.*, 1990).

In addition to gaseous C losses from soils (as CO_2 and CH_4), large amounts of particulate C (as SOM) can also be lost by topsoil erosion, principally of arable land (Harden *et al.*, 1999). For example, in Iceland, a country with large areas of very organic soils (Histosols and Andosols) and extensive erosion, Óskarsson *et al.* (2004) have estimated that current annual losses of SOM exceed 200,000 t. Much of the chemically more resistant SOM removed by erosion is sequestered by redeposition in water bodies, such as rivers and lakes (Smith *et al.*, 2001) or even the ocean, though some is probably decomposed and returned to the atmosphere as CO_2 (Raymond and Bauer, 2001). Soil erosion by water and wind is therefore another mechanism by which atmospheric CO_2 is increased, and this provides another reason for taking measures to decrease erosion (Rosenzweig and Hillel, 2000). However, Lal (1999, 2001) has pointed out that, once organic matter has been lost in this way, the reclamation of eroded soils has considerable potential for sequestering C. In Iceland, if the current loss of C to the atmosphere because of soil erosion were eliminated, and an equivalent amount sequestered by reclamation of the eroded areas, the total would be >60 per cent of the CO_2 released anthropogenically by that country (Óskarsson *et al.*, 2004).

In summary, despite some experimental evidence for the effects of increasing temperature and atmospheric CO_2 content on C uptake by plants, it is unlikely that increased afforestation and changes in agriculture can fully mitigate the anthropogenic increase in CO_2. With its ability to adapt more rapidly, agriculture could have a greater effect than afforestation, especially if biofuel crops are grown extensively, but at present it seems unlikely that even the combined effects of several techniques to increase C sequestration in vegetation and soils will reverse the inexorable rise in atmospheric CO_2 resulting from a range of human activities and various feedback mechanisms. Accumulation of organic matter in lakes and wetlands has considerable potential for C sequestration, and even soil erosion may have some potential, either by burial of SOM in new sediments or by accumulation during reclamation of the eroded soil. Perhaps the most useful change would be to decrease the currently high rate of deforestation in humid tropical regions, especially where this involves burning.

9.3.6 Methane emissions from soils

In anaerobic soils such as rice paddies, CH_4 is produced by methanogenic bacteria, some of which reduce CO_2 in the presence of H_2 and others decompose fatty acids, such as propionate or acetate, generated by anaerobic microbial degradation of SOM (Conrad, 2002). The intermediate decomposition products (fatty acids, CO_2 and H_2) are produced by anaerobic fermentation of sugars, which, in turn, result from hydrolysis of polysaccharides by bacteria such as *Clostridia*. As the initial degradation of SOM and the activity of methanogenic bacteria both require anaerobic conditions, CH_4 production is broadly dependent on soil redox potential (Eh) and decreases in periods when rice paddies are drained and partially aerated. However, methanogens can survive in aerobic environments even though their activity is temporarily inhibited. As a result, CH_4 production increases fairly rapidly when paddy fields are reflooded because a large methanogenic population is already present. It also increases with use of organic manures or incorporation of straw, which provide a substrate for the methanogens.

To minimize CH_4 emissions from rice paddies, the length of the flooding period should therefore be decreased as far as possible consistent with productivity. However, there is much variation in CH_4 emission rates from rice paddies. Even with uniform treatment (continuous flooding and no organic manures), Wassmann *et al.* (2000) reported a range of 15–200 kg $CH_4\,ha^{-1}\,yr^{-1}$. So factors other than Eh and organic matter may be important.

Other aspects of the behaviour of methanogens suggest ways of decreasing CH_4 emissions from rice paddies (Conrad, 2002). Those inhabiting the root surfaces of rice plants are sensitive to P, so use of P fertilizer can decrease CH_4 production as well as improve rice yields. Because of competition with denitrifying bacteria, which can utilize acetic acid and H_2 more efficiently than methanogens, CH_4 production is also decreased by use of N fertilizers. Urease inhibitors such as hydroquinone can also inhibit CH_4 production (Xu *et al.*, 2002), and applications of materials containing electron acceptors, such as Fe-containing slag from iron and steel furnaces (Furukawa and Inubushi, 2002), can have a similar effect. CH_4 production is also less in soils of low pH and high electrical conductivity (Rosenzweig and Hillel, 2000).

For various reasons, the rate at which CH_4 is emitted from rice paddies increases with increasing temperature. Measured emission rates are much less in the cooler parts of northern China and northern India than closer to the equator (Wassmann *et al.*, 2000). Temperature positively influences the rate of all microbial reactions but, in addition, the various microbial populations operate at different temperatures. Consequently, a positive feedback is likely to develop, with CH_4 production from rice paddies increasing if global warming continues. However, this prediction could be complicated by other factors influencing emissions, such as changes in precipitation and soil moisture contents and land use aspects.

In contrast to poorly drained soils, which generate CH_4, well-drained aerated soils have some capacity for oxidation of atmospheric CH_4, though they generate more CO_2 in the process. Worldwide, aerobic soils may act as a sink for up to 15 per cent of the CH_4 produced annually (Powlson *et al.*, 1997). However, this capacity is influenced by land use; for example, it is considerably decreased by repeated applications of ammonium-N fertilizers (Hütsch *et al.*, 1994), possibly because the nitrifying bacteria that convert ammonium to nitrate compete with the methanotrophic micro-organisms responsible for oxidation of CH_4. A similar inhibitory effect of ammonium fertilizers on methanotrophs has been noted in flooded rice paddies (Dubey, 2003). The CH_4 oxidation potential of mineral soils is also decreased by a decrease in pH (Hütsch, 2001), by herbicide or pesticide applications (Boeckx *et al.*, 1998) and by tillage and irrigation (Kessavalou *et al.*, 1998). Both mineral soils and drained organic soils have greater capacity for CH_4 oxidation under woodland than under agriculture, though the capacity of both is decreased by soil drying in summer, which probably stresses the CH_4-oxidizing micro-organisms (Maljanen *et al.*, 2003). The current conversion of forest and grassland to arable agriculture in many parts of the world is therefore decreasing the capacity of aerobic soils to oxidize CH_4 (Mosier *et al.*, 1997). This effect is likely to continue in the future because, even after reafforestation, previously cultivated soil has a lower CH_4 oxidation potential (Prieme *et al.*, 1997).

To feed the rapidly increasing population of Asian countries, such as China, India and Indonesia, land utilized for rice production is expected to increase by 65–70 per cent in the next few decades. This could lead to a similar increase in CH_4 emissions (Neue, 1997), unless paddy rice is extensively replaced by dryland rice cultivars, which have a much lower potential for CH_4 emission (Gupta *et al.*, 2002), or paddy fields are drained in the later part of the cultivation cycle for production of wheat or other staple cereal crops.

With more advanced global warming, an increase in atmospheric CH_4 could also result from melting of permafrost soils and destabilization of parts of the ocean floor, in which large amounts of methane hydrate and clathrate are stored (Section 9.6.2). Harvey and Huang (1995) calculated that the potential release from ocean floor sources is small, but emissions from the melting of permafrost could be much larger, though they would probably occur over a long period (MacDonald, 1990).

9.3.7 Methane emissions from fuel sources, landfill sites and ruminants

One of the main causes of the increase in atmospheric CH_4 during the industrial period has probably been losses from coal mines and oilwells, and more recently from leaks during extraction and distribution of natural gas (Wuebbles and Hayhoe, 2002). These could be minimized by more careful mining practices and improving the control of gas leakage.

Landfill sites used for disposal of waste materials, wastewater and slurry lagoons are another major source of CH_4, estimated to contribute up to $70 \times 10^6 \, t \, yr^{-1}$ worldwide (Bogner *et al.*, 1997) compared with emissions of $9–25 \times 10^6 \, t \, yr^{-1}$ from rice paddies (Sass *et al.*, 2002). The CH_4 is usually mixed with CO_2, especially in the early life of a landfill site, and traces of H_2, CO and H_2S can also occur. These emissions depend on the moisture and organic content of the waste, but are also affected by temperature, so they could increase with global warming. If the topsoil used to cover the sites is well aerated, methanotrophic micro-organisms can oxidize most of the CH_4 emitted, but their effect can be severely decreased by poor restoration practices, such as soil compaction leading to anaerobic conditions. At many landfill sites the emissions are channelled through a system of horizontal collector pipes into vertical wells, which allow the emissions to be monitored and then flared off at the surface or used as fuel for generation of electricity. With efficient capture, this can decrease CH_4 loss by >90 per cent (Borjesson and Svensson, 1997). Because the resulting CO_2 has less global warming potential than CH_4, either of these procedures is preferable to simply venting the CH_4 to the atmosphere.

Ruminants, such as buffalos, goats, sheep, cattle and other domesticated animals, emit CH_4 because of incomplete digestion of food (Johnson *et al.*, 2000). Cole *et al.* (1997) suggested that better quality cattle feed and dietary supplements could decrease these emissions as well as improve milk and meat production.

9.3.8 Nitrous oxide emissions from soils

Soils in agricultural use are the largest current source of N_2O emissions to the atmosphere. N_2O emissions result mainly from the microbiological processes of nitrification and denitrification, but are also increased by biomass burning and applications of organic manures. Growth of leguminous crops also increases N_2O emissions, as the *Rhizobia* responsible for N fixation can also produce N_2O by denitrification (Freney, 1997). Increased use of N fertilizers in the twentieth century is thought to be the main cause of the relatively small increase in atmospheric N_2O since the pre-industrial period (from about 270 to 310 parts per billion by volume, ppbv), though increased decomposition of SOM, for example by ploughing up old grassland, could have contributed. An average of about 2 per cent of fertilizer N is lost to the atmosphere as N_2O, and total emissions from agricultural land have increased by >60 per cent since the 1960s (Mosier and Kroeze, 1998).

Recently incorporated crop residues are often an important source of N_2O, especially after rainfall (Hou and Tsuruta, 2003). However, N_2O emissions from soils under cereal crops are less than those under grassland and non-cereal crops (Smith, 1999). Deforestation increases N_2O emissions, and soils under pasture produce about three times the N_2O generated by adjacent forest soils (Freney, 1997). Measured N_2O emissions from paddy fields are less than those from dryland rice production, especially if the paddies are flooded some weeks before N fertilizer is applied and kept flooded during crop growth.

Many soil factors, including organic matter content, particle size distribution, structure, air capacity and water holding capacity, temperature, pH, tillage and irrigation practices, fertilizer use and vegetation type, affect N_2O emissions and estimates are subject to considerable uncertainty. Brown *et al.* (2002) estimated that emissions of N_2O from UK agricultural sources total approximately $51 \times 10^3\,t\,yr^{-1}$ ($32 \times 10^3\,t\,yr^{-1}$ from soils, $6 \times 10^3\,t\,yr^{-1}$ from animals and $13 \times 10^3\,t\,yr^{-1}$ from other sources) and range from $<1.0\,kg\,N_2O\text{-}N\,ha^{-1}\,yr^{-1}$ in northern Scotland to $>6.5\,kg\,N_2O\text{-}N\,ha^{-1}\,yr^{-1}$ in areas of intensive animal production, such as southwest England, or intensive arable production, such as Norfolk. Sozanska *et al.* (2002) broadly agreed with this type of distribution, but calculated a much larger total UK annual emission rate ($127 \times 10^3\,t\,yr^{-1}$).

As with losses of nitrate by leaching (Section 5.4.1), gaseous N_2O losses from soils can be minimized by the use of nitrification inhibitors (Bremner, 1997; Merino *et al.*, 2001; Weiske *et al.*, 2001) or slow release N fertilizers (both rather expensive), by application of the minimum amounts of N fertilizer to achieve satisfactory crop yields, by careful timing of fertilizer applications in relation to crop demand and weather patterns, by better placement of fertilizer close to the roots of crops with low root density and by avoiding soil compaction.

9.4 EFFECTS OF FUTURE CLIMATIC CHANGE ON SOIL HYDROLOGY AND SEA LEVEL

Some aspects of predicted climatic change are likely to influence soil hydrological regimes considerably. With a global increase in temperature and localized decrease in rainfall, many soils will become drier, especially in continental interiors and in summer. With warming there is greater evaporation from land areas than oceans, and evaporation from land increases faster than precipitation, especially in summer.

Greater winter rainfall and earlier spring snowmelt at higher latitudes and altitudes could partially offset summer drying. However, a poleward shift in the mid-latitude cyclonic belts of high winter rainfall could negate the effect of winter rain on the equatorward side of these belts, leading here to increasingly frequent failures to return to field capacity in winter and the need for more extensive summer irrigation to maintain agricultural production. In other regions where rainfall increases, more efficient field and arterial drainage schemes to remove excess winter rainfall will be required.

The effects on plant growth in areas subject to an increasingly arid climate are likely to be greater than in those experiencing increased winter rainfall. With greater aridity, soil erosion is likely to increase by both wind and the less frequent but higher intensity rainfall events. Also, increasing evaporation could lead to greater capillary rise of groundwater and increasing soil salinization (Yeo, 1999). The combined effects of increased desiccation, soil erosion and salinization may well increase the rate of desertification in areas where it is currently already on the increase as a result of unsuitable agricultural activities (overgrazing and intensive cultivation).

Changes in soil moisture are also important because they can lead to feedbacks involving water vapour, which is an important 'greenhouse' gas and a source of precipitation. Higher temperatures increase evaporation from the oceans and wet soils, and the larger amounts of water vapour in the atmosphere and a denser cover of low clouds would then accentuate winter warming. However, in drier land areas the higher temperatures would lead to lower rates of evapotranspiration because of the smaller soil moisture reserves, and this would probably result in less water vapour in the atmosphere, a decreased low cloud cover, less precipitation but more solar radiation reaching the surface, and thus further soil drying. This would probably occur most strongly on the equatorward side of the mid-latitude storm belts, thus shifting the axis of maximum rainfall poleward and accentuating the changes in summer soil moisture.

Areas where rainfall is likely to decrease with global warming, such as eastern China, eastern and southern USA and southern Europe, are likely to see a decline in the areas of natural wetlands, many of which have already decreased considerably through drainage for arable agriculture and peat exploitation. Although organic-rich peat soils in wetland areas emit CH_4, they also sequester C and N from the atmosphere because of rapid plant growth and decreased mineralization in waterlogged anaerobic conditions. So although drainage and exploitation of peat soils for agriculture and horticulture have decreased CH_4 emissions slightly, they have probably contributed to global warming in the twentieth century by increasing CO_2 and N_2O emissions. As N_2O has about 206 times the global warming potential of CO_2, the overall effect of draining mires is to increase their global warming potential about 15 times. In Sweden cultivated peat soils represent <10 per cent of the agricultural land, but account for 25 per cent of N_2O emissions and as much as 10 per cent of the country's total anthropogenic CO_2 emissions (Eriksson, 1991). So, by increasing the rate of organic matter oxidation in drained mires, global warming will increase their emissions of CO_2 and N_2O even further.

After corrections where necessary for tectonic and isostatic movements of the Earth's crust, tide gauge records indicate a mean global sea-level rise over the period AD 1880–1980 of about 1.8 mm yr^{-1} (Douglas, 1997). Most of this probably resulted from melting of glaciers, especially the Antarctic and Greenland ice caps, and expansion of the oceans with increasing water temperature during the twentieth century, though other anthropogenic effects, such as permanent removal of water from aquifers and increased runoff from urban and deforested areas, may have also contributed.

IPCC (2001) estimated that with continued global warming the future sea-level rise would exceed 0.1 m by AD 2040 and reach about 0.5 m by 2100. Approximately 25 per cent of this expected rise will probably result from increased melting of glaciers and snowfields. The remainder would result from thermal expansion of the oceans, though this effect is slower because of the large thermal capacity of the oceans. Low-lying coastal and estuary areas are then likely to see more frequent and more extensive marine inundation and rising soil water tables, all of which could cause land to be abandoned or could adversely influence forest and agricultural production because of salinization (Taylor and Sanderson, 2002). Also, coastal areas with cliffs cut in soft sediments, such as much of eastern England, are likely to see increasing rates of coastal erosion and loss of high quality agricultural land. More importantly, coastal flooding will be a major problem in countries such as Egypt, where much of the industrial infrastructure lies close to sea level. By 2080 marine flooding is estimated to affect up to 200 million people living in low-lying coastal areas, unless coast protection measures are increased.

9.5 CLIMATIC CHANGE DURING THE QUATERNARY PERIOD AND ITS CAUSES

The evidence for past climatic change during the Quaternary period (2.6 million years (Ma) ago to present) and its likely causes throws some doubt on the assumption that the twentieth-century increase in temperature has resulted entirely from the anthropogenic increase in 'greenhouse' gases.

Neither of these increases are in doubt, but a simple causal relationship between them, which is the basis of climatic change predictions, looks unlikely in the light of the various natural factors that affected climate in the past. Future changes may well result from the same natural factors, and any proposals to influence future climatic change by management of the environment, including soils, must be viewed against the moving background of natural change, which is almost certainly beyond human control.

9.5.1 Glacial-interglacial cycles during the Quaternary

The Quaternary period is divided into two very unequal parts, the Pleistocene (2.6 Ma to 11,500 years ago) and the Holocene (11,500–present). Sedimentological and palaeontological studies of Pleistocene deposits in many areas have shown that there were numerous distinct cold periods producing a succession of glacials, and that these were separated by warm periods (interglacials) often approximately similar in climate and duration to the Holocene (Ehlers, 1996). As the Holocene seems to be just another interglacial, in the distant future it will probably give way to yet another glaciation.

The best evidence for glacial–interglacial cycles comes from cores of deep ocean sediments dated by palaeomagnetic methods, in which climatic fluctuations can be traced from changes in the ratio of $^{18}O/^{16}O$ in the carbonate shells of fossil foraminifera. The ratio is determined partly by water temperature (colder water and the carbonate deposited from it are slightly enriched in ^{18}O because the lighter isotope ^{16}O is preferentially lost by evaporation of $H_2{}^{16}O$ to the atmosphere), and partly by fluctuations in the volume of the Earth's glaciers, in which much of the $H_2{}^{16}O$ lost from the sea by evaporation is semi-permanently retained on land in glacier ice. The oceanic cores indicate approximately 50 glacials, each separated by warmer interglacials, over the last 2.6 Ma (Figure 9.4).

A similarly complex succession of major cold and warm stages during the Pleistocene is indicated by studies of the loess deposits in central Asia. The loess accumulated principally during cold stages when the climate of these regions was dominated by the cold, dry northwesterly winds of the winter monsoon. Unlike today, these winds persisted for much of the year, so that the warm, moist southeasterly winds of the summer monsoon had little effect on the overall climate. In contrast, in interglacials and the Holocene only small quantities of loess were deposited by occasional winter dust storms, as the summer monsoon then dominated the overall climate. Because of the dominantly warmer and wetter conditions, the interglacials of the loess areas are mainly represented by ancient buried soils (palaeosols), each formed in the upper part of the loess deposited in the preceding cold stage and buried by the loess of the succeeding cold stage. The palaeosols are often argillic brown earths (Luvisols) containing pollen and molluscs indicating formation under woodland.

The glacial–interglacial cycles indicated by the deep oceanic sediments and loess sequences are strongly correlated over time with variations in the amount of solar radiation reaching the Earth because of perturbations in the Earth's motion around the Sun. Three main cycles of perturbation affect radiation receipt (Lowe and Walker, 1997: chapter 1). These are:

- the eccentricity of the Earth's orbit, which varies by 0.5–6.0 per cent over a 100,000-year period;
- the obliquity of the ecliptic (tilt of the Earth's axis of rotation relative to the plane of the orbit), which varies from 21.5° to 24.5° over a 41,000-year period;
- the precession of the equinoxes (season of the year at which the Earth is nearest the Sun on its elliptical orbit), which affects the meridional temperature gradients in either hemisphere with periods lasting 23,000 and 19,000 years.

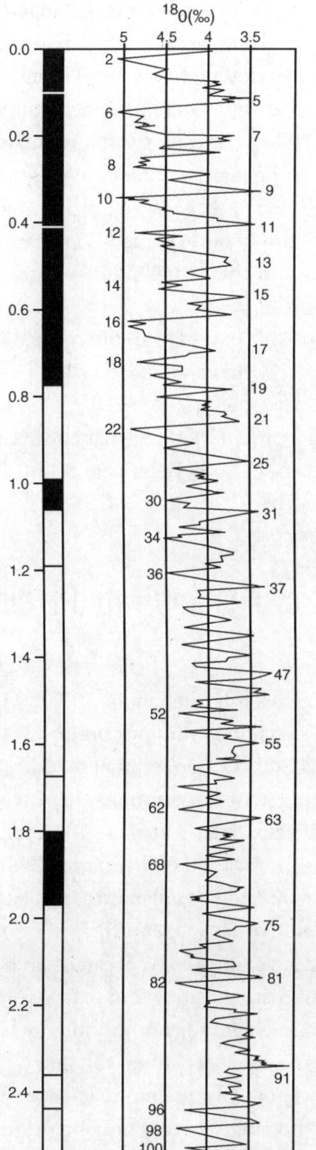

Figure 9.4 Record of climatic change during the Quaternary (since 2.6 Ma) based on oxygen isotope variation in foraminifera from deep oceanic sediments. The dating (left-hand column) is based on polarity reversals of the Earth's magnetic field (black, normal polarity; white, reversed polarity)
Source: Funnell (1995).

The radiation received by the Earth as a whole is mainly determined by the eccentricity cycle. In mid- and low latitudes the precession cycle also influences insolation, but in high latitudes the obliquity cycle has greater influence. Although each cycle is harmonic, when the three are superimposed they produce a disharmonic pattern of radiation receipt, because the cycles have different lengths and the strength of each varies over time. Changes in insolation received at various latitudes at different times over the past 1 Ma were originally calculated in the 1930s by the Serbian mathematician Milankovitch, and the cycles are known as Milankovitch cycles.

Despite strong statistical evidence that the main climatic fluctuations of the Quaternary are related to Milankovitch cycles (Hays *et al.*, 1976), the causative link remains obscure, because the variation in insolation values is too small to create the large temperature differences (>10°C) between glacials and interglacials. In mid-latitudes (40–65°), they give a range in radiation receipt from approximately 450 to 550 $W\,m^{-2}$, but the total variation worldwide is <1 per cent. It seems the cycles influenced the timing of major climatic changes, but their small effect on insolation was somehow amplified by other factors.

As they show a strong relationship with past glacial–interglacial fluctuations and can be accurately predicted from astronomical constants, Milankovitch cycles are probably the most reliable means of predicting future climate. However, the timescale (10^3–10^4 years) means that such forecasts are really of little immediate practical interest. Berger (1980) calculated that the first temperature minimum of the next 100,000-year glacial will be about 5000 years hence, but Loutre (2003) suggested that the Holocene may last much longer than this.

9.5.2 Stadial-interstadial cycles and rapid Pleistocene climatic change

Although the glacials were quite long, each lasting about 100,000 years over the last 0.9 Ma (Mudelse and Schulz, 1997), they were not periods of continuous uniformly cold conditions. Palaeontological and isotopic evidence from terrestrial and oceanic sediments indicates that they were often punctuated by short mild periods (interstadials), during which the winters were less cold than in the intervening stadials, and the summers almost as warm as in interglacials.

A very detailed climatic history for the last 500,000 years, including numerous interstadials, is provided by ice cores from the Antarctic and Greenland ice caps and other glaciers. The ice layers formed from thick winter snowfalls are lighter in colour than the summer layers, which are enriched in impurities because of summer thawing, so annual layers can often be dated by counting back from the present ice surface. Other methods of distinguishing and dating the layers include variations in dust, volcanic ash, electrical conductivity and chemical composition, including various stable isotopes (Budd *et al.*, 1989). Cores through the stacked sequence, often totalling >3 km in thickness, can provide a very detailed record of past climatic change. Temperature changes are indicated by oxygen isotope ($^{18}O/^{16}O$) ratios, the colder stadials showing relative enrichment in ^{16}O and the interstadials enrichment in ^{18}O (i.e., the reverse of the relationship in oceanic sediments). In the Greenland ice cores from approximately 115,000 to 14,000 years ago, 24 interstadials have been recognized each lasting 500–2000 years (Taylor *et al.*, 1997). They are named Dansgaard–Oeschger Events after two of the glaciologists who discovered them. Each began rapidly, the mean annual temperature rising by 5–10°C within a few decades; this was followed by a temperature plateau, and then each terminated only slightly less abruptly than it began. Dansgaard–Oeschger oscillations can also be traced in Antarctic ice cores, though they are more weakly expressed than in the Greenland cores (Hinnov *et al.*, 2002).

Six of the later cold (stadial) intervals in the ice cores correspond in time with influxes of coarse ice-rafted debris into the North Atlantic from the North American (Laurentide) and Fennoscandian (north European) ice sheets. During these so-called Heinrich Events, instability in the ice sheets led to the release of icebergs in large numbers, and as these melted they dropped coarse sediment onto the floor of the Atlantic (Bond and Lotti, 1995). Instability probably occurred when the ice sheets grew rapidly and reached a critical thickness, which caused a major ice surge into the ocean and a decrease in the total ice volume of about 10 per cent (Andrews, 1998). The events are designated H1–H6 (in order of increasing age).

The most likely cause of the Dansgaard–Oeschger and Heinrich Events were oscillations in the Atlantic thermohaline circulation. According to Ganopolski and Rahmstorf (2001), the rapid warming phase of each interstadial resulted from northward extension of the Gulf Stream into the Nordic Seas north of the Iceland–Scotland suboceanic ridge; this led to the temperature plateau when the circulation was in the warm mode. The post-interstadial cooling then occurred when NADW formation moved southwards out of the Nordic Seas; at this time the circulation moved into the cold mode. During Heinrich Events, NADW formation virtually ceased (Atlantic circulation in off mode), because the large volumes of fresh water released from icebergs decreased the salinity and density of the Gulf Stream (Keigwin *et al.*, 1994). With circulation in the cold or off modes, the loss of winter heat would have allowed sea ice to extend further south. As this has a greater albedo than the unfrozen ocean, more solar radiation would be reflected back into space, causing summers also to be cooler, so that annual snow covers would persist longer and glaciers could increase in size.

Evidence that increased freshwater inputs can affect the extent of sea ice and the North Atlantic climate has been provided by climatic modelling exercises (Stocker, 2000; Ganopolski and Rahmstorf, 2001) and the observation that a large expansion of ice on the East Greenland Sea in the mid-1960s resulted from an increase in river water inputs into the Arctic Ocean (Mysak and Powers, 1991). On a longer timescale, Clark *et al.* (2001) showed that variations in freshwater inputs from ice sheets around the North Atlantic were strongly correlated through the last glaciation with abrupt changes in NADW formation and North Atlantic climate. Consequently it is now widely accepted that switches in NADW formation played an important role in many rapid climatic changes of the past, and could do so in the future as well.

Rapid climatic changes are also indicated at the Pleistocene–Holocene boundary by terrestrial sequences, deep-sea sediments and ice cores. Fossil Coleopteran (beetle) assemblages suggest that the mean summer temperature in lowland Britain soon after 16,000 years ago rose as quickly as 7°C per century and that winter temperatures increased by approximately 20°C over a few centuries (Walker *et al.*, 1993). This warm period (the Late Glacial Interstadial) lasted until approximately 12,650 years ago, at which time renewed cold conditions led to the Younger Dryas Stadial, when temperatures were almost as low as before 16,000 years ago (Figure 9.5). The Younger Dryas resulted in major readvances of the Scandinavian and Alpine ice sheets, development of a 2000 km^2 ice sheet over the Grampian Highlands in western Scotland (Thorp, 1986) and expansion of the North American ice sheet into the St Lawrence lowland (La Salle and Shilts, 1993). As it shows some similarities to the earlier Heinrich Events (Manabe and Stouffer, 1997), it has been referred to as H0.

In the Greenland ice core records, the increase in temperature at the end of the Younger Dryas (beginning of the Holocene) occurred within about 40 years in three steps each lasting <5 years (Taylor *et al.*, 1997; Alley, 2000), and half the increase was accomplished within 15 years. As at other times of rapid change, there is some evidence that the climatic system 'flickered' at the Younger Dryas–Holocene transition (Taylor *et al.*, 1993), with some temperature-influencing variable

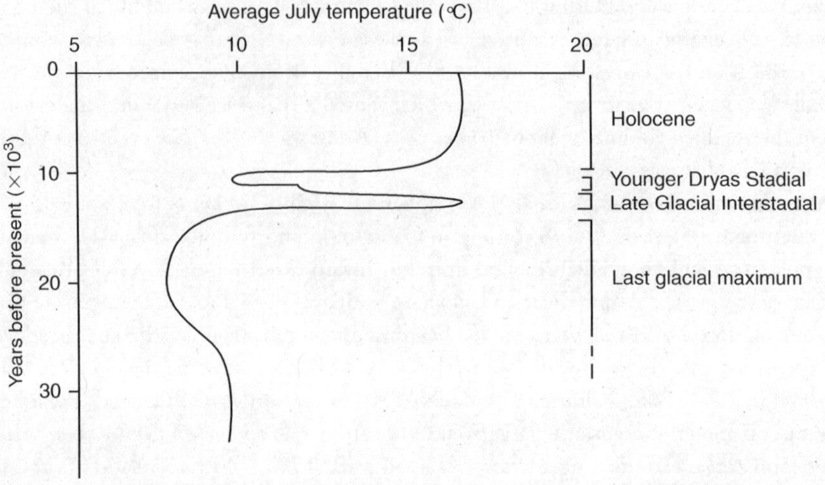

Figure 9.5 Variation in average summer temperature since the last glacial maximum, indicated by fossil Coleoptera (beetle) assemblages from dated sediments in lowland England
Source: Coope (1977).

oscillating rapidly between levels characteristic of cold (stadial) and warm (interstadial) conditions before stabilizing at the new level.

The various records of rapid Late Glacial climatic changes in the North Atlantic region derived from ice cores, ocean sediments and their foraminiferal assemblages, ice-rafting events and fossils from terrestrial sequences, all show remarkable synchroneity (Walker, 1995). This suggests that changes in regional sea surface temperatures, snow accumulation rates and seasonal continental temperatures occurred almost simultaneously in response to a single controlling mechanism. As originally suggested by Ruddiman and McIntyre (1973), the orchestration of events in these diverse environments almost certainly resulted from changes in the mode of Atlantic oceanic circulation.

The evidence that many temperature changes during the Quaternary occurred in sudden short jumps over a decade, or even less, suggests that the natural climatic system is quite delicately balanced. It probably alters in response to changes in oceanic circulation, which are then amplified by feedback mechanisms, such as changes in albedo (Rind and Overpeck, 1993). Therefore, if we are to predict future changes, we especially need to understand the initial causes of past rapid shifts and the relevant feedback mechanisms, but at present our understanding is rudimentary.

9.5.3 Holocene climatic change

In contrast to the last glacial, the Holocene seems to have been a period of more stable though not completely uniform climate. The most recent insolation change resulting from Milankovitch cycles, a gentle decrease in summer insolation in the northern hemisphere, probably explains a slow temperature decline since at least the mid-Holocene. However, as yet it seems to have had little effect on lowering global sea levels, which have been dominated by the decrease in global ice volume since about 16,000 years ago (Lambeck *et al.*, 2002), or on the oxygen isotope composition of the Greenland ice cap. This led Broecker (1998) to suggest that the Holocene climate is still locked in an interglacial operational mode but could change suddenly to a stadial mode.

Fluctuations in earlier parts of the Holocene are often indicated by the sediments that accumulated in lakes and mires, which were influenced mainly by variations in precipitation. For example, coarser layers in the Holocene sediments of lakes in New England indicate periods of increased natural soil erosion, which have been ^{14}C-dated at repeating intervals of approximately 3000 years (Noren *et al.*, 2002). A similar cyclicity is apparent in concentrations of sea-salt Na in Greenland ice cores, and has been attributed to increased storminess during periods of low index of the Northern Hemisphere Annular Mode (NAM). At such times, meridional northeasterly winds brought erosive winter storms to New England and sea-spray from the North Atlantic to the Greenland ice cap. Intervening periods, when finer sediments accumulated in the lakes and less sea-salt Na was transported to the Greenland ice cap, probably coincided with episodes of high NAM index, when the northern hemisphere atmospheric circulation was more zonal with a strong westerly air flow. The dated New England lake records suggest that winter storminess related to a low NAM index has increased gradually from a minimum about 600 years ago. With a half-cycle lasting approximately 1500 years, the storminess should therefore reach a peak in about 900 years. If the correlation with the NAM index is correct, this would mean a progressive increase in colder winters over most of the northern hemisphere over this period. Noren *et al.* (2002) pointed out that the increase in storminess and seasonal precipitation observed in the late twentieth century could be part of a natural trend related to the NAM cycle, rather than a result of increasing 'greenhouse' gases.

Late Holocene proxy-climatic data have also been obtained from the widths and densities of the rings of living and subfossil pine and larch trees growing in northern Sweden and Siberia, which cover the last 2000 years (Briffa, 2000). Where parts of these dendrochronological records overlap meteorological records in time and place, they have been calibrated against the instrumental data. Extrapolation has then allowed development of a detailed curve for mean annual temperature in the northern hemisphere since AD 1000 (Bradley, 2000). This shows:

- a long-term cooling trend (about 0.2°C overall) from AD 1000 to approximately AD 1900, probably a continuation of a general temperature decline since a mid-Holocene optimum;
- abrupt but minor fluctuations over periods lasting from a few decades to a century or more. 50-year means from the coldest part of the fifteenth century to the warmest period at the end of the twentieth century differed by only ~0.5°C;
- since about AD 1900 there has been a rapid reversal of the overall cooling trend. The mean northern hemisphere annual temperature for the twentieth century is 0.28°C higher than that for the nineteenth century and higher than for any period since AD 1000.

Despite the tree-ring evidence for fairly stable recent climatic conditions, there is good ice core (Dahl-Jensen *et al.*, 1998) and historical (Lamb, 1995: 187–240) evidence for a cold period within the last few centuries known as the 'Little Ice Age' (LIA). In Greenland ice cores, this began ca. AD 1500 but at lower latitudes other evidence suggests that it lasted from about AD 1300 to 1850 (De Menocal *et al.*, 2000).

Since the LIA, the progressive warming that became most evident in the twentieth century is also indicated by instrumental meteorological records. These cover approximately the last 250 years at a few recording stations, but they became sufficiently numerous for global averages for land, sea surface and marine air temperatures to be calculated only since the mid-nineteenth century. Analysis of global sea surface temperature and land surface air temperature records by the UK

Meteorological Office indicated an overall warming of 0.65 ± 0.15°C for the period 1861–1998, but occurring in two main steps: 1910–40 and 1975–98. In Europe the overall value for the twentieth century was 0.8°C (Bensiton and Tol, 1998). The instrumental data also confirm the concept of accelerating global warming, as the 1990s was the warmest decade since records began and contained the nine warmest years on record, with 1998 the hottest year yet recorded in many regions.

Warming during the twentieth century is also indicated by estimates of the retreat of valley glaciers in many mountain regions, which suggest an increase of 0.6–1.0°C (Oerlemans, 1994), and by a decrease of approximately 10 per cent in the extent of northern hemisphere snow cover between 1972 and 1992 (Groisman *et al.*, 1994). Between 1979 and 1997, there was also a gradual decrease in the extent of northern hemisphere sea ice (Maslanik *et al.*, 1996) and an increase in the annual number of days when melting of sea ice occurred (Smith, 1998). Remote sensing techniques also indicate that the arctic sea ice cover became thinner in the late twentieth century.

If the twentieth-century warming trend continues for whatever reason, there is likely to be an accelerated continuation of the melting of glaciers, polar sea ice and terrestrial snow cover. As Rahmstorf *et al.* (1996) pointed out, the resulting additional inputs of fresh water to the North Atlantic could weaken the thermohaline circulation and cause a southward shift of the North Atlantic polar front and Gulf Stream, bringing colder conditions to northwest Europe, especially in winter. Ensuing worldwide changes in ocean circulation and the increasing albedo resulting from greater snow cover in the northern hemisphere could lead in a few decades to a return to glacial conditions rather than increasing warmth. For the twenty-first century, this is perhaps a likely scenario, as Hansen *et al.* (2001) have already reported a weakening in the flow and salinity of water in the Faroe Bank Channel, which crosses the Greenland–Scotland ridge between the Nordic Seas and the North Atlantic, and is a major source of NADW.

The meteorological records for the last century or so also suggest the importance of global wind circulation in determining minor climatic fluctuations, and in the northern hemisphere this may be related in part to the NAM. In the first few and last few decades of the twentieth century, there was a pronounced increase in the frequency of westerly winds over the North Atlantic and an increased meridional pressure gradient between the Icelandic depression and the Azores anticyclone. As a result, depressions usually crossed the Atlantic on a more northerly course, bringing a generally milder southwesterly winter air flow over Britain and other parts of northwest Europe. However, between the late 1930s and 1970s there was a decline in westerly airflow, probably resulting from weakening of the mid-latitude circumpolar vortex, and for a few decades this temporarily halted the temperature increase of the early twentieth century.

Another factor thought to influence the climate of recent centuries are the changes in intensity and duration of sunspot activity, which can affect the amount of solar radiation reaching the Earth. When sunspot activity is high, the cycles of activity are short and the extended magnetic field of the Sun partially shields the Earth from cosmic rays (Svensmark and Friis-Christensen, 1997). When sunspot activity is low and the cycles longer, the increased cosmic ray flux cools the Earth by increasing high-level cloud cover. For example, two long periods of minimal sunspot activity with long cycles, the Spörer Minimum (AD 1400–1510) and the Maunder Minimum (AD 1645–1715), coincided approximately with the two coldest parts of the LIA. The length of sunspot cycles ranges from 9 to 14 years, with a mean of 11.1 years, and a similar periodicity has been detected in some long-term weather records, such as droughts in mid-western USA, northern hemisphere summer temperatures and the fluctuations of Pacific Ocean currents and atmospheric pressure giving rise to ENSO. In recent decades, there has been more sunspot activity than for several centuries, and the sunspot cycle has averaged 10.5 years. In the late nineteenth century and first decade or so of the twentieth century the cycle length was 11.5–12.0 years, and this change could explain at least part of the temperature increase recorded over the twentieth century (Laut and Gunderman, 1998).

9.6 THE ROLE OF 'GREENHOUSE' GASES IN QUATERNARY CLIMATIC CHANGE

9.6.1 Carbon dioxide

Analysis of air bubbles in Antarctic ice cores has shown that atmospheric CO_2 concentration varied closely in step with temperature change and global ice volumes during recent glacial–interglacial cycles (Petit *et al.*, 1999). During interglacials and the early Holocene, the atmospheric CO_2 concentration was typically 280 parts per million by volume (ppmv), but during glaciations it was 80–100 ppmv less. The current level of 370 ppmv is therefore an unprecedented value in the Quaternary, though as yet it has not led to unprecedented warmth, as palaeontological evidence suggests that the last (Eemian) interglacial was 3–4°C warmer than any part of the Holocene, including the last few decades.

As the ice core data also show a strong imprint of Milankovitch cycles, Petit *et al.* (1999) suggested that the weak effects of insolation changes were amplified first by CO_2 as a 'greenhouse' gas, and then by the feedback mechanism of changes in the albedo of the Earth's surface as ice and snow cover increased and decreased. However, despite considerable palaeoclimatic and oceanographic research in recent decades, the cause of the variations in CO_2 remains uncertain. One especially puzzling aspect is that any increase in atmospheric CO_2 in interglacials should have been at least partially counteracted by the increased growth of terrestrial and oceanic biomass resulting from the higher overall temperatures. During interglacials the atmosphere was therefore supplied with large amounts of CO_2 from some source other than the biomass, and this suggests that the CO_2 changes were an indirect result rather than a cause of the temperature changes.

Because the pool of inorganic C stored in the deep oceans is more than ten times greater than the sum of the total C pools in the atmosphere, soils, terrestrial biosphere and warm upper parts of the ocean, past changes in atmospheric CO_2 could have been related to oceanic circulation (Broecker, 1982a). Atmospheric CO_2 is absorbed initially by the surface layer of the oceans, where most of it reacts with water and carbonate ions to produce bicarbonate ions (HCO_3^-). Mixing of the upper 20–200 m of the ocean by winds and tides releases some CO_2 back to the atmosphere, especially in winter, but much of the bicarbonate is transferred to deeper layers by physical processes such as deep water formation and the slow sedimentation of organisms that have utilized dissolved bicarbonate in photosynthesis and secretion of carbonate.

As the solubility of CO_2 increases with decreasing water temperature, the lower temperature of the glacial oceans would have removed more CO_2 than usual from the atmosphere. The effect would have been greater in the warm tropical oceans than in high latitude seas, which were much closer to freezing point. However, estimates of a 5°C cooling in low latitude oceans and a 2–3°C decrease at high latitudes during glaciations give an atmospheric CO_2 decrease of only 30 ppmv (Keir, 1988), much less than that recorded in the ice cores. Also, the decrease would probably have been even less than this, because the increased salinity of the glacial ocean, resulting from large amounts of water stored in ice sheets, would have reduced the solubility of CO_2 in seawater. Consequently, changes in CO_2 solubility in the oceans do not seem to account for the large glacial–interglacial variations in atmospheric CO_2.

The main biological process removing CO_2 and bicarbonate from upper layers of the ocean is known as the 'biological pump'. This includes sequestration of C from dissolved CO_2 by photosynthesis in marine organisms growing in the sunlit upper layers of the ocean and the rain of a proportion of the dead organisms from these layers into the deeper ocean. It also includes removal of bicarbonate by secretion of $CaCO_3$ by marine organisms, mainly as calcite in foraminifera and coccolithophorids (algae). The Ca incorporated into the carbonate enters the ocean from terrestrial weathering of limestones and Ca-containing silicates in soils. Like the dead organic matter, the biogenic $CaCO_3$ sinks to lower levels in the ocean. However, approximately 75 per cent redissolves in deeper parts of the ocean below the depth of the 'calcite lysocline' (currently about 3.5 km), but 25 per cent is removed from the oceanic reservoir by sedimentation and burial on the sea floor where the ocean is shallower. Dissolution of calcite causes the deeper ocean to be supersaturated in CO_2, but some is returned to the atmosphere by localized upwelling and degassing (venting) of deep ocean water.

The overall effect of these various biological and physical processes is that short-term changes in atmospheric CO_2 are buffered by the oceans. However, this equilibrium could have been disturbed in the Pleistocene. For example, the lower weathering rates in glacial soils would have decreased Ca inputs to the ocean, and this would increase the depth of the lysocline, so that the proportion of calcite buried on the sea floor would increase and lower atmospheric CO_2. However, micropalaeontological studies of ocean sediment cores show that at the last glacial maximum, around 20,000 years ago, the depth of the lysocline was increased by <1 km compared with the Holocene, and this would have decreased atmospheric CO_2 by <25 ppmv (Sigman and Boyle, 2000). Therefore this mechanism also seems to have been insufficient to lower atmospheric CO_2 during glaciations to the concentrations recorded in the ice cores.

As biological production in surface layers of the oceans depends upon nutrients (mainly P, and N occurring as nitrate and ammonia) as well as C from CO_2, an increase in oceanic nutrient reserves could have led to a decrease in atmospheric CO_2 by increasing the efficiency of the 'biological pump' (Broecker, 1982b). Most of the increased production would occur in low-latitude oceans, because the biological productivity of polar oceans is limited by temperature and other non-nutrient factors (Chisholm and Morel, 1991). However, the observed amplitude of glacial–interglacial CO_2 change requires very large changes in the nutrient reservoirs; for example, Sigman *et al.* (1998) calculated that a massive 30 per cent increase in oceanic nutrients would decrease atmospheric CO_2 by only 30–45 ppmv.

Oceanic P concentrations are unlikely to have changed to such an extent, as the sole source of oceanic P is the input from land and there are no mechanisms for incorporating large amounts of P into oceanic sediments. However, the nitrate content of the ocean is influenced by at least two important microbial processes similar to those involved in N cycling in terrestrial soils, namely denitrification and N-fixation. The protein produced by N-fixation is decomposed to ammonia, which is then oxidized to nitrate either aerobically by microbial nitrification or anaerobically in the anammox reaction performed by a group of bacteria known as planctomycetes (Dalsgaard *et al.*, 2003).

Because of decreased activity of free-living micro-organisms at lower temperatures, denitrification was probably less effective in glacial periods. At the same time oceanic N-fixation could have been increased because of increased inputs of airborne dust, which contains Fe and other micronutrients important for N-fixation (Falkowski, 1997). Both of these processes could therefore have increased the availability of nitrate for biological production during glacial stages. However, the ratios of C:N:P in oceanic organisms are fairly constant (close to 106:16:1), and it is unlikely that marine organisms could have taken advantage of an increase in nitrate by varying their C:N:P ratios. Also, as in soils, a large increase in the N:P ratio would suppress N-fixation and limit marine productivity (Ganeshram *et al.*, 2002).

In summary, although the ice cores provide strong direct evidence for a positive correlation between atmospheric CO_2 content and temperature changes over the last 500,000 years, no known mechanism for increasing atmospheric CO_2 seems adequate to explain the full amplitude of observed temperature variation. Small changes in insolation resulting from perturbations in the Earth's orbit seem to be the 'pacemaker' of climatic change on the scale of 10^3–10^5 years, and the climatic response somehow involves CO_2, but little more can be said with certainty at present.

9.6.2 Methane

Methane also changed synchronously with climate during the later part of the Pleistocene. The ice cores show that during glaciations its abundance in the atmosphere was 320–350 ppbv, whereas in interglacials and the early Holocene it increased to 650–780 ppbv (Petit *et al.*, 1999). As the main natural CH_4 input to the atmosphere is from anaerobic wetland soils, in which its production by microbiological activity is temperature-dependent, the glacial–interglacial changes could have resulted from differences in soil microbial activity between the colder, drier soil environment of glacials and the warmer, wetter conditions of interglacials (Meeker *et al.*, 1997). Atmospheric CH_4 content also changed very rapidly; for example, after the Younger Dryas Stadial it reached its typical Holocene level within 150 years (Taylor *et al.*, 1997), probably as a result of increased emissions from tropical wetlands and mid-latitude peatlands (Maslin and Thomas, 2003).

As atmospheric CH_4 concentrations have always been much less than those of CO_2, they would have had a smaller effect on global temperatures (approximately 20 per cent that of CO_2), even though molecule for molecule the heat-trapping ability of CH_4 is about 66 times that of CO_2. Also CH_4 has a very short life in the atmosphere, about 7.9 years (Lelieveld *et al.*, 1998). Almost 90 per cent of it is rapidly oxidized to CO_2 and H_2O by reaction with hydroxyl radicals (OH), which are produced in the troposphere by photodissociation of ozone and water vapour, so any change in inputs to the atmosphere would have had a very short-lived effect on climate. Brook *et al.* (1996) suggested that sudden very large inputs of CH_4 from water–methane compounds termed methane hydrate and clathrate, which are stored in ocean floor sediments and deeper permafrost layers, could have led to some past episodes of rapid warming, but the ice cores do not record CH_4 increases of the magnitude required to generate the temperature increases. Also, during rapid climatic transitions, changes in CH_4 concentration in the ice cores lagged behind temperature changes (Brook *et al.*, 2000), so they seem to have been the result, rather than the primary cause, of climatic change.

9.6.3 Water vapour

As a 'greenhouse' gas, water vapour is actually more important than CO_2, accounting for approximately 60 per cent of the additional warming of the Earth resulting from the presence of the atmosphere (a total of about 33°C). Petit *et al.* (1999) estimated that the combined effect of the increases in CO_2, CH_4 and N_2O between glaciations and interglacials, as indicated by gas bubbles in ice cores, produced a global warming of only 0.95°C. The additional effect of water vapour probably increased this to approximately 2°C (Weaver *et al.*, 1998), because of the positive feedback mechanisms by which increases in global temperature increased evaporation from oceans and land surfaces and transpiration from an increased global vegetation cover. An increase in tropospheric water vapour content of 3–6 per cent per decade since the 1960s is indicated by measurements from satellites and the global networks of 'rawindsonde' and 'radiosonde' weather balloons (Elliott, 1995). It has been attributed to an increase in temperature at the tropopause over the tropics and the increase in stratospheric CH_4, which is oxidized to H_2O.

9.7 THE RELATIVE IMPORTANCE OF SOIL AND VEGETATION MANAGEMENT IN GLOBAL WARMING

Sections 9.5 and 9.6 show that numerous factors need to be taken into account in predicting future climatic change, and only a few of them (albedo and CO_2, CH_4, N_2O and water vapour as 'greenhouse' gases) are likely to be influenced by soils and vegetation. Prediction is difficult because many factors operate in ways that are at present uncertain and may be modified positively or negatively by a range of feedback mechanisms. Also, judging from the magnitude and speed of many past changes, the climate may behave chaotically, i.e., responding non-linearly with no simple proportional relationships between cause and effect. It seems to remain fairly stable for long periods and then 'flicker' or flip into another mode under the influence of various unknown (or at best partially understood) triggers and feedbacks. So it is unlikely that soil and vegetation management can strongly influence future climate; at best it may help decrease atmospheric CO_2, CH_4 and N_2O, but these effects could well be overruled by stronger, faster-acting natural factors.

9.7.1 Weaknesses of atmosphere-ocean global circulation models (AOGCMs) and the Kyoto Protocol

Although the latest AOGCMs have been expanded to take into account many of the factors known to influence climate, they still cannot cope adequately with water vapour, clouds and ocean currents. Also they do not adequately simulate either the NAM and ENSO cycles or the aspects of chaotic climatic behaviour, such as the rapid changes that were so frequent before the Holocene. Consequently some climate scientists have little faith in current AOGCM predictions, and suggest that the accumulated errors surrounding them amount to almost total uncertainty. If correct, this means that political resolutions, such as the Kyoto Protocol to limit CO_2 emissions in particular, are unjustified at present. These doubts have been expressed most forcefully by the Committee on Global Change Research of the US National Academy of Sciences (National Academy of Sciences, 2001).

A weakness of AOGCMs, which has led to the emphasis on soils and vegetation for limiting the effects of global warming, is the assumption that increasing atmospheric CO_2 will inevitably result in increasing warmth in the twenty-first century. The twentieth-century increase in mean surface temperature (0.65°C, Section 9.5.3) and the increase in atmospheric CO_2 since at least the middle of the century (Keeling, 1960) are established facts, and there is no doubt that increasing CO_2 can lead to warming. What is in doubt is that the CO_2 increase alone explains the 0.65°C increase and that further increases in the twenty-first century will definitely lead to yet higher temperatures. The uncertain cause of the glacial to interglacial temperature increase in atmospheric CO_2, and the unusually high temperature of the Eemian, even though the atmospheric CO_2 content was no greater then than in the pre-industrial Holocene, suggest that this assumption may not be correct. Also, the twentieth-century warming occurred in two separate periods (1910–40 and 1975–98) with little change between, whereas burning of fossil fuels, other industrial activities and atmospheric CO_2 concentrations would have increased progressively over the whole century. The twentieth-century temperature variation is perhaps more plausibly explained by changes in NAM index or sunspot activity (Section 9.5.3).

IPCC judged that the twentieth-century temperature increase resulted from human activities rather than natural factors, because AOGCM simulations including both natural and anthropogenic forcings fitted the temperature record over the century as a whole better than simulations using known natural forcings alone. However, the list of natural variables considered was far from complete. Until more is known about climatic forcing by natural as well as anthropogenic factors, the AOGCM predictions must remain uncertain.

Paradoxically, the most likely alternative outcome of continuing global warming is global cooling resulting from a weakening or shutdown in NADW-formation and transmission of the effect worldwide by various feedback mechanisms. This would have a completely different range of implications for soils and agriculture, probably posing greater challenges to agriculture in terms of feeding an increasing world population than the global warming predicted by AOGCMs.

In addition to the doubts over whether increasing CO_2 will increase temperature, climatic modelling exercises have suggested that decreasing CO_2 under the Kyoto Protocol will have a very small, almost worthless, effect on global warming in the twenty-first century. For example, Parry *et al.* (1998) calculated that by 2100 the temperature increase under the Kyoto agreement would be only 0.15°C less than under a 'business as usual' scenario, which means that the temperature expected under 'business as usual' will be the same in 2094 as that expected under the Kyoto Protocol only six years later in 2100.

Doubt is also cast on the value of the Kyoto Protocol by combined climate–economic models (Hamaide and Boland, 2000), such as the Regional Integrated Climate–Economy (RICE) Model (Nordhaus and Boyer, 1999). These include economic assessments of both climatic change and the restrictions on CO_2 emissions demanded by the Kyoto Protocol. They show that, although the total cost of global warming over the next century is likely to be the very large sum of approximately US \$5 $\times 10^{12}$ it will be much more expensive to restrict CO_2 emissions to something like present levels than adapt to the increase in temperature and other effects of global warming. In addition, the world probably faces greater problems at present. For example, the cost of implementing the Kyoto Protocol in the USA alone would be greater than the more philanthropic project of providing the entire world with clean drinking water and sanitation.

This type of argument is currently used by the world's two largest producers of CO_2 from fossil fuels, the USA and Russia, to justify not signing the Kyoto Protocol. However, purely financial considerations may, in the event, be overridden by social, political and security problems, such as increasing stress leading to upheaval, revolution or war in areas where life becomes increasingly intolerable because of insufficient water, inability to grow sufficient food, increasing storm frequency or flooding because of rising sea level. Impressed by AOGCM predictions, many climate scientists and politicians predict dire consequences such as these if action under the Kyoto Protocol is not implemented in the near future. However, it would be a pity to spend many trillions of dollars on decreasing atmospheric CO_2 only to find it had no effect on global warming because the modelling was inadequate or some other factor resulted in global cooling.

Summary

Assuming it is desirable to control the CO_2 content of the atmosphere very precisely, probably the most important initial strategy is to decrease rates of fossil fuel burning by further development of renewable energy sources (wind, wave, solar and nuclear power) and the production of 'CO_2-neutral' energy crops. Equally important is preventing further destruction of tropical rainforests and the drainage and cultivation of peatlands, which are contributing to atmospheric CO_2 at increasing rates. In addition, tropical forests seem to have ecological, genetic, and potential medicinal, values, which are still not perfectly understood. Attempts to increase carbon sequestration in forest plantations, agricultural crops and associated soils are limited in their effectiveness, and research with this aim could be raising false hopes and distracting attention from the other more effective strategies. This has led to the suggestion, which is gathering increasing support, that carbon sequestration in vegetation and soils should be given little or no credit against anthropogenic emissions of CO_2 under the Kyoto Protocol. Nevertheless, it is clear that the various agricultural, sea-level and other implications of impending climatic change are so severe for the future of humankind and wildlife that monitoring of the climatic system and studies of the various effects of climatic change are justified in increasing detail. We simply need to be aware that the very complex climatic system may produce unexpected changes at unexpectedly rapid rates.

FURTHER READING

Barry, R.G. and Chorley, R.J., 1998, *Atmosphere, Weather and Climate,* 7th edn, London: Routledge.

Ehlers, J., 1996, *Quaternary and Glacial Geology*, Chichester: John Wiley.

Harvey, L.D.D., 2000, *Global Warming. The Hard Science*, Harlow: Pearson.

Intergovernmental Panel on Climate Change, 2001, *Climate Change 2001: The Scientific Basis*, Cambridge: CUP. Also available at: http://www.ipcc.ch (accessed 4 March 2004).

Lal, R., Kimble, J., Levine, E. and Stewart, B.A. (eds), 1995, *Soils and Global Change*, Boca Raton FL: CRC Press.

Lamb, H.H., 1995, *Climate, History and the Modern World,* 2nd edn, London: Routledge.

Rosenzweig, C. and Hillel, D., 1998, *Climate Change and the Global Harvest*; *Potential Impacts of the Greenhouse Effect on Agriculture*, New York: OUP.

Wild, A., 1973, *Soils and the Environment: An Introduction*, Cambridge: CUP.

10 Prospects for the twenty-first century

We do not inherit the earth from our ancestors: we borrow it from our children.

Kenyan proverb

Introduction

This chapter reviews some of the challenges in managing soils that we face as a global society in the twenty-first century. In the future the increasing world population will place ever greater demands on soils for production of sufficient food. This poses special problems in a climatic environment that is changing at an increasing pace but perhaps ultimately in an uncertain direction. The growing population will also lead inevitably to expansion of urban areas, which cover and consequently sterilize some soil resources or through pollution render others less suitable for crop production. At the same time, the quality of urban life can be improved by the careful management of urban soils to create particular habitats that are more favourable to wildlife, provide amenity sites for recreation or limit pollution.

As a natural resource soil is strongly linked to the two other major aspects of environment: water and air. At national and international (especially European Union, EU) levels, policy-makers have focused unequally on these, with the result that there is much legislation on air and water quality, but little on strategies to protect and enhance soil quality. However, the balance is slowly being redressed. For example, within the EU, the Common Agricultural Policy (CAP) will soon place responsibilities on farmers to protect soil from erosion, to maintain organic matter levels and to improve its structural quality as conditions for receiving subsidies. Implementation of this policy will require a practical agenda supported by suitable campaigns to make soil users aware of the magnitude of the problems and how they can be overcome. This chapter summarizes management techniques that may be required to address some of the problems likely to arise in the next few decades.

10.1 MEETING THE CHALLENGES POSED BY FUTURE CLIMATIC CHANGE

In many parts of the world climatic change may well be the most significant problem affecting soils and their management in the twenty-first century. However, as explained in Chapter 9, it remains an ill-defined problem. There are three main ways in which it could develop.

1| The prediction favoured by 'The United Nations Intergovernmental Panel on Climate Change'(IPCC, 2001) and consequently by most politicians and climate scientists, in which increasing concentrations of 'greenhouse' gases will continue to cause increasing warmth, especially in mid- and high-latitudes, greater aridity in many mid-continental areas and greater winter rainfall in mid-latitude regions of the northern hemisphere.

2| Slight initial warming because of increasing 'greenhouse' gases, followed by a rapid transition to much colder conditions, initially in the northwest European winters as North Atlantic Deep Water (NADW) formation moves southwards or even ceases, but later extending to other parts of the world under the influence of feedback mechanisms such as increasing albedo.

3| Neither slow progressive warming nor sudden cooling, but a continuation of minor decadal temperature fluctuations under the influence of NAM (Northern Hemisphere Annular Mode) or sunspot cycles, despite continuing increases in 'greenhouse' gas emissions.

At present Atmosphere–Ocean General Circulation Models (AOGCMs) are really too poorly constrained to indicate which of these is the most likely. One particular unresolved question is whether the warming effect of increasing 'greenhouse' gases can overcome cooling resulting from changes in NADW formation, NAM cycles or sunspot activity. If it can, then the first prediction is perhaps inevitable; if it cannot, the choice between the other two remains uncertain, at least until there is strong evidence for a southward shift in NADW formation.

The implications for soil management are as extreme and uncertain as the three possible climatic predictions. The increased warming favoured by the IPCC (IPCC, 2001) primarily implies adopting all possible measures to minimize industrial 'greenhouse' gas emissions, as recommended under the Kyoto Protocol. Any additional attempts to increase the sequestration of C in soils and vegetation must be regarded as subordinate to these measures. The most useful soil management strategies are growth of biofuel crops and establishment of carefully managed permanent woodland on all surplus arable land in areas with enough rainfall to sustain tree growth, but these cannot be implemented at the expense of food production for an increasing world population. So their total impact on global warming is likely to be small. Some regions, such as northern Europe, favoured by improved conditions for crop production (Kenny *et al.*, 1993), will probably need to grow surplus crops to help feed the population of less favoured areas. This has major trading and political implications, which are beyond the scope of this book.

At the other extreme, a future climate with cooling initially in northwest Europe and perhaps subsequently in other mid- and high-latitude regions would present quite different problems for soil management and food production. Parts of Europe that currently produce large amounts of cereals and other crops would suffer decreasing yields, and higher latitude regions, which could have seen considerable yield increases under global warming, would instead become much less productive. Much is already known about the effects of lower temperatures on crop production from field and greenhouse experiments under observed or controlled environmental conditions. For example, the review by Porter and Gawith (1999) suggested that winter wheat could not be grown in northern Europe in a much colder climate, as the minimum air temperature for successful germination is between 2.4°C and 4.6°C, root growth ceases if the soil temperature is <2°C and an air temperature of -20°C is lethal to wheat roots. It is possible that future cultivars could be developed to survive harsher winter conditions, but cereal production in northern Europe would probably depend increasingly on the lower-yielding spring-sown varieties. Even their yields could decrease in response to lower summer temperatures, despite the beneficial effect of greater atmospheric CO_2 concentrations.

The third way in which climate may develop in the twenty-first century would probably involve least adaptation in terms of management for maintaining overall food production and least need to redistribute food resources. Any reductions in crop yields resulting from slightly cooler periods would be small and probably compensated by the CO_2 fertilization effect or improved crop husbandry.

The need for farmers and other soil users to adapt rapidly to whichever of these three possible situations arises will require the support of national and international policies that facilitate and encourage flexibility of land use. Increasing food production with any change in climate will need research to develop new crop varieties and novel farming systems. At the same time soil management in any of the three scenarios will need to consider implementing existing or new strategies to decrease 'greenhouse' gas emissions and avoid further anthropogenic climatic change. Within Europe all these problems can be linked to the agri-environment strategies of the CAP, but the effective global action that is required will demand wider international cooperation through the United Nations Food and Agriculture Organization (FAO).

10.2 URBANIZATION, INDUSTRY AND SOILS

Urbanization has effectively sterilized extensive areas of soil, much of it very fertile. This is a major issue in western Europe and North America, where rapid urban expansion began in the nineteenth century. In recent years approximately 120 ha day^{-1} have been converted to urban use in Germany, 35 ha day^{-1} in Austria and The Netherlands and 10 ha day^{-1} in Switzerland (Commission of the European Communities (CEC), 2001). Urban expansion is now especially problematic along the north Mediterranean coast. In 1996, 43 per cent of the coastal zone in Italy was completely occupied by urban areas and only 29 per cent was completely free from buildings and roads (CEC, 2002). In particular, airports are generally located on flat, well-drained peri-urban land, often with high soil fertility and agricultural potential. For instance, the soils derived from loess over Thames terrace gravels beneath Heathrow Airport were very fertile. In addition, urban areas may have great impacts on soils, changing water flow and drainage patterns and fragmenting areas of valuable biodiversity.

Currently, urban expansion is particularly rapid in the developing world. Increasing urban slums pose particular problems. For instance, the expanding urban slums or 'favelas' of Brazil are encroaching on steep land at the margins of urban areas. This puts them at risk from flash floods and gully erosion, which are serious threats to life, health and property (Guerra, 1995; Guerra and Favis-Mortlock, 1998). Construction sites in particular are at risk of high erosion rates (Wolman, 1967).

Urban soils are not usually mapped in detail and many scientists have argued they deserve more attention (e.g., Bullock and Gregory, 1991; Effland and Pouyat, 1997). For instance, on the Soil Survey of England and Wales (1983) 1:250,000 map, they are represented as 'unsurveyed, mainly urban and industrial areas' and are shaded grey. However, urban soils are very diverse and include gardens, parks, cemeteries, allotments, grass verges along roads, playing fields and sometimes much derelict and industrial land. The industrial land includes disposal sites, demolition and building sites, waste and derelict land, rubbish tips, spoil heaps, canal and railway land, collieries, docklands, power station land, shipbuilding land, scrap-yards, dried-out industrial lagoons, sewage works and land associated with mining, smelting and manufacture.

Craul (1985) identified eight environmentally important characteristics of urban soil:

1| great vertical and spatial variability, reflecting their often complex history of construction, management and modification;

2| compaction by trafficking, walking and vibration. This is reflected in increased bulk density and low porosity. The low organic content of many urban soils increases their susceptibility to compaction. Craul (1985) quoted bulk density values as high as 2.18 g cm^{-3}. High bulk densities (>1.7 g cm^{-3}) severely restrict root development and limit nitrogen fixation by legumes, which places urban plants under high stress (Gilbert, 1989);

3| the presence of a surface crust on bare soil, which is usually water-repellent;

4| modified pH. Incorporation of building rubble containing mortar rich in calcium carbonate ($CaCO_3$) often increases the pH (Gilbert, 1989). Use of calcium or sodium chloride for de-icing roads or use of Ca-rich irrigation water can have the same effect. Craul (1985) reported pH values of 8–9 in US urban soils. Brick rubble also contains high concentrations of phosphorus, potassium and magnesium, which may be released as the brick rubble weathers. Plaster is a major source of sulphur, derived from the dissolution of gypsum ($CaSO_4.2H_2O$);

5| restricted aeration and drainage;

6| interrupted nutrient cycling and decreased activity of soil organisms. The lack of annual organic matter inputs (e.g., autumn leaf-fall) decreases the organic content;

7| the presence of anthropogenic materials and contaminants, such as glass, wood, metal, plastic, asphalt, masonry, concrete, cloth, polythene, cardboard, tiles, cinders, ash, soot, bricks and rubble (Bridges, 1991b);

8| generally increased temperatures because of the urban heat island effect.

In the Soil Classification System of England and Wales, 'Man-made soils' are recognized as one of the ten Major Soil Groups (Avery, 1980; Section 1.2). They are defined as 'soils formed in material modified or created by human activity. They result from abnormal management practices such as the addition of earth containing manures or refuse, unusually deep cultivation, or the restoration of soil material following mining or quarrying' (Avery, 1980; Soil Survey of England and Wales, 1983). The man-made soils are subdivided into two groups, 'thick man-made A horizons, including "Plaggen" soils' (Section 7.4) and 'soils in restored open cast mines and quarries'. The latter group occupies 76 km^2, some 0.04 per cent of the surveyed area of England and Wales.

For mining, soils are often stripped away for storage and returned as part of reclamation projects. A major problem is compaction and the deterioration of soil structure during storage. Effects on the stored soil are dependent on soil type, moisture content, handling processes and the equipment used, and generally increase with the depth of the stockpile (Rimmer, 1991). Such soils often lose organic matter, soil fauna (especially earthworms and microbial biomass) and may become nitrogen deficient. These combined effects can make reclamation difficult (Bloomfield *et al.*, 1981; Section 7.7).

Industrial sites often have pollution problems, with very varied pollutants, such as heavy metals and hydrocarbons (Bridges, 1991b; Section 5.3.1). Industrial sites are also major sources of eroded sediment, as these soils are often poorly vegetated and steep. Old mine and colliery mounds are major sediment sources, as in South Wales (Bridges and Harding, 1971; Haigh 1979, 1992; Higgitt *et al.*, 1994), South Yorkshire (Haigh and Sansom, 1999) and the USA (Haigh, 1988). However, they can be visually striking and it has been argued that such disturbed sites should be preserved for their historical and landscape heritage interest (Quinn, 1988, 1992).

10.3 SOILS AND HUMAN HEALTH

There are many impacts of soils on human health (Oliver, 1997). As soils are sources of the nutrients required for production of plant and animal crops they are the starting point for nutrients required in human nutrition. However, although many elements present in the soil are beneficial in small or 'trace' amounts, they are toxic in larger quantities. For instance, selenium (Se) is beneficial in trace amounts for both humans and animals, but is toxic to brain tissues in larger amounts. These complex relationships also apply to copper (Cu), iron (Fe) and zinc (Zn).

Macronutrient deficiencies can cause human health problems, such as the established link between Ca deficiency and osteoporosis (decrease in bone density). Micronutrients necessary for human health include manganese (Mn), copper (Cu), nickel (Ni), zinc (Zn), molybdenum (Mo), boron (B), chlorine (Cl), fluorine (F), iron (Fe), iodine (I), cobalt (Co), chromium (Cr), vanadium (V) and thallium (TI) (FitzPatrick, 1974; Oliver, 1997). Se is not required by plants, but is needed for the health of both humans and animals, particularly in maintaining bone health. Macronutrients in soils and crops are usually sufficient from the human health perspective because they can easily be supplemented by NPK fertilizers, but there are concerns that under continuous intensive arable agriculture soils may become depleted in micronutrients. In part, this concern has contributed to the development of organic farming.

The chemical pollution of soils (Chapter 5) can damage human health. However, identifying precise relationships is difficult, as these are complicated by many other factors (e.g., age, dietary and lifestyle habits, general health, genetic factors, reproductive health and social class). In addition, the geographical source of foods is important. Affluent urban communities obtain their food supplies from a geographically wide range of sources, and are much less likely to suffer nutrient deficiencies than poor rural populations dependent on food grown locally on soils that may lack certain micronutrients. In terms of soil–health interactions, we must distinguish between clinical and subclinical effects. Clinical effects are evident as visual symptoms of nutrient deficiency; with subclinical effects there are no visual symptoms of deficiency, but health improves with nutrient additions.

There are multiple pathways by which soil can enter the human body and cause problems, including eating soil, either deliberately or accidentally (geophagia), dust inhalation and absorption through skin lesions (Oliver, 1997). This last pathway is particularly effective in introducing diseases and is associated with elephantiasis, hookworm and tetanus.

Soil erosion can be damaging to human health. Water erosion can transport poisonous chemicals such as pesticides downslope to streams, lakes and reservoirs. Moreover, dust carried by the wind is a major lung irritant (Piper, 1989) and major respiratory problems were associated with the US Dust Bowl of the 1930s (Hurt, 1981).

Some combinations of elements derived from soil may produce synergistic toxicities (e.g., Cd and Pb acting together are especially damaging to brain tissues). Heavy metals are particularly damaging to brain tissues (Alloway, 1995). The biomedical effects of individual metals are reviewed below.

- *Aluminium*: Al usually enters the body in water and, since it is soluble at low pH (<5.5), concentrations are particularly high in waters draining acid soils. The World Health Organization (WHO) recommend a maximum concentration of 0.2 mg l^{-1} (WHO, 1993). High Al concentrations have been linked with Alzheimer's disease, but this is subject to continued research (Section 6.3.5).

- *Arsenic*: As is poisonous and carcogenic and has been linked with many forms of cancer (bladder, kidney, liver, lung and skin). It is particularly associated with metalliferous mining and industrial activity and the main pathways are inhalation and ingestion of contaminated particles.

- *Cadmium*: Cd is toxic, even in trace amounts, and has been linked with itai-itai disease, cancer and many renal disorders. In pregnant women it is retained by the placenta and severely restricts the transfer of Cu and Zn to the embryo (WHO, 1996). The highest concentrations occur in soils derived from black shales or sites contaminated by smelting, fossil fuel combustion, industrial waste disposal and sewage sludge. Cd may also be present as an impurity in phosphatic fertilizers. Food, especially cereals, is the main pathway into the body.

- *Lead*: Pb is particularly damaging to the brain motor functions, especially in children. It can also damage neurological and renal systems and reproductive health. Damage can occur before birth, as Pb can easily access placenta tissues. Intakes greater than 500 μg day^{-1} are considered damaging to health. Industrial contamination is a major source of Pb. Previous additions to petrol caused problems, but this is now decreasing because of the greater use of lead-free petrol.

- *Mercury*: industrial activities account for most Hg contamination in soils. The old saying 'mad as a hatter' related to the mental health problems of hat manufacturers, who used Hg to stiffen top hats in Victorian times. The acute effects of Hg pollution were demonstrated in Minamata, a fishing village on the west coast of Kyushu, Japan, after people consumed fish contaminated with Hg from industrial pollution. In 1932 the Chisso Corporation began to manufacture acetaldehyde, used to produce plastics. Mercury from the production process began to spill into Minamata Bay and methyl mercury chloride entered the food chain. At the time, Minamata residents relied almost exclusively on fish and shellfish from the bay for protein. People developed impaired brain motor functions, evidenced by stumbling while walking, not being able to write or coordinate actions, having difficulty hearing or swallowing and trembling uncontrollably. Mercury was identified as the source of the illness in 1956 and it is estimated that 10,000 victims exist currently and that at least 3000 have died, with over US $611 million paid to victims in compensation. For further details visit: http://www1.umn.edn./ships/ethics/minamata.htm (accessed 5 March 2004).

- *Zinc*: Zn deficiency is a major global problem (Alloway, 1995) associated with various medical disorders (e.g., dwarfism, anaemia, failure of wounds to heal, skin lesions and birth defects). The recommended daily intake is 15–45 μg day^{-1}. The main sources are red meat, legumes, whole grains, pulses and unpolished rice. Excess Zn (>150 μg day^{-1}) is associated with disorders such as damage to reproductive health. Sources of excess Zn include pollution, sewage sludge, fertilizers, pesticides and mining.

Sometimes the difference in concentration between beneficial and toxic amounts is small. This applies to B, Cl, Co, Cr, Cu, Fe, F, Na and Zn, and especially I and Se. The roles of some of these elements are reviewed below.

- *Chromium (Cr)*: beneficial intake of Cr is 50–200 μg day^{-1}, but damaging effects occur above 400 mg kg^{-1}. However, Cr toxicity is rare and is only likely to be a potential problem in acid clay-rich soils. Sewage sludge is the main source of contamination (Oliver, 1997).
- *Copper*: Cu deficiency can be a serious problem. In humans an intake of 2 mg Cu day^{-1} is recommended, with a safe upper limit of 12 mg day^{-1}. Symptoms of deficiency include osteoporosis and damage to white blood cells. In cattle and sheep, symptoms of deficiency include unthriftiness (general lack of response to food), poor fertility and poor milk yields. The bioavailability of Cu is strongly related to soil pH, being very low at high soil pH.
- *Fluorine (F)*: the recommended intake of F is 0.2–2.0 mg day^{-1}, water being the main pathway into the body. Deficiency is linked to dental caries, but excess can lead to bony outgrowths and, in extreme cases, to fluorosis (calcification of ligaments). Excess is a particular problem in developing semi-arid countries, especially India and Sri Lanka, with extensive irrigation because of rapid evaporation and high drinking water consumption (Oliver, 1997).
- *Iodine*: I is an essential constituent of the thyroid hormone, thyroxine, which is essential for growth and development. An intake of 100–150 μg day^{-1} is recommended. Coastal areas tend to have higher concentrations of I and seaweeds are an important source. Concurrent I and Se deficiencies are particularly damaging to health.
- *Selenium*: Se deficiency can cause the diseases Kashin-Beck (associated with bone disorders and chronic arthritis) and Keshan (a heart disease) and has been implicated in cancer. The main pathway into the body is in food, so intake is a function of diet and the geographical origin of the food. The recommended daily intake is 50–200 μg day^{-1}. However, if taken in excess (>9 mg day^{-1}) it can result in adverse clinical effects, including morbidity (Oliver, 1997).

10.4 SOILS AND HABITAT CREATION

Changes in farming practices since the Second World War have led to catastrophic losses of species-rich meadows in the UK and elsewhere. The decline in floristically rich grasslands has altered the visual impression of the British countryside, particularly in the spring and summer months. The remaining species-rich hay meadows are of rich botanical and ecological interest and are now the focus of conservation initiatives (Atkinson *et al.*, 1995; Jones *et al.*, 1995). Only a fraction of this once extensive environment remains, existing as isolated pockets. Although habitat creation cannot hope to fully compensate for these losses, attempts at reconstruction provide a means of reversing trends that have continued for several decades.

In urban areas, creation of hay meadows can improve recreational and amenity sites by increasing diversity of structure and species, which makes landscaping schemes more interesting, visually attractive and results in management economies. These hay meadows make established multi-functional contributions to biodiversity, landscape aesthetics and education and particularly benefit inner city residents. They not only improve the aesthetic value of the area, but also provide environmental and educational resources. Many of these sites are developed on mineral soils deficient in nutrients, which might encourage wildflower assemblages that require low soil fertility. Since these soils are also often deficient in organic matter, they have considerable potential to store soil organic matter and so act as carbon sinks. This should help promote urban hay meadows as part of national environmental development policies.

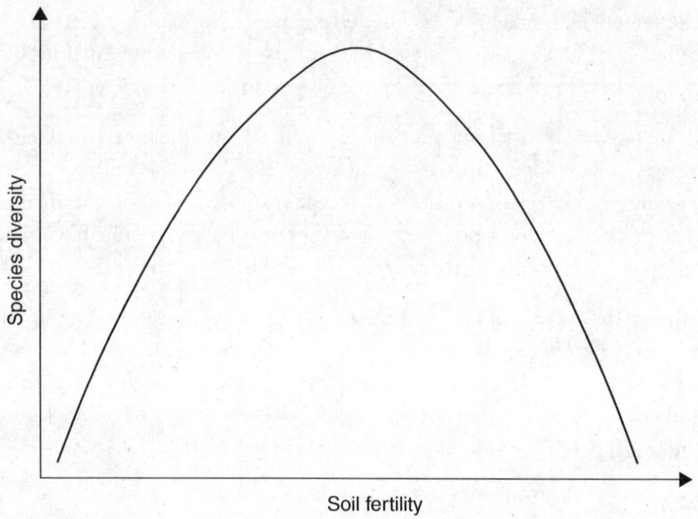

Figure 10.1 The 'hump-back model', relating soil fertility to species diversity within grasslands
Source: adapted from Grime (1973).

Between 1983 and 1989, 17 species-rich meadows were established in various locations within and around the Wolverhampton and Dudley conurbations of the West Midlands of England. These studies have formed the basis of a growing corpus of knowledge on habitat creation and management (Besenyei and Trueman, 2001; Trueman and Millett, 2003). These include studies on site establishment (Jones, 1993), soil fertility (McCrea, 1999; Vaz, 2001) and monitoring and management (Besenyei, 2000). Management records at the various sites have been used to investigate the causes of change in the vegetation (Besenyei, 2000; Besenyei *et al.*, 2002) and the minimum management input needed to retain or increase species diversity.

The habitat creation technique in the Wolverhampton area is to strew hay cut in old floristically diverse rural meadows, such as along the Welsh Marches (Trueman and Millett, 2003). After several years, vegetation comparable with that at the donor site establishes itself in the new urban/industrial environment. Desirable species include key indicators of old grassland, such as Crested dog's tail (*Cynosurus cristatus*) and Lesser Knapweed (*Centaurea nigra*) and attractive meadow species, such as Oxeye Daisy (*Leucanthum vulgaris*), Yellow Rattle (*Rhinanthus minor*) and Green-winged Orchid (*Orchis morio*). Soil conditions affect the success of habitat creation schemes. Grime (1973) proposed the 'hump-back model' (Figure 10.1) for the relationship between soil fertility and species diversity. If soil fertility is too low, introduced species cannot establish themselves. If fertility is too high, then a few species dominate and exclude others from the site, as found in the Rothamsted Park Grass Hay Experiment (Catt and Henderson, 1993). Tall grasses are particularly effective in dominating such sites. This means that creation of species-rich grassland can only be successful in areas of moderate to low soil fertility (Marrs, 1993).

Attempts are in progress to define 'envelopes of soil fertility'. If soil conditions fall within the envelope, then habitat creation is likely to be successful. If conditions fall outside, then the scheme is likely to fail. Table 10.1 shows the range of fertility conditions in which habitat creation for hay meadows is likely to be successful (McCrea, 1999; McCrea *et al.*, 2001a). Fertility depletion studies have shown that arable crops can deplete available nutrients and allow rapid establishment of a diverse sward. Barley is particularly efficient in depleting nutrients compared with potatoes, maize or tobacco (Jones, 1993; McCrea *et al.*, 2001b).

Table 10.1 Soil fertility conditions suitable for species-rich hay meadows

Soil property	Optimum
Extractable phosphorus (mg 100 g^{-1})	<7.0
Extractable potassium (mg 100 g^{-1})	10–30
Soil pH	>5.2

Source: McCrea (1999); McCrea *et al.* (2001a).

10.5 SOIL QUALITY

Because of increasing recognition that soils can profoundly influence crop production, human and animal health and many aspects of the environment, there is at present growing national and international pressure to establish standards of soil quality, sometimes alternatively termed soil health. This is strongly advocated by policy makers as well as environmental pressure groups and members of the public. However, most scientists recognize the considerable difficulties involved in deciding standards applicable to all soil types and all properties that affect fertility and the environment. So, although some proposals have been made, principally for standardization of soil description, sampling and analytical methods (Hortensius and Nortcliff, 1991), no standards have yet been agreed for national or international use.

Early suggestions for assessing soil quality were based on comparisons with 'pristine' soil, that is, soil derived from the same parent material and with similar subsurface horizons to the one evaluated, but under long-established natural vegetation and not significantly affected by human activities. However, very few pristine soils remain, especially in countries such as the UK where almost all areas have been influenced by a range of human activities for several millennia. Also, even if pristine soils can be found, they are not necessarily of greater value for some purpose than those which have been carefully improved for those purposes.

It is now widely recognized that a more pragmatic approach is to define quality according to a soil's agreed function (i.e., fitness for a selected purpose). This has the advantage of focusing on the limited range of soil properties required for that purpose. However, it may ignore others, which could be important for future uses, perhaps leading to unforeseen problems when management is changed.

A list of the general soil ecosystem characteristics that should be considered when evaluating soil quality would include:

- the levels of pollutants entering soil and leaving it as a result of plant uptake, leaching, erosion and gaseous emissions, including nitrate, ammonia, nitrous oxide, radon, phosphorus, heavy metals, pesticides and industrial organic chemicals;
- trends in amounts of total organic matter or easily mineralizable organic matter resulting from land use change;
- the size of the total microbial biomass or of selected functional groups;
- the populations of some important or sensitive faunal components, such as earthworms, millipedes or isopods;
- the ability to generate or oxidize methane;

- populations of pathogenic micro-organisms;
- structural quality and its resilience to changes in cropping, cultivation techniques and climate, as reflected in hydraulic conductivity, bulk density, erodibility and the need for field drainage or irrigation;
- levels of acidifying inputs from the atmosphere or the soil itself (e.g., pyrite);
- the soil's pH buffering capacity.

Unfortunately, knowledge of most if not all of these characteristics is probably insufficient at present for soil scientists to agree on more detailed lists of individual properties important for selected soil functions, or on their relative importance, the values to be set as limits or targets and the best strategies to achieve them. Further problems arise from the heterogeneity of soil, which means that standards and strategies chosen for one site may not be applicable even to others close by.

However, at least soil scientists are aware of the multi-functional role of soil, which puts them in the best position to mediate the often opposing demands of the various soil user groups and ensure that any strategies chosen do not create more problems than they solve. This is likely to develop into a major role for soil scientists in the twenty-first century.

Summary

As we progress into the twenty-first century, many problems exist in terms of our soil resources. These problems present both challenges and opportunities. For instance, if increased atmospheric CO_2 and other 'greenhouse gases' are contributing to global warming, there are opportunities to ameliorate these changes by increasing the carbon stored in soil. This is unlikely to solve the problem, but may provide time while we develop the appropriate technologies and policies. Increased carbon sequestration offers opportunities to increase soil organic content, thus also improving the structure and fertility of soils. The spread of urban areas also poses challenges, in terms of both the quantity of land lost to urban cover and soil quality. Reclamation of old urban-industrial sites is possible, but great care must be exercised in both avoiding chemical contamination from reclaimed land and assessing suitability to the newly designated land use. Another series of problems are posed by interactions between soils and human health. These relationships are poorly understood and soil scientists and biomedical scientists need to collaborate to understand their interaction and thus develop solutions to improve human health. The question of habitat creation illustrates how we can begin to tackle many of these problems. Uninteresting, urban derelict land can be reclaimed to make floristically diverse grassland. These recreated meadows are visually attractive, valuable as educational and recreational resources and may well act as carbon sinks. Thus, by taking holistic views of soil management problems and cooperating with other disciplines, soil scientists can arrive at practical, viable and realistic solutions to the environmental challenges we face in the twenty-first century.

FURTHER READING

Alloway, B.J., 1995, *Heavy Metals in Soils*, 2nd edn, Glasgow: Blackie Academic and Professional.

Bullock, P. and Gregory, P.J. (eds), 1991, *Soils in the Urban Environment*, Oxford: OUP.

Gilbert, O.L. and Anderson, P., 1998, *Habitat Creation and Repair*, Oxford: OUP.

Bibliography

Adams, J., Maslin, M. and Thomas, E., 1999, 'Sudden climatic transitions during the Quaternary', *Progress in Physical Geography,* 23, 1–36.

Adams, M.J., 1990, 'Epidemiology of fungally-transmitted viruses', *Soil Use and Management,* 6, 184–9.

Adams, W.A., 1986, 'Practical aspects of sportsfield drainage', *Soil Use and Management,* 2, 51–4.

Addiscott, T.M., 1988, 'Long-term leakage of nitrate from bare unmanured soil', *Soil Use and Management,* 4, 91–5.

Addiscott, T.M. and Thomas, D., 2000, 'Tillage, mineralization and leaching: phosphate', *Soil & Tillage Research,* 53, 255–73.

Addiscott, T.M. and Whitmore, A.P., 1991, 'Simulation of solute leaching in soils of differing permeabilities', *Soil Use and Management,* 7, 94–102.

Agnew, C. and Anderson, E., 1992, *Water Resources in the Arid Realm,* London: Routledge.

Agricultural Development and Advisory Service, 1982, *The Use of Sewage Sludge on Agricultural Land,* Ministry of Agriculture, Fisheries and Food Publication 2409, Alnwick: MAFF.

Al-Dabbagh, A., Forman, D., Bryson, D., Stratton, I. and Doll, E.C., 1986, 'Mortality of nitrate fertilizer workers', *British Journal of Industrial Medicine*, 43, 507–15.

Alich, J.A., Jr and Inman, R.E., 1975, 'Utilization of plant biomass as an energy feedstock', in Jewell, W.J. (ed.), *Energy, Agriculture, and Waste Management*, Ann Arbor MI: Ann Arbor Scientific Publications Inc., 453–66.

Alley, R.B., 2000, 'The Younger Dryas cold interval as viewed from central Greenland', *Quaternary Science Reviews,* 19, 213–26.

Alloway, B.J., 1995, *Heavy Metals in Soils,* 2nd edn, Glasgow: Blackie Academic and Professional.

Aluko, O.B. and Koolen, A.J., 2000, 'The essential mechanics of capillary crumbling of structured agricultural soils', *Soil & Tillage Research,* 55, 117–26.

Amelung, W., Zech, W., Zhang, X., Follett, R.F., Tiessen, H., Knox, E. and Flach, K.W., 1998, 'Carbon, nitrogen, and sulfur pools in particle-size fractions as influenced by climate', *Soil Science Society of America Journal*, 62, 172–81.

Amoozegar, A. and Warrick, A.W., 1986, 'Field measurements of saturated hydraulic conductivity', in A. Klute (ed.), *Methods of Soil Analysis. Part 1. Physical and Mineralogical Methods,* 2nd edn, Madison WI: American Society of Agronomy Monograph, 9, 735–70.

Anderson, C.W.N., Brooks, R.R., Stewart, R.B. and Simcock, B., 1998, 'Harvesting a crop of gold in plants', *Nature,* 395, 1185–6.

Anderson, D.W., 1995, 'The role of non-living organic matter in soils', in R.G. Zepp and C. Sonntag (eds), *The Role of Non-living Organic Matter in the Earth's Carbon Cycle*, Chichester: J. Wiley and Sons, 81–92.

Andrews, J., 1998, 'Abrupt changes (Heinrich Events) in Late Quaternary North Atlantic marine environments: a history and review of data and concepts', *Journal of Quaternary Science,* 13, 3–16.

Angulo-Jaramillo, R., Vandervaere, J.-P., Roulier, S., Thony, J.-T., Gaudet, J.-P. and Vauclin, M., 2000, 'Field measurement of surface hydraulic properties by disc and ring infiltrometers. A review and recent developments', *Soil & Tillage Research,* 55, 1–29.

Arden-Clarke, C. and Hodges, D., 1987, 'Soil erosion: the answer lies in organic farming', *New Scientist,* 12 February 1987, 42–3.

Argabright, M.S., 1991, 'Evolution in use and development of the wind erosion equation', *Journal of Soil and Water Conservation,* 46, 104–105.

Arvidsson, J., 2001, 'Subsoil compaction caused by heavy sugarbeet harvesters in southern Sweden. I. Soil physical properties and crop yield in six experiments', *Soil & Tillage Research,* 60, 67–78.

Ashman, M.R. and Puri, G., 2002, *Essential Soil Science,* Oxford: Blackwell.

Atkinson, M.D., Trueman, I.C., Millett, P., Jones, G.H. and Besenyei, L., 1995, 'The use of hay strewing to create species rich grasslands (ii) monitoring the vegetation and the seed bank', *Land Contamination and Reclamation,* 3, 108–10.

Aubertot, J.N., Dürr, C., Kieu, K. and Richard, G., 1999, 'Characterization of sugar beet (*Beta vulgaris* L.) seedbed structure', *Soil Science Society of America Journal,* 63, 1377–84.

Aubrèville, A., 1949, *Climats, Forêts et Désertification de l'Afrique Tropicale,* Paris: Societé d'Editions Géographiques Maritimes et Colonials.

Aura, E., 1999, 'Effects of shallow tillage on physical properties of clay soil and growth of spring cereals in dry and moist conditions in southern Finland', *Soil & Tillage Research,* 50, 169–76.

Avery, A.A., 1999, 'Infantile methaemoglobinaemia: re-examining the role of drinking water nitrates', *Environmental Health Perspectives,* 107, 583–6.

Avery, B.W., 1980, *Soil Classification for England and Wales (Higher Categories)*, Soil Survey Technical Monograph 14, Harpenden: Soil Survey of England and Wales.

Aylmore, L.A.G., 1993, 'The use of computer assisted tomography in studying water movement around plant roots', *Advances in Agronomy,* 49, 1–54.

Båårth, E., 1989, 'Effects of heavy metals in soil on microbial processes and populations (a review)', *Water, Air and Soil Pollution,* 47, 335–79.

Backes, C.A., Pulford, I.D. and Duncan, H.J., 1993, 'Seasonal variation of pyrite oxidation rates in colliery spoil', *Soil Use and Management,* 9, 30–4.

Bailey, R.J. and Spackman, E., 1996, 'A model for estimating soil moisture changes as an aid to irrigation scheduling and crop water-use studies: I. Operational details and description', *Soil Use and Management,* 12, 122–8.

Bailey, R.J., Groves, S.J. and Spackman, E., 1996, 'A model for estimating soil moisture changes as an aid to irrigation scheduling and crop water-use studies: II. Field test of the model', *Soil Use and Management,* 12, 129–33.

Baker, S.W., 1989, 'Soil physical conditions of the root zone layer and the performance of winter games' pitches', *Soil Use and Management,* 5, 116–22.

Baksiene, E., 2002, 'Changes in the properties of light textured soils in relation to lake sediment fertilization', in *Sustainable Utilization of Global Soil and Water Resources. Technology and Method of Soil and Water Conservation,* Proceedings of the 12th International Soil Conservation Organization Conference, 26–31 May, Beijing, China, Beijing: Tsinghua University Press, Vol III, 291–5.

Baldock, J.A. and Nelson, P.N., 2000, 'Soil organic matter', in M.E. Sumner (ed.), *Handbook of Soil Science,* Boca Raton FL: CRC Press, B25–B84.

Balesdent, J., Chenu, C. and Balabane, M., 2000, 'Relationship of soil organic matter dynamics to physical protection and tillage', *Soil & Tillage Research,* 53, 215–30.

Ball, B.C., Campbell, D.J., Douglas, J.T., Henshall, J.K. and O'Sullivan, M.F., 1997, 'Soil structural quality, compaction and land management', *European Journal of Soil Science,* 48, 593–601.

Ball, D.F., 1964, 'Loss-on-ignition as an estimate of organic matter and organic carbon in non-calcareous soils', *Journal of Soil Science,* 15, 84–92.

Balling, R.C., Jr, Michaels, P.J. and Knappenberger, C., 1998, 'Analysis of winter and summer warming rates in gridded temperature time series', *Climate Research,* 9, 175–82.

Barber, D.A. and Martin, J.K., 1976, 'The release of organic substances by cereal roots into soil', *New Phytologist,* 76, 69–80.

Barker, R.D., 1981, 'The offset system of electrical resistivity sounding and its use with a multicore cable', *Geophysical Prospecting,* 29, 128–43.

Barnhisel, R.I. and Hower, J.M., 1997, 'Coal surface mine reclamation in the eastern United States: the revegetation of disturbed lands to hayland/pasture or cropland', *Advances in Agronomy,* 61, 233–75.

Barry, L., Craig, G.C. and Thuburn, J., 2002, 'Poleward heat transport by the atmospheric heat engine', *Nature,* 415, 774–7.

Barry, R.G. and Chorley, R.J., 1998, *Atmosphere, Weather and Climate,* 7th edn, London: Routledge.

Bar-Yosef, B., 1999, 'Advances in fertigation', *Advances in Agronomy*, 65, 1–77.

Beck, A.J., Alcock, R.E., Wilson, S.C., Wang, M.-J., Wild, S.R., Sewart, A.P. and Jones, K.C., 1995, 'Long-term persistence of organic chemicals in sewage sludge-amended agricultural land: a soil quality perspective', *Advances in Agronomy,* 55, 345–91.

Bennett, H.H., 1939, *Soil Conservation,* New York: McGraw-Hill.

Bensiton, M. and Tol, R.S.J., 1998, 'Europe', in R.T. Watson, M.C. Zinyowera and R.H. Moss (eds), *The Regional Impacts of Climatic Change: An Assessment of Vulnerability,* Special Report of the IPCC Working Group II, Cambridge: Cambridge University Press.

Beresford, S.A., 1985, 'Is nitrate in drinking water associated with gastric cancer in the urban UK?', *International Journal of Epidemiology,* 14, 57–63.

Berger, A., 1980, 'The Milankovitch astronomical theory of paleoclimates – a modern review', *Vistas in Astronomy,* 24, 103–22.

Berntsen, R. and Berre, B., 2002, 'Soil fragmentation and the efficiency of tillage implements', *Soil & Tillage Research,* 64, 137–47.

Berry, P.M., Sylvester-Bradley, R., Philipps, L., Hatch, D.J., Cuttle, S.P., Rayns, F.W. and Gosling, P., 2002, 'Is the productivity of organic farms restricted by the supply of available nitrogen?', *Soil Use and Management*, 18, 248–55.

Besenyei, L., 2000, The Management of Artificially Created Species-rich Meadows in Urban Landscaping Schemes, Unpublished Ph.D. Thesis, The University of Wolverhampton, 275 pp.

Besenyei, L. and Trueman, I.C., 2001, 'Creating species-rich grasslands using traditional hay cutting and removal techniques (England)', *Ecological Restoration*, 19, 114–15.

Besenyei, L., Trueman, I.C., Atkinson, M.D., Jones, G.H. and Millett, P., 2002, 'Retaining diversity in hay meadows by traditional management', in Durand, J.L., Emile, J.C., Huyghe, C. and Lemaire, G. (eds), *Multi-Function Grassland – Quality Forages, Animal Products and Landscapes,* Grassland Science in Europe, Volume 7, 762–3.

Beyer, L., 1996, 'The chemical composition of soil organic matter in classical humic compound fractions and in bulk samples – a review, *Zeitschrift für Pflanzenernährung und Bodenkunde*, 159, 527–39.

Bhogal, A., Shepherd, M.A., Hatch, D.J., Brown, L. and Jarvis, S.C., 2001, 'Evaluation of two N cycle models for the prediction of N mineralization from grassland soils in the UK', *Soil Use and Management,* 17, 163–72.

Bhogal, A., Young, S.D. and Sylvester-Bradley, R., 1997, 'Straw incorporation and immobilization of spring-applied nitrogen', *Soil Use and Management,* 13, 111–16.

Bibby, J.S. (ed.), 1991, *Land Use Capability Classification for Agriculture,* Aberdeen: Macaulay Land Use Research Institute.

Bird, J.A., Horwath, W.R., Eagle, A.J. and Van Kessel, C., 2001, 'Immobilization of fertilizer nitrogen in rice: effects of straw management practices', *Soil Science Society of America Journal*, 65, 1143–52.

Biswas, A.S., 1990, 'Conservation and management of water resources', in A.S. Goudie (ed.), *Techniques for Desert Reclamation*, Chichester: J. Wiley, 251–65.

Blake, L., Goulding, K.W.T., Mott, C.J.B. and Johnston, A.E., 1999, 'Changes in soil chemistry accompanying acidification over more than 100 years under woodland and grass at Rothamsted Experimental Station, UK', *European Journal of Soil Science*, 50, 401–12.

Blake, L., Hesketh, N., Fortune, S. and Brookes, P.C., 2002, 'Assessing phosphorus 'Change-Points' and leaching potential by isotopic exchange and sequential fractionation', *Soil Use and Management,* 18, 199–207.

Blake, L., Johnston, A.E. and Goulding, K.W.T., 1994, 'Mobilization of aluminium in soil by acid deposition and its uptake by grass cut for hay – a chemical time bomb', *Soil Use and Management*, 10, 51–5.

Blavet, D., Mathe, E. and Leprun, J.C., 2000, 'Relations between soil color and waterlogging duration in a representative hillside of the West African granito-gneissic bedrock', *Catena*, 39, 187–210.

Blaylock, M.J., Salt, D.E., Dushenkov, S., Zhakarova, O., Gussman, C., Kapulnik, Y., Emsley, B.D. and Raskin, I., 1997, 'Enhanced accumulation of Pb in Indian mustard by soil-applied chelating agents', *Environmental Science and Technology,* 31, 860–5.

Bloomfield, C. and Zahari, A.B., 1982, 'Acid sulphate soils', *Outlook on Agriculture*, 11, 48–54.

Bloomfield, H.E., Handley, J.F. and Bradshaw, A.D., 1981, 'Top soil quality', *Landscape Design*, 135, 32–4.

Boardman, J., Ligneau, L., de Roo, A. and Vandaele, K., 1994, 'Flooding of property by runoff from agricultural land in northwestern Europe', *Geomorphology,* 10, 183–96.

Boast, C.W. and Langbartel, R.G., 1984, 'Shape factor for seepage into pits', *Soil Science Society of America Journal*, 48, 10–15.

Boeckx, P., Van Cleemput, O. and Meyer, T., 1998, 'The influence of land use and pesticides on methane oxidation in some Belgian soils', *Biology and Fertility of Soils,* 27, 293–8.

Boeuf-Tremblay, V., Plantureux, S. and Guckert, A., 1993, 'Mechanical constraint and organic matter losses of young maize roots under sterile conditions', in H.F. Cook and H.C. Lee (eds), *Soil Management in Sustainable Agriculture,* Wye: Wye College Press, 325–30.

Bogner, J., Meadows, M. and Czepiel, P., 1997, 'Fluxes of methane between landfills and the atmosphere: natural and engineered controls', *Soil Use and Management,* 13, 268–77.

Bolker, B.M., Pacala, S.W., Bazzaz, F.A., Canham, C.D. and Levin, S.A., 1995, 'Species-diversity and ecosystem response to carbon dioxide fertilization – conclusions from a temperate forest model', *Global Change Biology,* 1, 373–81.

Bollinne, A., 1978, 'Study of the importance of splash and wash on cultivated loamy soils in Hesbaye (Belgium)', *Earth Surface Processes,* 3, 71–84.

Bolt, G.H. (ed.), 1982, *Soil Chemistry B: Physico-chemical Models*, 2nd edn, Amsterdam: Elsevier.

Bolton, J., 1977, 'Changes in soil pH and exchangeable calcium in two liming experiments on contrasting soils over 12 years', *Journal of Agricultural Science, Cambridge*, 89, 81–6.

Bond, G. and Lotti, R., 1995, 'Iceberg discharge into the North Atlantic on millennial time scales during the last glaciation', *Science,* 267, 1005–10.

Borjesson, G. and Svensson, B., 1997, 'Effects of a gas extraction interruption on emissions of methane and carbon dioxide from a landfill, and on methane oxidation in the cover soil', *Journal of Environmental Quality,* 26, 1182–90.

Børreson, T., 1999, 'The effect of straw management and reduced tillage on soil properties and crop yields of spring-sown cereals on two loam soils in Norway', *Soil & Tillage Research*, 51, 91–102.

Boudot, J.P., Bel Hadi Brahim, A. and Chone, T., 1988, 'Dependence of carbon and nitrogen mineralization rates upon amorphous metallic constituents and allophanes in highland soils', *Geoderma*, 42, 245–60.

Bower, A.S., Le Cann, B., Rossby, T., Zenk, W., Gould, J., Speer, K., Richardson, P.L., Prater, M.D. and Zhang, H.-M., 2001, 'Directly measured mid-depth circulation in the northeastern North Atlantic Ocean', *Nature,* 419, 603–607.

Brade-Birks, S.G., 1944, *Good Soil,* London: English University Press.

Bradley, R.S., 2000, 'Past global changes and their significance for the future', *Quaternary Science Reviews,* 19, 391–402.

Brady, N.C. and Weil, R.R., 1999, *The Nature and Properties of Soils,* Upper Saddle River NJ: Prentice Hall.

Brandsma, R.T., Fullen, M.A. and Hocking, T.J., 1999a, 'Soil conditioner effects on soil structure and erosion', *Journal of Soil and Water Conservation,* 54, 485–9.

Brandsma, R.T., Fullen, M.A., Hocking, T.J. and Allen, J.R., 1999b, 'An X-ray scanning technique to determine soil macroporosity by chemical mapping', *Soil & Tillage Research,* 50, 95–8.

Bransden, B.E., 1991, 'Soil protection as a component of gravel raising', *Soil Use and Management,* 7, 139–45.

Bremner, J.M., 1997, 'Sources of nitrous oxide in soils', *Nutrient Cycling in Agroecosystems,* 49, 7–16.

Bresson, J.M. and Boiffin, J., 1990, 'Morphological characterization of soil crust development stages on an experimental field', *Geoderma*, 47, 301–25.

Brevik, E.C., Fenton, T.E. and Moran, L.P., 2002, 'Effect of soil compaction on organic carbon amounts and distribution, south-central Iowa', *Environmental Pollution,* 116, S137–S141.

Bridges, E.M., 1991a, 'Dealing with contaminated soils', *Soil Use and Management,* 7, 151–8.

Bridges, E.M., 1991b, 'Waste materials in urban soils', in P. Bullock and P.J. Gregory (eds), *Soils in the Urban Environment*, Oxford: Blackwell Scientific Publications, 28–46.

Bridges, E.M. and Harding, D.M., 1971, 'Micro-erosion processes and factors affecting slope development in the Lower Swansea Valley', in D. Brunsden (ed.), *Slopes: Form and Process*, IBG Special Publication No. 3, Oxford: Alden & Mowbray, 65–79.

Briffa, K.R., 2000, 'Annual climate variability in the Holocene: interpreting the message of ancient trees', *Quaternary Science Reviews,* 19, 87–105.

Broecker, W.S., 1982a, 'Glacial to interglacial changes in ocean chemistry', *Progress in Oceanography,* 2, 151–97.

Broecker, W.S., 1982b, 'Ocean chemistry during glacial time', *Geochimica et Cosmochimica Acta,* 46, 1689–1706.

Broecker, W.S., 1998, 'The end of the present interglacial: how and when?', *Quaternary Science Reviews,* 17, 689–94.

Broecker, W.S. and Denton, G.H., 1990, 'The role of ocean–atmosphere reorganizations in glacial cycles', *Quaternary Science Reviews,* 9, 305–41.

Bronswijk, J.J.B., Groenenberg, J.E., Ritsema, C.J., Van Wijk, A.L.M. and Nugroho, K., 1995, 'Evaluation of water management strategies for acid sulphate soils using a simulation model: a case study in Indonesia', *Agricultural Water Management*, 27, 125–42.

Brook, E., Harder, S., Severinghaus, J., Steig, E. and Sucher, C., 2000, 'On the origin and timing of rapid changes in atmospheric methane during the last glacial period', *Global Biogeochemical Cycles,* 14, 559–72.

Brook, E.J., Sowers, T. and Orchado, J., 1996, 'Rapid variations in atmospheric methane concentration during the past 110 000 years', *Science,* 273, 1087–91.

Brookes, P.C., 1995, 'The use of microbial parameters in monitoring soil pollution by heavy metals', *Biology and Fertility of Soils,* 19, 269–79.

Brown, L.R., 1984, 'The global loss of topsoil'. *Journal of Soil and Water Conservation,* 39, 162–5.

Brown, L.R., 1991, 'The global competition for land', *Journal of Soil and Water Conservation,* 46, 394–7.

Brown, L., Syed, B., Jarvis, S.C., Sneath, R.W., Phillips, V.R., Goulding, K.W.T. and Li, C., 2002, 'Development and application of a mechanistic model to estimate emission of nitrous oxide from UK agriculture', *Atmospheric Environment,* 36, 917–28.

Bruckler, L., 1983, 'Rôle des propriétés physiques du lit de semences sur l'imbibition et la germination. II. Contrôle expérimental d'un modèle d'imbibition et possibilités d'application', *Agronomie,* 3, 223–32.

Brundtland, G.H. (Chairman), 1987, *Our Common Future,* Report of the World Commission on Environment and Development, presented to the Chairman of the Intergovernmental Intersessional Preparatory Committee, UNEP Governing Council, Oxford: Oxford University Press.

Budd, W.F., Andrews, J.T., Finkel, R.C., Fireman, E.L., Graf, W., Hammer, C.U., Jouzel, J., Raynaud, D.P., Reeh, N., Shoji, H., Stauffer, B.R. and Weertman, J., 1989, 'Group report: how can an ice core chronology be established?', in H. Oeschger and C.C. Langway Jr (eds), *The Environmental Record in Glaciers and Ice Sheets,* Chichester: J. Wiley, 177–92.

Budyko, M.I., Nayefimova, N.A., Aubenok, L.I. and Strokhina, L.A., 1962, 'The heat balance of the surface of the earth', *Soviet Geography,* 3(5), 3–16.

Bullock, P. and Gregory, P.J., 1991, 'Soils: a neglected resource in urban areas', in P. Bullock and P.J. Gregory (eds), *Soils in the Urban Environment,* Oxford: Blackwell Scientific Publications, 1–4.

Bulson, H.A.J., Welsh, J.P., Stopes, C.E. and Woodward, L., 1996, 'Agronomic viability and potential economic performance of three organic four year rotations without livestock, 1988–1995', *Aspects of Applied Biology,* 47, 277–86.

Burger, N., Lebert, M. and Horn, R., 1988, 'Prediction of the compressibility of arable land', *Catena Supplementband,* 11, 141–51.

Burke, I.C., Elliott, E.T. and Cole, C.V., 1995, 'Influence of macroclimate, landscape position, and management on soil organic matter in agroecosystems', *Ecological Applications,* 5, 124–31.

Burkholder, J.A. and Glasgow, H.B., Jr, 1997, '*Pfeisteria piscicidia* and other Pfeisteria-dinoflagellates behaviors, impacts and environmental controls', *Limnology and Oceanography,* 42, 1052–75.

Burt, T. and Labadz, J., 1990, 'Blanket peat erosion in the Southern Pennines', *Geography Review,* 3, 31–5.

Burton, R.G.O. and Hodgson, J.M. (eds), 1987, *Lowland Peat in England and Wales*, Soil Survey Special Survey 15, Harpenden: Soil Survey of England and Wales.

Cajuste, L.J., Laird, R.J., Cajuste, L. and Cuevas, B.G., 1996, 'Citrate and oxalate influence on phosphate, aluminium, and iron in tropical soils', *Communications in Soil Science and Plant Analysis*, 27, 1377–86.

Caldwell, T.H. and Richardson, S.J., 1975, 'Field behaviour of lowland peats and organic soils', in *Soil Physical Conditions and Crop Production*, Ministry of Agriculture, Fisheries and Food Technical Bulletin 29, London: Her Majesty's Stationery Office, 94–111.

Cannell, R.Q. and Hawes, J.D., 1994, 'Trends in tillage practices in relation to sustainable crop production with special reference to temperate climates', *Soil & Tillage Research,* 30, 245–82.

Cassaro, F.A.M., Tominaga, T.T., Bacchi, O.O.S., Reichardt, K., de Oliveira, J.C.M. and Timm, L.C., 2000, 'The use of a surface gamma-neutron gauge to explore compacted soil layers', *Soil Science,* 165, 665–76.

Cassel, D.K. and Nielsen, D.R., 1986, 'Field capacity and available water capacity', in A. Klute, (ed.), *Methods of Soil Analysis. Part 1. Physical and Mineralogical Methods*, Madison WI: American Society of Agronomy Monograph, 9, 901–26.

Catt, J.A. and Henderson, I.F., 1993, 'Rothamsted Experimental Station – 150 years of agricultural research. The longest continuous scientific experiment?', *Interdisciplinary Science Reviews*, 18, 365–78.

Catt, J.A., Howse, K.R., Christian, D.G., Lane, P.W., Harris, G.L. and Goss, M.J., 1998a, 'Strategies to decrease nitrate leaching in the Brimstone Farm Experiment, Oxfordshire, UK, 1988–1993: the effects of winter cover crops and unfertilised grass leys', *Plant and Soil,* 203, 57–69.

Catt, J.A., Howse, K.R., Christian, D.G., Lane, P.W., Harris, G.L. and Goss, M.J., 1998b, 'Strategies to decrease nitrate leaching in the Brimstone Farm Experiment, Oxfordshire, UK, 1988–1993: the effect of straw incorporation', *Journal of Agricultural Science, Cambridge,* 131, 309–19.

Catt, J.A., Howse, K.R., Christian, D.G., Lane, P.W., Harris, G.L. and Goss, M.J., 2000, 'Assessment of tillage strategies to decrease nitrate leaching in the Brimstone Farm Experiment, Oxfordshire, UK', *Soil & Tillage Research,* 53, 185–200.

Catt, J.A., Howse, K.R., Farina, R., Brockie, D., Todd, A., Chambers, B.J., Hodgkinson, R., Harris, G.L. and Quinton, J.N., 1998c, 'Phosphorus losses from arable land in England', *Soil Use and Management,* 14, 168–74.

Ceuppens, J. and Wopereis, M.C.S., 1999, 'Impact of non-irrigated rice cropping on soil salinization in the Senegal River delta', *Geoderma*, 92, 125–40.

Chalmers, A.G., Bacon, E.T.G. and Clarke, J.H., 2001, 'Changes in soil mineral nitrogen during and after 3-year and 5-year set-aside and nitrate leaching losses after ploughing out the 5-year plant covers in the UK', *Plant and Soil,* 228, 157–77.

Chambers, B., Smith, K. and Pain, B., 1999, 'Strategies to encourage better use of nitrogen in animal manures', in *Tackling Nitrate from Agriculture Strategy from Science,* London: MAFF Publications, 27–36.

Chaney, K., 1990, 'Effect of nitrogen fertilizer rate on soil nitrate content after harvesting winter wheat', *Journal of Agricultural Science, Cambridge,* 114, 171–6.

Cheshire, M.V., Bedrock, C.N., Williams, B.L., Chapman, S.J., Solntseva, I. and Thomsen, I., 1999, 'The immobilization of nitrogen by straw decomposing in soil', *European Journal of Soil Science,* 50, 329–41.

Chiou, C.T., 1990, 'Roles of organic matter, minerals and moisture in sorption of non-ionic compounds and pesticides by soil', in P. MacCarthy, C.E. Clapp, R.L. Malcolm and P.R. Bloom (eds), *Humic Substances in Soil and Crop Sciences: Selected Readings*, Madison WI: Soil Science Society of America, 111–60.

Chisholm, S.W. and Morel, F.M.M., 1991, 'What controls phytoplankton production in nutrient-rich areas of the open sea?', *Limnology and Oceanography,* 36, U1507–U1511.

Christensen, B.T., 1992, 'Physical fractionation of soil and organic matter in primary size particles and density separates', *Advances in Soil Science*, 20, 1–90.

Christensen, B.T., 1996, 'Carbon in primary and secondary organomineral complexes', in M.R. Carter and B.A. Stewart (eds), *Structure and Organic Matter Storage in Agricultural Soils,* Boca Raton FL: CRC Press, 97–165.

Christensen, B.T. and Johnston, A.E., 1997, 'Soil organic matter and soil quality: lessons learned from long-term field experiments at Askov and Rothamsted', in E.G. Gregorich and M.R. Carter (eds), *Soil Quality for Crop Production*, Amsterdam: Elsevier, 399–430.

Christian, D.G. and Bacon, E.T.G., 1988, 'A comparison of tine cultivation and ploughing with two methods of straw disposal on the growth, nutrient uptake and yield of winter wheat grown on a clay soil', *Soil Use and Management*, 4, 51–7.

Christian, D.G. and Bacon, E.T.G., 1990, 'A long-term comparison of ploughing, tine cultivation and direct drilling on the growth and yield of winter cereals and oilseed rape on clayey and silty soils', *Soil & Tillage Research,* 18, 311–31.

Christian, D.G. and Bacon, E.T.G., 1991, 'The effects of straw disposal and depth of cultivation on the growth, nutrient uptake and yield of winter wheat on a clay and a silt soil', *Soil Use and Management*, 7, 217–22.

Christian, D.G. and Ball, B.S., 1994, 'Reduced cultivation and direct drilling for cereals in Great Britain', in M.R. Carter (ed.), *Conservation Tillage in Temperate Agroecosystems,* Boca Raton FL: CRC Press Inc., 117–40.

Christian, D.G. and Miller, D.P., 1986, 'Straw incorporation by different tillage systems and the effect on growth and yield of winter oats', *Soil & Tillage Research,* 8, 239–52.

Christian, D.G., Bacon, E.T.G., Brockie, D., Glen, D., Gutteridge, R.J. and Jenkyn, J.F., 1999, 'Interactions of straw disposal methods and direct drilling or cultivations on winter wheat (*Triticum aestivum*) grown on a clay soil', *Journal of Agricultural Engineering Research*, 73, 297–309.

Clark, P.U., Marshall, S.J., Clarke, G.K.C., Hostetler, S.W., Licciardi, J.M. and Teller, J.T., 2001, 'Freshwater forcing of abrupt climate change during the Last Glaciation', *Science,* 293, 283–7.

Clayden, B. and Hollis, J.M., 1984, *Criteria for Differentiating Soil Series,* Soil Survey Technical Monograph, 17, Harpenden: Soil Survey of England and Wales.

Cole, C., Duxbury, J., Freney, J., Heinemeyer, O., Minami, K., Mosier, A., Paustian, K., Rosenberg, N., Sampson, N., Sauerbeck, D. and Zhao, Q., 1997, 'Global estimates of potential mitigation of greenhouse gas emissions by agriculture', *Nutrient Cycling in Agroecosystems,* 49, 221–8.

Cole, L., Bardgett, R.D., Ineson, P. and Adamson, J.K., 2002, 'Relationships between enhytraeid worms (Oligochaeta), climate change, and the release of dissolved organic carbon from blanket peat in Northern England', *Soil Biology and Biochemistry*, 34, 599–607.

Coleman, K. and Jenkinson, D.S., 1996, 'RothC-26.3 – a model for the turnover of carbon in soil', in D.S. Powlson, P. Smith and J.U. Smith (eds), *Evaluation of Soil Organic Matter Models*, Berlin: Springer Verlag, 237–46.

Collis-George, N., 2001, 'The application of double-layer theory to drainage, drying and wetting, and the Gapon exchange constant in a soil with mono- and divalent ions', *European Journal of Soil Science,* 52, 1–12.

Commission of the European Communities (CEC), 2001, *The Soil Protection Communication – DG ENV Draft,* Brussels: CEC (26.10.2001).

Commission of the European Communities (CEC), 2002, *Towards a Thematic Strategy for Soil Protection,* Communication from the Commission to the Council, The European Parliament, The Economic and Social Committee and the Committee of the Regions COM(2002), 179 final, Brussels: CEC (16.4.2002).

Connor, R.F., Chmura, G.L. and Beecher, C.B., 2001, 'Carbon accumulation in Bay of Fundy salt marshes; implications for restoration of reclaimed marshes', *Global Biogeochemical Cycles,* 15, 943–54.

Conrad, R., 2002, 'Control of microbial methane production in wetland rice fields', *Nutrient Cycling in Agroecosystems,* 64, 59–69.

Conry, M.J., 1974, 'Plaggen soils: a review of man-made raised soils', *Soils and Fertilizers*, 37, 319–26.

Conry, M.J. and MacNaeidhe, F., 1999, 'Comparative nutrient status of a peaty gleyed podzol and its plaggen counterpart on the Dingle Peninsula in the south-west of Ireland', *European Journal of Agronomy*, 11, 85–90.

Cook, R.J., Boosalis, M.G. and Doupnik, B., 1978, 'Influence of crop residues on plant diseases', in W.R. Oschwald (ed.), *Crop Residue Management Systems*, Madison WI: American Society of Agronomy Special Publication, 31, 147–63.

Cooke, G.W., 1982, *Fertilizing for Maximum Yield*, 3rd edn, London: Granada.

Coope, G.R., 1977, 'Fossil Coleopteran assemblages as sensitive indicators of climatic changes during the Devensian (Last) cold stage', *Philosophical Transactions of the Royal Society of London*, B280, 313–40.

Cornelis, W.M., Ronsyn, J., Van Meirvenne, M. and Hartmann, R., 2001, 'Evaluation of pedotransfer functions for predicting the soil moisture retention curve', *Soil Science Society of America Journal*, 65, 638–48.

Craul, P.J., 1985, 'A description of urban soils and their desired characteristics', *Journal of Arboriculture*, 11, 330–9.

Crawford, J.W., Verrall, S. and Young, I.M., 1997, 'The origin and loss of fractal scaling in simulated soil aggregates', *European Journal of Soil Science*, 48, 643–50.

Crestana, S., Cesareo, R. and Mascarenhas, S., 1986, 'Using a computer assisted tomography miniscanner in soil science', *Soil Science*, 142, 56–61.

Critical Loads Advisory Group, 1994, *Critical Loads of Acidity in the United Kingdom: Summary Report*, London: Department of the Environment.

Cruickshank, J.G., 1972, *Soil Geography*, Newton Abbot: David & Charles.

Cunningham, S.D., Anderson, T.A., Schwab, A.P. and Hsu, F.C., 1996, 'Phytoremediation of soils contaminated with organic pollutants', *Advances in Agronomy*, 56, 56–114.

Curmi, P., Merot, P., Roger-Estrade; J. and Caneill, J., 1996, 'Use of environmental isotopes for field study of water infiltration in the ploughed soil layer', *Geoderma*, 72, 203–17.

Curtin, D. and Syers, J.K., 2001, 'Lime-induced changes in indices of soil phosphate availability', *Soil Science Society of America Journal*, 65, 147–52.

Curtin, D., Campbell, C.A. and Messer, D., 1996, 'Prediction of titratable acidity and soil sensitivity to pH change', *Journal of Environmental Quality*, 25, 1280–4.

Da Silva, A.P. and Kay, B.D., 1997, 'Estimating the least limiting water range of soils from properties and management', *Soil Science Society of America Journal*, 61, 877–83.

Dacey, P.W. and Colbourn, P., 1979, 'An assessment of methods for the determination of iron pyrite in coal mine spoil', *Reclamation Revue,* 2, 113–21.

Dahl-Jensen, D., Mosegaard, K., Gundestrup, N., Clow, G.D., Johnsen, S.J., Hansen, A.W. and Balling, N., 1998, 'Past temperatures directly from the Greenland ice sheet', *Science,* 282, 268–71.

Dalsgaard, T., Canfield, D.E., Petersen, J., Thamdrup, B. and Acuña-González, J., 2003, 'N_2 production by the anammox reaction in the anoxic water of the Golfo Dulce, Costa Rica', *Nature,* 422, 606–608.

Dampney, P.M.R., 1985, 'A trial to determine the lime requirement for reseeded grassland on a peaty hill soil', *Soil Use and Management,* 1, 95–100.

Dancer, W.S. and Jansen, I.J., 1981, 'Greenhouse evaluation of solum and substratum materials in the southern Illinois coalfield: I. Forage crops', *Journal of Environmental Quality,* 10, 396–400.

Darby, H.C. (ed.), 1976, *A New Historical Geography of England after 1600,* Cambridge: Cambridge University Press.

Darcy, H., 1856, *Les Fontaines Publique de la Ville de Dijon,* Paris: Dalmont.

Datnoff, L.E., Nemec, S. and Pernezny, K., 1995, 'Biological control of fusarium crown and root rot of tomato in Florida using *Trichoderma harzianum* and *Glomus intraradices*', *Biological Control,* 5, 427–31.

Davies, D.B., 2000, 'The nitrate issue in England and Wales', *Soil Use and Management,* 16, 142–4.

Davies, D.B., Eagle, D.J. and Finney, J.B., 1982, *Soil Management,* Ipswich: Farming Press Ltd.

Davies, M.G., Vinten, A.J.A. and Smith, K.A., 1996, 'The mineralization and fate of nitrogen following the incorporation of grass and grass/clover swards', in D. Younie (ed.), *Legumes in Sustainable Farming Systems,* Reading: British Grassland Society, 133–4.

De Laune, R.D., Patrick, W.H., Lindau, C.W. and Smith, C.J., 1990, 'Nitrous oxide and methane emissions from Gulf Coast wetlands', in A.F. Bouwman (ed.), *Soils and the Greenhouse Effect,* New York: J. Wiley, 497–502.

De Menocal, P., Ortiz, J., Gilderson, T. and Sarnthein, M., 2000, 'Coherent high- and low-latitude climate variability during the Holocene warm period', *Science,* 288, 2198–202.

De Ploey, J. and Gabriels, D., 1980, 'Measuring soil loss and experimental studies', in M.J. Kirkby and R.P.C. Morgan (eds), *Soil Erosion,* Chichester: John Wiley, 63–108.

De Souza, M.P., Chu, D., Zhao, M., Zayed, A.M., Ruzin, S.E., Schichnes, D. and Terry, N., 1999, 'Rhizosphere bacteria enhance selenium accumulation and volatilization by Indian Mustard', *Plant Physiology,* 119, 565–73.

De Varennes, A. and Torres, M.O., 1999, 'Remediation of a long-term copper-contaminated soil using a polyacrylate polymer', *Soil Use and Management,* 15, 230–2.

Dean, W.E. and Gorham, E., 1998, 'Magnitude and significance of carbon burial in lakes, reservoirs and peatlands', *Geology,* 26, 535–8.

Dekker, L.W. and de Weerd, M.D., 1973, 'The value of soil survey for archaeology', *Geoderma,* 10, 169–78.

Dekker, L.W. and Ritsema, C.J., 1997, 'Effect of maize canopy and water repellency on moisture patterns in a Dutch black plaggen soil', *Plant and Soil,* 195, 339–50.

Del Campillo, M.C., Van Der Zee, S.E.A.T.M. and Torrent, J., 1999, 'Modelling long-term phosphorus leaching and changes in phosphorus fertility in excessively fertilized acid sandy soils', *European Journal of Soil Science*, 50, 391–9.

Department for Environment, Food and Rural Affairs (DEFRA), 2002, 'Ammonia in the UK', London: DEFRA Publications. Also available at http://www.defra.gov.uk/environment/airquality/ammonia (accessed 18 February 2004).

Derenne, S. and Largeau, C., 2001, 'A review of some important families of refractory macromolecules: composition, origin and fate in soils and sediments', *Soil Science*, 166, 833–47.

Dexter, A.R., 1986, 'Model experiments on the behaviour of roots at the interface between a tilled seed-bed and a compacted subsoil. II. Entry of pea and wheat roots into sub-soil cracks', *Plant and Soil,* 95, 135–47.

Di, H.J. and Cameron, K.C., 2002, 'Nitrate leaching in temperate agroecosystems: sources, factors and mitigating strategies', *Nutrient Cycling in Agroecosystems,* 64, 237–56.

Dick, W.A., Blevins, R.L., Frye, W.W., Peters, S.E., Christensen, D.R., Pierce, F.J. and Vitosk, M.L., 1998, 'Impacts of agricultural management practices on carbon sequestration in forest-derived soils of the eastern Corn Belt', *Soil & Tillage Research*, 47, 235–44.

Dick, W.A., McCoy, E.L., Edwards, W.M. and Lal, R., 1991, 'Continuous application of no-tillage to Ohio soils', *Agronomy Journal,* 83, 65–73.

Dixon, R.K., Brown, S., Houghton, R.A., Solomon, A.M., Trexler, M.C. and Wisniewski, J., 1994, 'Carbon pools and flux of global forest ecosystems', *Science,* 263, 185–90.

Djodjic, F., Bergström, L. and Ulén, B., 2002, 'Phosphorus losses from a structured clay soil in relation to tillage practices', *Soil Use and Management,* 18, 79–83.

Douglas, B.C., 1997, 'Global sea rise: a redetermination', *Surveys in Geophysics,* 18, 279–92.

Douglas, C.J., Jr, Allmaras, R.R., Rasmussen, P.E., Ramig, R.E. and Roager, N.C., Jr, 1980, 'Wheat straw composition and placement effects on decomposition in dry land agriculture of the Pacific Northwest', *Soil Science Society of America Journal,* 44, 833–7.

Douglas, J.T., Aitken, M.N. and Smith, C.A., 2003, 'Effects of five non-agricultural organic wastes on soil composition, and on the yield and nitrogen recovery of Italian ryegrass', *Soil Use and Management*, 19, 135–8.

Douglas, J.T., Koppi, A.J. and Crawford, C.E., 1998, 'Structural improvement in a grassland soil after changes to wheel-traffic systems to avoid soil compaction', *Soil Use and Management*, 14, 14–18.

Downing, T.E., Barrow, E.M., Brooks, R.J., Butterfield, R.E., Carter, T.R., Harrison, P.A., Hulme, M., Oleson, J.E., Porter, J.R., Schellberg, J., Semenov, M.A., Vinther, F.P., Wheeler, T.R. and Wolf, J., 2000, 'Quantification of uncertainty in climate change impact assessment', in T.E. Downing, P.A. Harrison, R.E. Butterfield and K.G. Lonsdale (eds), *Climate Change, Climatic Variability and Agriculture in Europe: an Integrated Assessment*, Oxford: Environmental Change Institute Research Report 21, 415–34.

Dubey, S.K., 2003, 'Spatio-kinetic variation of methane oxidizing bacteria in paddy soil at mid-tillering: effect of N-fertilizers', *Nutrient Cycling in Agroecosystems*, 65, 53–9.

Dunker, R.E. and Jansen, I.J., 1987, 'Corn and soybean response to topsoil and rooting medium replacement after surface mining', in D.H. Graves (ed.), *Proceedings of the 1987 National Symposium on Mining, Hydrology, Sedimentology, and Reclamation*, Lexington KT: University of Kentucky, 83–9.

Dürr, C., Aubertot, J.N., Richard, G., Dubrulle, P., Duval, Y. and Boiffin, J., 2001, 'SIMPLE: a model for simulation of plant emergence predicting the effects of soil tillage and sowing operations', *Soil Science Society of America Journal*, 65, 414–23.

Dushenkov, V., Kumar, P.B.A.N., Motto, H. and Raskin, I., 1995, 'Rhizofiltration: the use of plants to remove heavy metals from aqueous streams', *Environmental Science and Technology*, 29, 1239–45.

Dykhuisen, R.S., Frazer, R., Duncan, C., Smith, C.C., Golden, M., Benjamin, N. and Leifert, C., 1996, 'Antimicrobial effect of acidified nitrite on gut pathogens; importance of dietary nitrate in host defence', *Antimicrobial Agents and Chemotherapy*, 40, 1422–5.

Easterling, D.R., Evans, J.L., Groisman, P.Y., Karl, T.R., Kunkel, K.E. and Ambenje, P., 2000, 'Observed variability and trends in extreme climate events: a brief review', *Bulletin of the American Meteorological Society*, 81, 417–25.

Ebbs, S., Kochian, L., Lasat, L., Pence, N. and Jiang, T., 2000, 'An integrated investigation of the phytoremediation of heavy metal and radionuclide contaminated soils: from the laboratory to the field', in D.L. Wise, D.J. Trantolo, E.J. Cichon, H.I. Inyang and U. Stottmeister (eds), *Bioremediation of Contaminated Soils*, New York: Marcel Dekker, 745–69.

Edmeades, D.C., 2003, 'The long-term effect of manures and fertilisers on soil productivity and quality: a review', *Nutrient Cycling in Agroecosystems*, 66, 165–80.

Edwards, W.M., Norton, L.D. and Redmond, C.E., 1988, 'Characterizing macropores that affect infiltration into non-tilled soil', *Soil Science Society of America Journal,* 52, 483–7.

Effland, W.R. and Pouyat, R.V., 1997, 'The genesis, classification and mapping of soils in urban areas', *Urban Ecosystems,* 1, 217–18.

Ehlers, J., 1996, *Quaternary and Glacial Geology,* Chichester: J. Wiley & Sons.

Elliott, L.F. and Stott, D.E., 1997, 'Influence of no-till cropping systems on microbial relationships', *Advances in Agronomy*, 60, 121–47.

Elliott, L.F., McCalla, T.M. and Waiss, A., Jr, 1978, 'Phytotoxicity associated with residue management', in W.R. Oschwald (ed.), *Crop Residue Management Systems,* Madison WI: American Society of Agronomy Special Publication, 31, 131–46.

Elliott, W.P., 1995, 'On detecting long-term changes in atmospheric moisture', *Climatic Change,* 31, 349–67.

Ellis, S. and Mellor, A., 1995, *Soils and Environment,* London: Routledge.

El-Swaify, S.A., Arunin, S.S. and Abrol, I.P., 1983, 'Soil salinization: development of salt-affected soils', in R.A. Carpenter (ed.), *Natural Systems for Development,* London: Macmillan, 162–228.

Epstein, E., Alpert, J.E. and Calvert, C.C., 1978, 'Alternative uses of excess crop residues', in W.R. Oschwald (ed.), *Crop Residue Management Systems*, American Society of Agronomy Special Publication 31, Madison WI: American Society of Agronomy, 219–29.

Eriksen, J. and Kristensen, K., 2001, 'Nutrient excretion by outdoor pigs: a case study of distribution, utilization and potential for environmental impact', *Soil Use and Management,* 17, 21–9.

Eriksson, H., 1991, 'Sources and sinks of carbon dioxide in Sweden', *Ambio,* 20, 146–50.

Fairchild, M.A., Coyne, M.S., Grove, J.H. and Thom, W.O., 1999, 'Denitrifying bacteria stratify above fragipans', *Soil Science*, 164, 190–6.

Falkowski, P.G., 1997, 'Evolution of the nitrogen cycle and its influence on the biological sequestration of CO_2 in the ocean', *Nature,* 387, 272–5.

Falloon, P. and Smith P., 2002, 'Simulating SOC changes in long-term experiments with RothC and CENTURY: model evaluation for a regional scale application', *Soil Use and Management,* 18, 101–11.

Favis-Mortlock, D.T. and Guerra, A.J.T., 1999, 'The implications of general circulation model estimates of rainfall for future erosion: a case study from Brazil', *Catena,* 37, 329–54.

Fearnehough, W., Fullen, M.A., Mitchell, D.J., Trueman, I.C. and Zhang Jixian, 1998, 'Aeolian deposition and its effect on soil and vegetation changes on stabilised desert dunes in northern China', *Geomorphology,* 23, 171–82.

Ferrier, R.C., Edwards, A.C., Dutch, J., Wolstenholme, R. and Mitchell, D.S., 1996, 'Sewage sludge as a fertilizer of pole stage forests: short-term hydrochemical fluxes and foliar response', *Soil Use and Management,* 12, 1–7.

Field, C.B., Jackson, R.B. and Mooney, H.A., 1995, 'Stomatal responses to increased CO_2: implications from the plant to the global scale', *Plant, Cell and Environment,* 18, 1214–25.

Fiès, J.C. and Braund, A., 1998, 'Particle packing and organisation of the textural porosity in clay-silt-sand mixtures', *European Journal of Soil Science,* 49, 557–68.

Fiès, J.C. and Stengel, P., 1981, 'Textural density of natural soils. I. Method of measurements', *Agronomie,* 1, 651–8.

FitzPatrick, E.A., 1956, 'An indurated soil horizon formed by permafrost', *Journal of Soil Science,* 7, 248–54.

FitzPatrick, E.A., 1986, *An Introduction to Soil Science,* 2nd edn, London: Longman.

FitzPatrick, R.W., Merry, R.H., Williams, J., White, I., Bowman, G. and Taylor, G., 1998, *Acid Sulphate Soil Assessment: Coastal, Inland and Minespoil Conditions,* National Land and Water Resources Audit Methods Paper, Canberra: National Land and Water Resources Department.

Food and Agriculture Organization/United Nations Educational, Scientific and Cultural Organization (FAO/UNESCO), 1974, *Soil Map of the World* (1:5,000,000), Vol. 1 (Legend) and Vols II–X (Maps). Paris: UNESCO.

Forman, D., Al-Dabbagh, A. and Doll, E.C., 1985, 'Nitrate, nitrite and gastric cancer in Great Britain', *Nature,* 313, 620–5.

Foster, I.D.L., Fullen, M.A., Brandsma, R.T. and Chapman, A.S., 2000, 'Drip-screen rainfall simulators for hydro- and pedo-geomorphological research: the Coventry experience', *Earth Surface Processes and Landforms,* 25, 691–707.

Fournier, F., 1960, *Climat et Érosion: La Relation entre l'erosion du Sol par l'eau et les Precipitations Atmosphériques,* Paris: Presses Universitaires de France.

Francis, G.S., Cameron, K.C. and Swift, R.S., 1987, 'Soil physical conditions after six years of direct drilling or conventional cultivation on a silt loam soil in New Zealand', *Australian Journal of Soil Research,* 25, 517–29.

Freney, J.R., 1997, 'Emission of nitrous oxide from soils used for agriculture', *Nutrient Cycling in Agroecosystems,* 49, 1–6.

Fullen, M.A., 1985, 'Compaction, hydrological processes and soil erosion on loamy sands in east Shropshire, England', *Soil & Tillage Research,* 6, 17–29.

Fullen, M.A., 2003, 'Soil erosion and conservation in northern Europe', *Progress in Physical Geography,* 27, 331–58.

Fullen, M.A. and Mitchell, D.J., 1991, 'Taming the shamo dragon', *The Geographical Magazine,* 63, 26–9.

Fullen, M.A. and Mitchell, D.J., 1994, 'Desertification and reclamation in north-central China', *Ambio (The Journal of the Royal Swedish Academy of Sciences)*, 23, 131–5.

Fullen, M.A., Fearnehough, W., Mitchell, D.J. and Trueman, I.C., 1995, 'Desert reclamation using Yellow River irrigation water in Ningxia, China', *Soil Use and Management,* 11, 77–83.

Fullen, M.A., Mitchell, D.J., Barton, A.P., Hocking, T.J., Liu Liguang, Wu Bo Zhi, Zheng Yi and Xia Zheng Yuan, 1999, 'Soil erosion and conservation in Yunnan Province, China', *Ambio (The Journal of the Royal Swedish Academy of Sciences),* 28, 125–9.

Funnell, B.M., 1995, 'Global sea-level and the (pen)insularity of late Cenozoic Britain', in R.C. Preece (ed.), *Island Britain: a Quaternary Perspective,* London: Geological Society of London Special Publication, 96, 3–13.

Furukawa, Y. and Inubushi, K., 2002, 'Feasible suppression technique of methane emission from paddy soil by iron amendment', *Nutrient Cycling in Agroecosystems,* 64, 193–202.

Ganeshram, R.S., Pedersen, T.F., Calvert, S.E. and François, R., 2002, 'Reduced nitrogen fixation in the glacial ocean inferred from changes in marine nitrogen and phosphorus inventories', *Nature,* 415, 156–9.

Ganopolski, A. and Rahmstorf, S., 2001, 'Rapid changes of glacial climate simulated in a coupled climate model', *Nature,* 409, 153–8.

Garland, J.L., 1996, 'Analytical approaches to the characterisation of samples of microbial communities using patterns of potential C source utilisation', *Soil Biology and Biochemistry*, 28, 213–21.

Garland, J.L. and Mills, A.L., 1991, 'Classification and characterization of heterotrophic microbial communities on the basis of patterns of community-level sole-carbon-source utilization', *Applied and Environmental Microbiology*, 57, 2351–9.

Garten, C.T. and Wullschleger, S.D., 1999, 'Soil carbon inventories under a bioenergy crop (Switchgrass): measurement limitations', *Journal of Environmental Quality*, 28, 1359–65.

Gerrard, J., 1992, *Soil Geomorphology: An Integration of Pedology and Geomorphology,* London: Chapman and Hall.

Gianfreda, L. and Bollag, J.M., 1996, 'Influence of natural and anthropogenic factors on enzyme activity in soils', in G. Stotzky and J.M. Bollag (eds), *Soil Biochemistry*, Volume 9, New York: Marcel Dekker, 123–93.

Gilbert, O.L., 1989, 'Soils in urban areas', in O.L. Gilbert (ed.), *The Ecology of Urban Habitats,* London: Chapman and Hall, 41–54.

Gildon, A. and Rimmer, D.L., 1993, 'The use of soil in colliery spoil reclamation', *Soil Use and Management,* 9, 148–52.

Glendining, M.J., Powlson, D.S., Poulton, P.R., Bradbury, N.J., Palazzo, D. and Li, X., 1996, 'The effects of long-term applications of inorganic nitrogen fertilizer on soil nitrogen in the Broadbalk wheat experiment', *Journal of Agricultural Science, Cambridge*, 127, 347–63.

Glentworth, R., 1944, 'Studies on the soils developed on basic igneous rocks in central Aberdeenshire', *Transactions of the Royal Society of Edinburgh,* 61, 149–70.

Goenadi, D.H., Siswanto and Sugiarto, Y., 2000, 'Bioactivation of poorly soluble phosphate rocks with a phosphorus-solubilizing fungus', *Soil Science Society of America Journal,* 64, 927–32.

Golchin, A., Baldock, J.A. and Oades, J.M., 1998, 'A model linking organic matter decomposition, chemistry and aggregate dynamics', in R. Lal, J.M. Kimble, R.F. Follett and B.A. Stewart (eds), *Soil Processes and the Carbon Cycle*, Boca Raton FL: CRC Press Inc., 245–66.

Gooddy, D.C., Hughes, A.G., Williams, A.T., Armstrong, A.C., Nicholson, R.J. and Williams, J.R., 2001, 'Field and modelling studies to assess the risk to UK groundwater from earth-based stores for livestock manure', *Soil Use and Management,* 17, 128–37.

Goossens, R., Ghabour, T.K., Ongena, T. and Gad, A., 1994, 'Waterlogging and soil salinity in the newly reclaimed areas of the western Nile Delta of Egypt', in A.C. Millington and K. Pye (eds), *Effects of Environmental Change in Drylands,* Chichester: J. Wiley, 365–77.

Goss, M.J., Carvalho, M.J.G.P.R., Cosimini, V. and Fearnhead, M.L., 1992, 'An approach to the identification of potentially toxic concentration of manganese in soils', *Soil Use and Management,* 8, 40–4.

Goss, M.J., Howse, K.R., Lane, P.W., Christian, D.G. and Harris, G.L., 1993, 'Losses of nitrate-nitrogen in water draining from under autumn-sown crops established by direct drilling or mouldboard ploughing', *Journal of Soil Science,* 44, 35–48.

Goudie, A.S., 1990, 'Desert reclamation', in A.S. Goudie (ed.), *Techniques for Desert Reclamation,* Chichester: J. Wiley, 1–33.

Goudie, A.S., 1994, 'Deserts in a warmer world', in A.C. Millington and K. Pye (eds), *Effects of Environmental Change in Drylands,* Chichester: J. Wiley, 1–24.

Goulding, K.W.T., 2000, 'Nitrate leaching from arable and horticultural land', *Soil Use and Management,* 16, 151–4.

Goulding, K.W.T., McGrath, S.P. and Johnston, A.E., 1989, 'Predicting the lime requirement of soils under permanent grassland and arable crops', *Soil Use and Management*, 5, 54–8.

Goulding, K.W.T., Poulton, P.R., Thomas, V.H. and Williams, R.J.B., 1986, 'Atmospheric deposition at Rothamsted, Saxmundham and Woburn Experimental Stations, England, 1969–1984', *Water, Air and Soil Pollution,* 29, 27–49.

Goulding, K.W.T., Poulton, P.R., Webster, C.P. and Howe, M.T., 2000, 'Nitrate leaching from the Broadbalk Wheat Experiment, Rothamsted, UK, as influenced by fertilizer and manure inputs and the weather', *Soil Use and Management,* 16, 244–50.

Goulding, K.W.T., Bailey, N.J., Bradbury, N.J., Hargreaves, P., Howe, M., Murphy, D.V., Poulton, P.R. and Willison, T.W., 1998a, 'Nitrogen deposition and its contribution to nitrogen cycling and associated soil processes', *New Phytologist,* 139, 49–58.

Goulding, K.W.T., Bailey, N.J. and Bradbury, N.J., 1998b, 'A modelling study of nitrogen deposited to arable land from the atmosphere and its contribution to nitrate leaching', *Soil Use and Management,* 14, 70–7.

Govers, G., 1985, 'Selectivity and transport capacity of thin flows in relation to rill erosion', *Catena,* 12, 35–46.

Govers, G., Vandaele, K., Desmet, P., Poesen, J. and Bunte, K., 1994, 'The role of tillage in soil redistribution on hillslopes', *European Journal of Soil Science,* 45, 469–78.

Goyal, M.R., Leland, O.D. and Carpenter, T.G., 1982, 'Analytical prediction of seedling emergence force', *Transactions of the American Society of Agricultural Engineers,* 25, 38–41.

Greacen, E.L., 1981, *Soil Water Assessment by the Neutron Method,* Canberra: CSIRO.

Gregory, J.M., 1982, 'Soil cover prediction with various amounts and types of crop residue', *Transactions of the American Society of Agricultural Engineers,* 25, 1333–7.

Grelle, C., Fabre, M-C., Leprêtre, A. and Decamps, M., 2000, 'Myriapod and isopod communities in soils contaminated by heavy metals in northern France', *European Journal of Soil Science,* 51, 425–33.

Grieve, I.C., 1979, *Soil Aggregate Stability Tests for the Geomorphologist,* British Geomorphological Research Group (BGRG) Technical Bulletin no. 25, Norwich: Geoabstracts.

Grime, J.P., 1973, 'Competitive exclusion in herbaceous vegetation', *Nature,* 242, 344–7.

Groenevelt, P.H. and Grunthal, P.E., 1998, 'Utilisation of crumb rubber as a soil amendment for sports turf', *Soil & Tillage Research,* 47, 169–72.

Groisman, P.Y., Karl, T.R., Knight, R.W. and Stenchikov, G.L., 1994, 'Changes of snow cover, temperature, and radiative heat balance over the Northern Hemisphere', *Journal of Climate,* 7, 1633–56.

Grove, A.T., 1977, 'Desertification', *Progress in Physical Geography,* 1, 296–310.

Guerra, A.J.T., 1995, 'Catastrophic events in Petrópolis City (Rio de Janeiro State), between 1940 and 1990', *GeoJournal*, 37, 349–54.

Guerra, A.J.T. and Favis-Mortlock, D., 1998, 'Land degradation in Brazil', *Geography Review,* 12, 18–23.

Gupta, P.K., Sharma, C., Bhattacharya, S. and Mitra, A.P., 2002, 'Scientific basis for establishing country greenhouse gas estimates for rice-based agriculture: an Indian case study', *Nutrient Cycling in Agroecosystems,* 64, 19–31.

Gysi, M., 2001, 'Compaction of a Eutric Cambisol under heavy wheel traffic in Switzerland: field data and a critical state soil mechanics model approach', *Soil & Tillage Research,* 61, 133–42.

Gysi, M., Klubertanz, G. and Vulliet, L., 2000, 'Compaction of a Eutric Cambisol under heavy wheel traffic in Switzerland – field data and modelling, *Soil & Tillage Research,* 56, 117–29.

Haeberli, W. and Burn, C.R., 2002, 'Natural hazards in forests: glacier and permafrost effects as related to climate change', in R.C. Sidle (ed.), *Environmental Changes and Geomorphic Hazards in Forests,* Wallingford: CABI Publishing, 167–202.

Haigh, M.J., 1979, 'Ground retreat and slope evolution on plateau-type colliery spoil mauls at Bleenavon, Gwent', *Transactions of the Institute of British Geographers*, 4, 321–8.

Haigh, M.J., 1988, 'Slope evolution on coal-mine disturbed land', in A.S. Balasubramanian, S. Chanda, D.T. Bergado, and Prinya Natalaya (eds), *Environmental Geotechnics and Problematic Soils and Rocks,* Rotterdam: A.A. Balkema, 3–13.

Haigh, M.J., 1992, 'Degradation of "reclaimed" lands previously disturbed by coal mining in Wales: causes and remedies', *Land Degradation and Rehabilitation*, 3, 169–80.

Haigh, M.J. and Sansom, B., 1999, 'Soil compaction, runoff and erosion on reclaimed coal-lands (UK)', *International Journal of Surface Mining, Reclamation and Environment*, 13, 135–46.

Håkansson, I., 1985, 'Swedish experiments on subsoil compaction by vehicles with high axle load', *Soil Use and Management,* 1, 113–16.

Håkansson, I., Steinberg, M. and Rydberg, T., 1998, 'Long-term experiments with different depths of mouldboard ploughing in Sweden', *Soil & Tillage Research,* 46, 209–23.

Hakimata, T., Matsumoto, N., Ikeda, H. and Nakane, K., 1997, 'Do plant and soil systems contribute to global carbon cycling as a sink of CO_2?', *Nutrient Cycling in Agroecosystems,* 49, 287–93.

Hall, D.G.M., Reeve, M.J., Thomasson, A.J. and Wright, V.F., 1977, *Water Retention, Porosity and Density of Field Soils,* Soil Survey Technical Monograph, 9, Harpenden: Rothamsted Experimental Station.

Hamaide, B. and Boland, J.J., 2000, 'Benefits, costs and cooperation in greenhouse gas abatement', *Climatic Change,* 47, 239–58.

Hammer, D. and Keller, C., 2003, 'Phytoextraction of Cd and Zn with *Thlaspi caerulescens* in field trials', *Soil Use and Management,* 19, 144–9.

Han, F.X., Kingery, W.L., Selim, H.M. and Gerard, P.D., 2000, 'Accumulation of heavy metals in a long-term poultry waste-amended soil', *Soil Science,* 165, 260–8.

Hansen, B., Turrell, W.R. and Østerhus, S., 2001, 'Decreasing overflow from the Nordic Seas into the Atlantic Ocean through the Faroe Bank Channel since 1950', *Nature,* 411, 927–30.

Hansen, D., Duda, P.J., Zayed, A.M. and Terry, N., 1998, 'Selenium removal by constructed wetlands: role of biological volatilisation', *Environmental Science and Technology,* 32, 591–7.

Harden, J.W., Sharpe, J.M., Parton, W.J., Ojima, D.S., Fries, T.L., Huntington, T.G. and Dabney, S.M., 1999, 'Dynamic replacement and loss of soil carbon on eroding cropland', *Global Biogeochemical Cycles,* 13, 885–901.

Hare, F.K., 1966, *The Restless Atmosphere,* 4th edn, London: Hutchinson University Library.

Hargreaves, P., Leidi, A., Grubb, H.J., Howe, M.T. and Mugglestone, M.A., 2000, 'Local and seasonal variations in atmospheric nitrogen dioxide levels at Rothamsted, UK, and relationships with meteorological conditions', *Atmospheric Environment*, 34, 843–53.

Hargrove, W.L. and Thomas, G.W., 1984, 'Extraction of aluminium from aluminium-organic matter in relation to titratable acidity', *Soil Science Society of America Journal*, 48, 1458–60.

Harris, G.L., Howse, K.R. and Pepper, T.J., 1993, 'Effects of moling and cultivation on soil-water and runoff from a drained clay soil', *Agricultural Water Management*, 23, 161–80.

Harris, J.A. and Birch, P., 1989, 'Soil microbial activity in opencast coal mine restorations', *Soil Use and Management,* 5, 155–60.

Harvey, L.D.D., 2000, *Global Warming. The Hard Science,* Harlow: Pearson Education Ltd.

Harvey, L.D.D. and Huang, Z., 1995, 'Evaluation of the potential impact of methane clathrate destabilization on future global warming', *Journal of Geophysical Research,* 100, 2905–26.

Hatcher, P.G., Dria, K.J., Kim, S. and Frazier, S.W., 2001, 'Modern analytical studies of humic substances', *Soil Science*, 166, 770–94.

Hayes, M.H.B. and Clapp, C.E., 2001, 'Humic substances: considerations of composition, aspects of structure, and environmental influences', *Soil Science*, 166, 723–37.

Haynes, R.J., 1997, 'Fate and recovery of ^{15}N derived from grass/clover residues when incorporated into a soil and cropped with spring or winter wheat for two succeeding seasons', *Biology and Fertility of Soils*, 25, 130–5.

Haynes, R.J. and Swift, R.S., 1990, 'Stability of soil aggregates in relation to organic constituents and soil water content', *Journal of Soil Science*, 41, 73–84.

Haynes, R.J., Swift, R.S. and Stephen, R.C., 1991, 'Influence of mixed cropping pasture–arable rotations on organic matter content, water stable aggregation and clod porosity in a group of soils', *Soil & Tillage Research*, 19, 77–87.

Hays, J.D., Imbrie, J. and Shackleton, N.J., 1976, 'Variations in the earth's orbit: pacemaker of the ice ages', *Science*, 194, 1121–32.

He, D.H., Liao, X.L., Xing, T.X., Zhou, W.J., Fang, Y.J. and He, L.H., 1995, 'The fate of nitrogen from ^{15}N labelled straw and green manure in soil crop domestic animal system', *Soil Science*, 158, 65–73.

Heathwaite, A.L., Fraser, A.I., Johnes, P.J., Hutchins, M., Lord, E. and Butterfield, D., 2003, 'The Phosphorus Indicators Tool: a simple model of diffuse P loss from agricultural land to water', *Soil Use and Management*, 19, 1–11.

Heckrath, G., Brookes, P.C., Poulton, P.R. and Goulding, K.W.T., 1995, 'Phosphorus leaching from soils containing different phosphorus concentrations in the Broadbalk Experiment', *Journal of Environmental Quality*, 24, 904–10.

Hempfling, A., Schulten, H. and Horn, R., 1990, 'Relevance of humus composition for the physical/mechanical stability of agricultural soils: a study by direct pyrolysis-mass spectrometry', *Journal of Analytical and Applied Pyrolysis*, 17, 275–81.

Henderson, W.C. and Farr, E., 1992, 'Field drainage in temperate climates', in P. Smart, and J.G. Herbertson, (eds), *Drainage Design*, Glasgow: Blackie & Sons, 61–89.

Hesketh, N. and Brookes, P.C., 2000, 'Development of an indicator for risk of phosphorus leaching', *Journal of Environmental Quality*, 29, 105–10.

Higgitt, D.L., Walling, D.E. and Haigh, M.J., 1994, 'Estimating rates of ground retreat on mining spoils using caesium-137', *Applied Geography*, 14, 294–307.

Hills, R.C., 1970, 'The determination of the infiltration capacity of field soils using the cylinder infiltrometer', *British Geomorphological Research Group Technical Bulletin*, 3, Norwich: Geoabstracts.

Hinnov, L.A., Schulz, M. and Yiou, P., 2002, 'Interhemispheric space-time attributes of the Dansgaard-Oeschger oscillations between 100 and 0 ka', *Quaternary Science Reviews*, 21, 1213–28.

Hodgson, J.M. (ed.), 1974, *Soil Survey Field Handbook*, Soil Survey Technical Monograph 5, Harpenden: Rothamsted Experimental Station.

Hodgson, J.M., 1976, *Soil Survey Field Handbook*, Describing and sampling soil profiles, Soil Survey Technical Monograph No. 5, Harpenden: Lawes Agricultural Trust.

Hodgson, J.M. and Thompson, D., 1985, 'Uncovering the secrets of soil', *New Scientist,* 14 November 1985, 44–7.

Hogg, W.H., 1967, *Atlas of Long-term Irrigation Needs for England and Wales,* London: Ministry of Agriculture, Fisheries and Food.

Høgh-Jensen, H., 1996, 'Symbiotic N_2 fixation in grass-clover mixtures and transfer from clovers to the accompanying grass', in J. Raupp (ed.), *Symbiotic Nitrogen Fixation in Crop Rotations with Manure Fertilization*, Darmstadt: Institute for Biodynamic Research, 7–31.

Holden, J. and Burt, T.P., 2002, 'Piping and pipeflow in a deep peat catchment', *Catena*, 48, 163–99.

Holford, I.C.R. and Crocker, G.J., 1997, 'A comparison of chickpeas and pasture legumes for sustaining yields and nitrogen status of subsequent wheat', *Australian Journal of Agricultural Research*, 48, 305–15.

Hons, F.M. and Hossner, L.R., 1980, 'Soil nitrogen relationships in spoil material generated by the surface mining of lignite coal', *Soil Science,* 129, 222–8.

Horgan, G.W., 1998, 'Mathematical morphology for analysing soil structure from images', *European Journal of Soil Science,* 49, 161–74.

Hortensius, D. and Nortcliff, S., 1991, 'International standardization of soil quality measurement procedures for the purpose of soil protection', *Soil Use and Management*, 7, 163–6.

Hou, A.X. and Tsuruta, A., 2003, 'Nitrous oxide and nitric oxide fluxes from an upland field in Japan: effect of urea type, placement, and crop residues', *Nutrient Cycling in Agroecosystems,* 65, 191–200.

Howes, B.L., Dacey, J.W.H. and Teal, J.M., 1985, 'Annual carbon mineralization and belowground production of *Spartina alterniflora* in a New England salt marsh', *Ecology* 66, 595–605.

Huang, J.W., Blaylock, M.J., Kapulnik, Y. and Ensley, B.D., 1998, 'Phytoremediation of uranium-contaminated soils: role of organic acids in triggering uranium hyperaccumulation in plants', *Environmental Science and Technology,* 32, 2004–2008.

Huber, D.M., 1990, 'Fertilizers and soil-borne diseases', *Soil Use and Management,* 6, 168–73.

Hurt, R.D., 1981, *The Dust Bowl: An Agricultural and Social History,* Chicago IL: Nelson-Hall.

Hutchings, T.R., Moffat, A.J. and French, C.J., 2002, 'Soil compaction under timber harvesting machinery: a preliminary report on the role of brash mats in its prevention', *Soil Use and Management*, 18, 34–8.

Hutchinson, H.B. and Richards, E.H., 1921, 'Artificial farmyard manure', *Journal of the Ministry of Agriculture*, 28, 567–72.

Hutchinson, J.N., 1980, 'The record of peat wastage in the East Anglian Fenlands at Holme Post, 1848–1978 A.D.', *Journal of Ecology*, 68, 229–49.

Hütsch, B.W., 2001, 'Methane oxidation in non-flooded soils as affected by crop production', *European Journal of Agronomy,* 14, 237–60.

Hütsch, B.W., Webster, S.P. and Powlson, D.S., 1994, 'Methane oxidation in soil as affected by land use, soil pH and N fertilization', *Soil Biology and Biochemistry,* 26, 1613–22.

Idso, K.E. and Idso, S.B., 1994, 'Plant responses to atmospheric CO_2 enrichment in the face of environmental constraints: a review of the past 10 years' research', *Agricultural and Forest Meteorology,* 69, 153–203.

Inter-Department Committee on the Redevelopment of Contaminated Land, 1987, *Guidance on the Assessment and Redevelopment of Contaminated Land, ICRCL 59/83,* London: Department of the Environment.

Intergovernmental Panel on Climate Change (IPCC), 1996, *Climate Change 1995: The Science of Climate Change,* Cambridge: Cambridge University Press.

Intergovernmental Panel on Climate Change (IPCC), 2001, *Climate Change 2001: The Scientific Basis,* Cambridge: Cambridge University Press. Also available at: http://www.ipcc.ch (accessed 4 March 2004).

Ismail, I., Blevins, R.L. and Fry, W.W., 1994, 'Long-term no-tillage effects on soil properties and continuous corn yields', *Soil Science Society of America Journal,* 58, 193–8.

Iversen, B.V., Schjønning, P., Poulsen, T.G. and Moldrup, P., 2001, 'In situ, on-site and laboratory measurements of soil air permeability: boundary conditions and measurement scale', *Soil Science,* 166, 97–106.

Jansen, I.J. and Melsted, S.W., 1988, 'Land shaping and soil construction', in L.R. Hossner (ed.), *Reclamation of Surface-mined Lands,* Boca Raton: CRC Press, 125–36.

Jenkinson, D.S., 1985, 'How straw incorporation affects the nitrogen cycle', in J. Hardcastle (ed.), *Straw, Soils and Science,* London: Agricultural and Food Research Council, 14–15.

Jenkinson, D.S., 2001, 'The impact of humans on the nitrogen cycle, with focus on temperate arable agriculture', *Plant and Soil,* 228, 3–15.

Jenkinson, D.S. and Coleman, K., 1994, 'Calculating the annual input of organic matter to soil from measurements of total organic carbon and radiocarbon', *European Journal of Soil Science,* 45, 167–74.

Jenkinson, D.S. and Ladd, J.N., 1981, 'Microbial biomass in soil: measurement and turnover', in E.A. Paul and J.N. Ladd (eds), *Soil Biochemistry, Volume 5,* New York: Marcel Dekker, 415–71.

Jenkyn, J.F., Christian, D.G., Bacon, E.T.G., Gutteridge, R.J. and Todd, A.D., 2001, 'Effects of incorporating different amounts of straw on growth, diseases and yield of consecutive crops of winter wheat grown on contrasting soil types', *Journal of Agricultural Science, Cambridge,* 136, 1–14.

Jetten, V. and de Roo, A.P.J., 2001, 'Spatial analysis of erosion and conservation measures with LISEM', in R. Harmon (ed.), *Landscape Erosion and Evolution Modelling,* Dordrecht: Kluwer Academic Press, 429–45.

Jiao, B., 1983, 'Utilization of green manure for raising soil fertility in China', *Soil Science,* 135, 65–9.

Johnson, D., Johnson, K., Ward, G.M. and Branine, M., 2000, 'Ruminants and other animals', in M. Khalil (ed.), *Atmospheric Methane: Its Role in the Global Environment,* New York: Springer-Verlag, 112–33.

Johnson, D.L. and Lewis, L.A., 1995, *Land Degradation: Creation and Destruction*, Cambridge MA: Blackwell.

Johnson, D.W., 1992, 'The effects of forest management on soil carbon storage', *Water, Air and Soil Pollution,* 64, 83–120.

Johnston, A.E., 1991, 'Soil fertility and soil organic matter', in W.S. Wilson (ed.), *Advances in Soil Organic Matter Research,* Cambridge: Royal Society of Chemistry, 351–63.

Johnston, A.E., 1997, 'The value of long-term field experiments in agricultural, ecological and environmental research', *Advances in Agronomy,* 59, 291–333.

Johnston, A.E., McEwan, J., Lane, P.W., Hewitt, M.V., Poulton, P.R. and Yeoman, D.P., 1994, 'Effects of one to six year old ryegrass-clover leys on soil nitrogen and on the subsequent yields and fertilizer nitrogen requirements of the arable sequence winter wheat, potatoes, winter wheat, winter beans (*Vicia faba*) grown on a sandy loam soil', *Journal of Agricultural Science, Cambridge,* 122, 73–89.

Jones, D.L., 1999, 'Potential health risks associated with the persistence of *Escherichia coli* O157 in agricultural environments', *Soil Use and Management,* 15, 76–83.

Jones, G.H., 1993, Factors Controlling the Establishment of Species-rich Grasslands in Urban Landscape Schemes, Unpublished Ph.D. Thesis, The University of Wolverhampton, 274 pp.

Jones, G.H., Trueman, I.C. and Millett, P., 1995, 'The use of hay strewing to create species-rich grasslands (i) general principles and hay strewing versus seed mixes', *Land Contamination and Reclamation,* 3, 104–107.

Jones, P.D., New, M., Parker, D.E., Martin, S. and Rigor, I.G., 1999, 'Surface air temperature and its changes over the last 150 years', *Reviews of Geophysics,* 37, 173–99.

Kaiser, K. and Guggenberger, G., 2000, 'The role of DOM sorption to mineral surfaces in the preservation of organic matter in soils', *Organic Geochemistry,* 31, 711–25.

Kamprath, E.J., 1970, 'Exchangeable Al as a criterion for liming leached mineral soils', *Soil Science Society of America Proceedings*, 34, 252–4.

Kaspar, T.C., Erbach, D.C. and Cruse, R.M., 1990, 'Corn response to seed-row residue removal', *Soil Science Society of America Journal,* 54, 1112–17.

Keeling, C.D., 1960, 'The concentration and isotopic abundances of carbon dioxide in the atmosphere', *Tellus,* 12, 200–203.

Keigwin, L.D., Curry, W.B., Lehman, S.J. and Johnsen, S., 1994, 'The role of the deep ocean in North Atlantic climate change between 70 and 130 kyr ago', *Nature,* 371, 323–6.

Keir, R.S., 1988, 'On the late Pleistocene ocean geochemistry and circulation', *Paleoceanography,* 3, 413–45.

Kennedy, A.C. and Gewin, V.L., 1997, 'Soil microbial diversity: present and future considerations', *Soil Science,* 162, 607–17.

Kenny, G.J., Harrison, P.A. and Parry, M.L. (eds), 1993, *The Effect of Climate Change on Agricultural and Horticultural Potential in Europe,* Oxford: University of Oxford Environmental Change Unit.

Kessavalou, A., Doran, J.W., Mosier, A.R. and Drijber, R.A., 1998, 'Greenhouse gas fluxes following tillage and wetting in a wheat-fallow cropping system', *Journal of Environmental Quality,* 27, 1105–16.

Kilbertus, G., 1980, 'Étude des microhabitats contenus dans les agrégates du sol. Leur relation avec la biomass bactérienne et la taille des procaryotes présents', *Revue d'Écologie et de Biologie du Sol,* 17, 543–7.

Killham, K., 1994, *Soil Ecology,* Cambridge: Cambridge University Press.

King, J.A., 1988, 'Some physical features of soil after opencast mining', *Soil Use and Management,* 4, 23–30.

King, J.A., Smith, K.A. and Pyatt, G.D., 1986, 'Water and oxygen regimes under conifer plantations and native vegetation on upland peaty gley soils and deep peat soils', *Journal of Soil Science,* 37, 485–97.

Klass, D.L., 1976, 'Making SNG from waste and biomass', *Hydrocarbon Processing,* 55, 76–82.

Kleinman, P.J.A., Bryant, R.B., Reid, W.S., Sharpley, A.N. and Pimentel, D., 2000, 'Using soil phosphorus behaviour to identify environmental thresholds', *Soil Science,* 165, 943–50.

Klute, A., 1986, 'Water retention: laboratory methods', in A. Klute (ed.), *Methods of Soil Analysis. Part 1. Physical and Mineralogical Methods*, 2nd edn, Madison WI: American Society of Agronomy Monograph, 9, 635–62.

Klute, A. and Dirksen, C., 1986, 'Hydraulic conductivity and diffusivity; laboratory methods', in A. Klute (ed.), *Methods of Soil Analysis. Part 1. Physical and Mineralogical Methods*, 2nd edn, Madison WI: American Society of Agronomy Monograph, 9, 687–734.

Koekkoek, E.J.W. and Booltink, H., 1999, 'Neural network models to predict soil water retention', *European Journal of Soil Science,* 50, 489–95.

Körner, C. and Miglietta, F., 1994, 'Long term effects of naturally elevated CO_2 on Mediterranean grassland and forest trees', *Oecologia,* 99, 343–51.

Kosaka, H. and Tyuma, I., 1987, 'Mechanism of autocatalytic oxidation of oxyhaemoglobin by nitrite', *Environmental Health Perspectives,* 78, 147–51.

Kotak, B.G., Kenefick, S.L., Fritz, D.L., Rousseaux, C.G., Prepas, E.E. and Hrudey, S.E., 1993, 'Occurrence and toxicological evaluation of cyanobacterial toxins in Alberta lakes and farm dugouts', *Water Research,* 27, 495–506.

Kumar, K. and Goh, K.M., 2000, 'Crop residues and management practices: effects on soil quality, soil nitrogen dynamics, crop yield, and nitrogen recovery', *Advances in Agronomy,* 68, 197–319.

La Salle, J. and Shilts, W.W., 1993, 'Younger-Dryas age readvance of Laurentide ice into the Champlain Sea', *Boreas,* 22, 25–37.

Ladd, J.N., Amato, M., Grace, P.R. and Van Veen, J.A., 1995, 'Simulation of ^{14}C turnover through the microbial biomass in soils incubated with ^{14}C-labelled plant residues', *Soil Biology and Biochemistry,* 27, 777–83.

Ladd, J.N., Amato, M. and Oades, J.M., 1985, 'Decomposition of plant materials in Australian soils. III. Residual organic and microbial biomass C and N from isotope-labelled legume materials and soil organic matter, decomposing under field conditions', *Australian Journal of Soil Research,* 23, 603–11.

Laird, D., 2001, 'Nature of clay-humic complexes in an agricultural soil: II. Scanning electron microscopy analysis', *Soil Science Society of America Journal,* 65, 1419–25.

Lal, R., 1990, 'Water erosion and conservation: an assessment of the water erosion problem and the techniques available for soil conservation', in A.S. Goudie (ed.), *Techniques for Desert Reclamation,* Chichester: J. Wiley, 161–98.

Lal, R., 1999, 'Soil management and restoration for C sequestration to mitigate the accelerated greenhouse effect', *Progress in Environmental Science,* 1, 307–26.

Lal, R., 2001, 'Potential of desertification control to sequester carbon and mitigate the greenhouse effect', *Climatic Change,* 51, 35–72.

Lal, R., Hall, G.F. and Miller, F.P., 1989, 'Soil degradation 1. Basic processes', *Land Degradation and Rehabilitation,* 1, 51–69.

Lal, R., Kimble, J., Levine, E. and Whitman, C., 1995, 'World soils and the greenhouse effect; an overview', in R. Lal, J. Kimble, E. Levine and B.A. Stewart (eds), *Soils and Global Change,* Boca Raton FL: CRC Press, 1–7.

Lamb, H.H., 1995, *Climate, History and the Modern World,* 2nd edn, London: Routledge.

Lambeck, K., Esat, T.M. and Potter, E-K., 2002, 'Links between climate and sea levels for the past three million years', *Nature,* 419, 199–206.

Langbein, W.B. and Schumm, S.A., 1958, 'Yield of sediment in relation to mean annual precipitation', *Transactions of the American Geophysical Union,* 39, 257–66.

Larney, F.J. and Fortune, R.A., 1986, 'Recompaction effects of mouldboard ploughing and seedbed cultivation on four deep loosened soils', *Soil & Tillage Research,* 8, 77–87.

Laut, P. and Gundermann, J., 1998, 'Solar cycle length hypothesis appears to support the IPCC on global warming', *Journal of Atmospheric and Solar-Terrestrial Physics,* 60, 1719–28.

Lawson, D.M., 1989, 'The principles of fertilizer use for sports turf', *Soil Use and Management,* 5, 122–7.

Lehrsch, G.A., 1998, 'Freeze-thaw cycles increase near-surface aggregate stability', *Soil Science,* 163, 63–70.

Leinweber, P., Meissner, R., Eckhardt, K.U. and Saeger, J., 1999, 'Management effects on forms of phosphorus in soil and leaching losses', *European Journal of Soil Science,* 50, 413–24.

Leinweber, P., Reuter, G. and Brozio, K., 1993, 'Cation exchange capacity of organo-mineral particle size fractions in soils from long-term experiments', *Journal of Soil Science,* 44, 111–19.

Leiros, M.C., Trasar-Cepeda, C., Seoane, S. and Gil-Sotres, F., 1999, 'Dependence of mineralization of soil organic matter on temperature and moisture', *Soil Biology and Biochemistry,* 31, 327–35.

Lelieveld, J., Crutzen, P.J. and Dentener, F.J., 1998, 'Changing concentration, lifetime and climate forcing of atmospheric methane', *Tellus,* 50B, 128–50.

Lewis, K.A., Tzilivakis, J., Skinner, J.A., Finch, J., Kaho, T. and Bardon, K.S., 1997, 'Scoring and ranking farmland conservation activities to evaluate environmental performance to encourage sustainable farming', *Sustainable Development,* 5, 71–8.

L'Hirondel, J. and L'Hirondel, J.-L., 2002, *Nitrate and Man – Toxic, Harmless or Beneficial?,* Wallingford: CABI Publishing.

Liski, J., Ilvesniemi, H., Makela, A. and Starr, M., 1998, 'Model analysis of the effects of soil age, fires and harvesting on the soil carbon storage of boreal forest soils', *European Journal of Soil Science,* 49, 406–16.

Logan, T.J., Lal, R. and Dick, W.A., 1991, 'Tillage systems and soil properties in North America', *Soil & Tillage Research,* 20, 241–70.

Loizeau, J.-L., Arbouille, D., Santiago, S. and Vernet, J.P., 1994, 'Evaluation of a wide range laser diffraction grain size analyser for use with sediments', *Sedimentology,* 41, 353–61.

Lord, E.I. and Anthony, S.G., 2000, 'MAGPIE: a modelling framework for evaluating nitrate losses at national and catchment scales', *Soil Use and Management,* 16, 167–74.

Loughran, R.J., 1989, 'The measurement of soil erosion', *Progress in Physical Geography,* 13, 216–33.

Loutre, M.F., 2003, 'Clues from MIS 11 to predict the future climate – a modelling point of view', *Earth and Planetary Science Letters,* 212, 213–24.

Lovell, C.J. and Youngs, E.G., 1984, 'A comparison of steady-state land-drainage equations', *Agricultural Water Management*, 9, 1–21.

Lowe, J.J. and Walker, M.J.C., 1997, *Reconstructing Quaternary Environments,* Harlow: Longman.

Loya, W.M., Pregitzer, K.S., Karberg, N.J., King, J.S. and Giardina, C.P., 2003, 'Reduction of soil carbon formation by tropospheric ozone under increased carbon dioxide levels', *Nature,* 425, 705–707.

Lundgren, D.G. and Silver, M., 1980, 'Ore leaching by bacteria', *Annual Review of Microbiology*, 34, 263–83.

Lupini, J.F., Skinner, A.E. and Vaughan, P.R., 1981, 'The drained residual strength of cohesive soils', *Géotechnique,* 31, 181–213.

Lyles, L. and Tatarko, J., 1986, 'Wind erosion effects on soil texture and organic matter', *Journal of Soil and Water Conservation,* 41, 191–3.

Macdonald, A.J., Poulton, P.R., Powlson, D.S. and Jenkinson, D.S., 1997, 'Effects of season, soil type and cropping on recoveries, residues and losses of ^{15}N-labelled fertilizer applied to arable crops in spring', *Journal of Agricultural Science, Cambridge,* 129, 125–54.

Macdonald, A.J., Powlson, D.S., Poulton, P.R. and Jenkinson, D.S., 1989, 'Unused fertiliser nitrogen in arable soils – its contribution to nitrate leaching', *Journal of the Science of Food and Agriculture,* 46, 407–19.

MacDonald, G.J., 1990, 'Role of methane clathrates in past and future climates', *Climatic Change,* 16, 247–81.

Maljanen, M., Liikanen, A., Silvola, J. and Martikainen, P.J., 2003, 'Methane fluxes on agricultural and forested boreal organic soils', *Soil Use and Management,* 19, 73–9.

Manabe, S. and Stouffer, R., 1997, 'Coupled ocean–atmosphere model response to freshwater input: comparison to Younger Dryas event', *Paleoceanography,* 12, 321–36.

Marinissen, J.C.Y., 1994, 'Earthworm populations and stability of soil structure in a silt loam soil of a recently reclaimed polder', *Agricultural Ecosystems and Environment,* 51, 75–87.

Markham, G., 1636, *The Inrichment of the Weald of Kent*, London: J. Harison.

Marrs, R.H., 1993, 'Soil fertility and nature conservation in Europe: theoretical considerations and practical management solutions', *Advances in Ecological Research,* 24, 241–300.

Marsden, R. and Allison, C., 1992, 'The answer's in the soil', *The Geographical Magazine (Geographical Analysis Supplement)*, 64, 1–6.

Martin, V.L., McCoy, E.L. and Dick, W.A., 1990, 'Allelopathy of crop residues influences corn seed germination and early growth', *Agronomy Journal*, 82, 555–60.

Martyn, D., 1992, *Climates of the World*, Amsterdam: Elsevier.

Maslanik, J.A., Serreze, M.C. and Barry, R.G., 1996, 'Recent decreases in Arctic summer ice cover and linkages to atmospheric circulation changes', *Geophysical Research Letters*, 23, 1677–80.

Maslin, M.A. and Thomas, E., 2003, 'Balancing the deglacial global carbon budget: the hydrate factor', *Quaternary Science Reviews*, 22, 1729–36.

Matula, S. and Kozáková, H., 1997, 'A simple pressure infiltrometer for determination of soil hydraulic properties by *in situ* infiltration measurements', *Rostlinná Výroba*, 43, 405–13.

McBride, M.B., 1994, *Environmental Chemistry of Soils*, Oxford: Oxford University Press.

McConnaughay, K.D.M., Bassow, S.L., Berntson, G.M. and Bazzaz, F.A., 1996, 'Leaf senescence and decline of end-of-season gas exchange in five temperate deciduous tree species grown in elevated CO_2 concentrations', *Global Change Biology*, 2, 25–33.

McCrea, A.R., 1999, 'Relationships between soil fertility and species-richness in created and semi-natural grassland in the English West Midlands', Unpublished Ph.D. Thesis, The University of Wolverhampton, 325 pp.

McCrea, A.R., Trueman, I.C., Fullen, M.A., Atkinson, M.D. and Besenyei, L., 2001a, 'Relationships between soil characteristics and species richness in two botanically heterogeneous created meadows in the English West Midlands', *Biological Conservation*, 97, 171–80.

McCrea, A.R., Trueman, I.C. and Fullen, M.A., 2001b, 'A comparison of the effects of four arable crops on the fertility depletion of a sandy silt loam for grassland habitat creation', *Biological Conservation*, 97, 181–7.

McDowell, R., Sharpley, A., Brookes, P. and Poulton, P., 2001, 'Relationship between soil test phosphorus and phosphorus release to solution', *Soil Science*, 166, 137–49.

McGrath, S.P., 1998, 'Phytoextraction for soil remediation', in R.R. Brooks (ed.), *Plants that Hyperaccumulate Heavy Metals*, Wallingford: CAB International, 305–11.

McGrath, S.P., Zhao, F.J. and Lombi, E., 2002, 'Phytoremediation of metals, metalloids and radionuclides', *Advances in Agronomy*, 75, 1–56.

McIlveen, R., 1986, *Basic Meteorology: A Physical Outline*, Wokingham: Van Nostrand Reinhold.

McNeil, A.M. and Wood, M., 1990, 'Fixation and transfer of nitrogen by white clover to ryegrass', *Soil Use and Management*, 6, 84–6.

McRae, S.G., 1989, 'The restoration of mineral workings in Britain – a review', *Soil Use and Management,* 5, 135–42.

McSweeney, K. and Jansen, I.J., 1984, 'Soil structure and associated rooting behavior in minesoils', *Soil Science Society of America Journal,* 48, 607–12.

McSweeney, K., Jansen, I.J., Boast, C.W. and Dunker, R.E., 1987, 'Row crop productivity of eight constructed minesoils', *Reclamation and Revegetation Research,* 6, 137–44.

Meeker, L.D., Mayewski, P.A., Twickler, M.S., Whitlow, S.I. and Meese, D., 1997, 'A 110 000 year history of change in continental biogenic emissions and related atmospheric circulation inferred from the Greenland Ice Sheet Project ice core', *Journal of Geophysical Research,* 102, 489–505.

Mele, P.M. and Carter, M.R., 1999, 'Impact of crop management factors in conservation tillage farming on earthworm density, age structure and species abundance in south-eastern Australia', *Soil & Tillage Research,* 50, 1–10.

Merino, P., Estavillo, J.M., Besga, G., Pinto, M. and González-Murua, C., 2001, 'Nitrification and denitrification derived N_2O production from a grassland soil under application of DCD and Actilith F2', *Nutrient Cycling in Agroecosystems,* 60, 9–14.

Merrill, S.A., Black, A.L. and Bauer, A., 1996, 'Conservation tillage affects root growth of dryland spring wheat under drought', *Soil Science Society of America Journal,* 60, 575–83.

Middleton, A.C., 1949, 'Clay marling: some historical notes', *Agriculture,* 56, 80–4.

Ministry of Agriculture, Fisheries and Food (MAFF), 1973, *Lime and Liming*, MAFF Reference Book 35, London: Her Majesty's Stationery Office.

Ministry of Agriculture, Fisheries and Food (MAFF), 1976, *Organic Manures,* MAFF Bulletin 210, London: Her Majesty's Stationery Office.

Ministry of Agriculture, Fisheries and Food (MAFF), 1977, *Water for Irrigation*, MAFF Reference Book 202, 2nd edn, London: Her Majesty's Stationery Office.

Ministry of Agriculture, Fisheries and Food (MAFF), 1981, *The Analysis of Agricultural Materials,* 2nd edn, MAFF Reference Book 427, London: Her Majesty's Stationery Office.

Ministry of Agriculture, Fisheries and Food (MAFF), 1982, *Irrigation*. MAFF Reference Book 138, 5th edn, London: Her Majesty's Stationery Office.

Ministry of Agriculture, Fisheries and Food (MAFF), 2000, *Fertiliser Recommendations for Agricultural and Horticultural Crops,* MAFF Reference Book 209, London: Her Majesty's Stationery Office.

Mitchell, D.J. and Fullen, M.A., 1994, 'Soil-forming processes on reclaimed desertified land in north-central China', in A.C. Millington and K. Pye (eds), *Effects of Environmental Change in Drylands,* Chichester: J. Wiley, 393–412.

Moffat, A.J. and Bending, N.A.D., 2000, 'Replacement of soil and soil-forming materials by loose tipping in reclamation to woodland', *Soil Use and Management,* 16, 75–81.

Montgomery, J.A., McCool, D.K., Busacca, A.J. and Frazier, B.E., 1999, 'Quantifying tillage translocation and deposition rates due to moldboard ploughing in the Palouse region of the Pacific Northwest, USA', *Soil & Tillage Research,* 51, 175–87.

Mooney, S.J., 2002, 'Three-dimensional visualization and quantification of soil macroporosity and water flow patterns using computed tomography', *Soil Use and Management,* 18, 142–51.

Moran, C.J. and McBratney, A.B., 1992, 'Image measurement and modeling of the two-dimensional spatial distribution of wheat straw', *Geoderma,* 53, 201–16.

Morgan, R.P.C., 1995, *Soil Erosion & Conservation,* 2nd edn, London: Longman.

Mosier, A., Delgado, J., Cochran, V., Valentine, D. and Parton, W., 1997, 'Impact of agriculture on soil consumption of atmospheric CH_4 and a comparison of CH_4 and N_2O flux in subarctic, temperate and tropical grasslands', *Nutrient Cycling in Agroecosystems,* 49, 71–3.

Mosier, A.R., 1998, 'Soil processes and global change', *Biology and Fertility of Soils,* 27, 221–9.

Mosier, A.R. and Kroeze, C., 1998, 'A new approach to estimate emissions of nitrous oxide from agriculture and its implications for the global N_2O budget', *Newsletter of International Global Atmospheric Chemistry Project of IGBP,* 12, 17–25.

Mudelse, M. and Schulz, M., 1997, 'The mid-Pleistocene climatic transition: onset of 100 ka cycle lags ice volume build up by 280 ka', *Earth and Planetary Science Letters,* 151, 117–23.

Mulholland, B. and Fullen, M.A., 1991, 'Cattle trampling and soil compaction on loamy sands', *Soil Use and Management,* 7, 189–93.

Mulvaney, R.L., Khan, S.A., Hoeft, R.G. and Brown, H.M., 2001, 'A soil organic nitrogen fraction that reduces the need for nitrogen fertilization', *Soil Science Society of America Journal,* 65, 1164–72.

Munk, W. and Wunsch, C., 1998, 'Abyssal recipes II: energetics of wind and tidal mixing', *Deep Sea Research,* 45, 1977–2010.

Muñoz-Carpena, R., Regalado, C.M., Álvarez-Benedi, J. and Bartoli, F., 2002, 'Field evaluation of the new Philip-Dunne permeameter for measuring saturated hydraulic conductivity', *Soil Science,* 167, 9–24.

Murphy, D.V., Bhogal, A., Shepherd, M.A., Goulding, K.W.T., Jarvis, S.C., Barraclough, D. and Gaunt, J.L., 1999, 'Comparison of ^{15}N labelling methods to measure gross nitrogen mineralisation', *Soil Biology and Biochemistry,* 31, 2015–24.

Murphy, D.V., Macdonald, A.J., Stockdale, E.A., Goulding, K.W.T., Fortune, S., Gaunt, J.L., Poulton, P.R., Wakefield, J.A., Webster, C.P. and Wilmer, W.S., 2000, 'Soluble organic nitrogen in agricultural soils', *Biology and Fertility of Soils,* 30, 374–87.

Mysak, L.A. and Powers, S.B., 1991, 'Greenland sea ice and salinity anomalies and interdecadal climate variability', *Climatological Bulletin,* 25, 81–91.

National Academy of Sciences, 2001, *Climate Change Science: An Analysis of Some Key Questions,* Washington DC: National Academy Press.

Neff, J.C., Townsend, A.R., Gleixner, G., Lehman, S.J., Turnbull, J. and Bowman, W.D., 2002, 'Variable effects of nitrogen additions on the stability and turnover of soil carbon', *Nature,* 419, 915–17.

Negri, M.C. and Hinchman, R.R., 2000, 'The use of plants for treatment of radionuclides', in I. Rankin and B.D. Ensley (eds), *Phytoremediation of Toxic Metals: Using Plants to Clean up the Environment,* New York: J. Wiley & Sons, 107–32.

Neill, C., Cerri, C.C., Melillo, J.M., Feigl, B.J., Steudler, P.A., Moraes, J.F.L. and Piccolo, M.C., 1998, 'Stocks and dynamics of soil carbon following deforestation for pasture in Rondonia', in R. Lal, J.M. Kimble, R.F. Follett and B.A. Stewart (eds), *Soil Processes and the Carbon Cycle,* Boca Raton FL: CRC Press, 9–28.

Nelson, D.W. and Sommers, L.E., 1996, 'Total carbon, organic carbon and organic matter', in D.L. Sparks, A.L. Page, P.A. Helmke, R.H. Loeppert, P.N. Soltanpour, M.A. Tabatabai, C.T. Johnston and M.E. Sumner (eds), *Methods of Soil Analysis Part 3. Chemical Methods,* Madison WI: Soil Science Society of America, 961–1010.

Nelson, P.N. and Oades, J.M., 1998, 'Organic matter, sodicity and soil structure', in M.E. Sumner, and R. Naidu (eds), *Sodic Soils: Distribution, Processes, Management and Environmental Consequences,* New York: Oxford University Press, 67–91.

Nero, A.V., Gadgil, A.J., Nazaroff, W.W. and Revzen, K.L., 1990, *Indoor radon and decay products; concentrations, causes and control strategies,* Report DOE/ER-0480P, Department of Energy, Office of Health and Environmental Research, Washington DC: DOE/ER.

Neue, H.U., 1997, 'Fluxes of methane from rice fields and potential for mitigation', *Soil Use and Management,* 13, 258–67.

Nguyen, M.L. and Goh, K.M., 1994, 'Sulphur cycling and its implications on sulphur fertilizer requirements of grazed grassland ecosystems', *Agriculture, Ecosystems and Environment,* 49, 173–206.

Nicks, L. and Chambers, M.F., 1995, 'Farming for metals', *Mining Engineering and Management,* 15–18 September, 1995.

Nimmo, J.R., 1997, 'Modeling structural influences on soil water retention', *Soil Science Society of America Journal,* 61, 712–19.

Noble, A.D., Randall, P.J. and James, T.R., 1995, 'Evaluation of two coal-derived organic products in ameliorating surface and subsurface soil acidity', *European Journal of Soil Science,* 46, 65–75.

Nordhaus, W. and Boyer, J., 1999, 'Requiem for Kyoto: an economic analysis of the Kyoto Protocol', *The Energy Journal,* 93–130.

Noren, A.J., Bierman, P.R., Steig, E.J., Lini, A. and Southon, J., 2002, 'Millennial-scale storminess variability in the northeastern United States during the Holocene epoch', *Nature,* 419, 821–4.

Oades, J.M., 1984, 'Soil organic matter and structural stability; mechanisms and implications for management', *Plant and Soil,* 76, 319–37.

Ochsner, T.E., Horton, R. and Ren, T., 2001, 'Simultaneous water content, air-filled porosity and bulk density measurements with thermo-time domain reflectometry', *Soil Science Society of America Journal*, 65, 1618–22.

Oerlemans, J., 1994, 'Quantifying global warming from retreat of glaciers', *Science,* 264, 243–5.

Oldeman, L.R., Hakkeling, R.T.A. and Sombroek, W.G., 1990, *World Map of the Status of Human-Induced Soil Degradation,* Wageningen (The Netherlands): International Soil Reference and Information Centre (ISRIC)/UNEP in cooperation with Winand Staring Centre-International Soil Science Society (ISSS)-FAO-ITC (The International Institute for Geo-Information Science and Earth Observation).

Oleson, J.E. and Bindi, M., 2002, 'Consequences of climate change for European agricultural productivity, land use and policy', *European Journal of Agronomy,* 16, 239–62.

Oliver, M.A., 1997, 'Soil and human health: a review', *European Journal of Soil Science*, 48, 573–92.

Oostwoud Wijdenes, D.J. and Poesen, J., 1999, 'The effect of soil moisture on the vertical movement of rock fragments by tillage', *Soil & Tillage Research,* 49, 301–12.

Oren, R., Ellsworth, D.S., Johnsen, K.H., Phillips, N., Ewers, B.E., Maier, C., Schäfer, K.V.R., McCarthy, H., Hendrey, G., McNulty, S.G. and Katul, G.G., 2001, 'Soil fertility limits carbon sequestration by forest ecosystems in a CO_2-enriched atmosphere', *Nature,* 411, 469–72.

Organisation for Economic Co-operation and Development (OECD), 1982, *Eutrophication of Waters: Monitoring, Assessment and Control*, Paris: OECD.

Óskarsson, H., Arnalds, O., Gudmundsson, J. and Gudbergsson, G., 2004, 'Organic carbon in Icelandic Andosols; geographical variation and impact of erosion', *Catena,* 56, 225–38.

Pachepsky, Y., Rawls, W.J. and Giménez, D., 2001, 'Comparison of soil water retention at field and laboratory scales', *Soil Science Society of America Journal*, 65, 460–2.

Pain, B.F., Misselbrook, T.H. and Rees, Y.J., 1994, 'Effect of nitrification inhibitor and acid addition to cattle slurry on nitrogen losses and herbage yields', *Grass and Forage Science,* 49, 209–15.

Palm, C., 1997, 'Nutrient management: combined use of organic and inorganic fertilizers for increasing soil phosphorus availability', in M.J. Swift (ed.), *The Biology and Fertility of Tropical Soils: Report of the Tropical Soil Biology and Fertility Programme,* Nairobi: TSBF/UNESCO, 4–7.

Pape, J.C., 1970, 'Plaggen soils in The Netherlands', *Geoderma,* 4, 229–56.

Parkes, M.E., Campbell, J. and Vinten, A.J.A., 1997, 'Practice to avoid contamination of drainflow and runoff from slurry spreading in spring', *Soil Use and Management,* 13, 36–42.

Parkinson, R. and Reid, I., 1987, 'Field drainage, soil water management and flood hazard', *Soil Use and Management,* 3, 133–8.

Parrington, J.R., Zoler, W.H. and Aras, N.K., 1983, 'Asian dust: seasonal transport to the Hawaiian Islands', *Science,* 220, 195–7.

Parry, M., Arnell, N., Hulme, M., Nicholls, R. and Livermore, M., 1998, 'Buenos Aires and Kyoto targets do little to reduce climate change impacts', *Global Environmental Change,* 8, 285–9.

Parry, M.L., Rosenzweig, C., Iglesias, A., Fisher, G. and Livermore, M., 1999, 'Climate change and world food security: a new assessment', *Global Environmental Change,* 9, 51–67.

Parton, W.J., 1996, 'The CENTURY model', in D.S. Powlson, P. Smith and J.U. Smith (eds), *Evaluation of Soil Organic Matter Models,* Berlin: Springer Verlag, 284–91.

Paterson, E., Clark, L. and Birnie, C., 1993, 'Sequential selective dissolution of iron, aluminium and silicon from soils', *Communications in Soil Science and Plant Analysis,* 24, 2015–23.

Paul, E.A., Follett, R.F., Leavitt, S.W., Halvorsen, A.D., Peterson, G.A. and Lyon, D., 1997, 'Radiocarbon dating for determination of soil organic matter pool sizes and dynamics', *Soil Science Society of America Journal,* 61, 1058–67.

Paustian, K., 1994, 'Modelling soil biology and biochemical processes for sustainable agricultural research', in C.E. Pankhurst, B.M. Doube, V.V.S.R. Gupta and Grace, P.R. (eds), *Soil Biota Management in Sustainable Farming Systems,* East Melbourne: CSIRO Information Services, 182–93.

Paustian, K., Collins, H.P. and Paul, E.A., 1997, 'Management controls on soil carbon', in E.A. Paul, E.T. Elliott, K. Paustian and C.V. Cole (eds), *Soil Organic Matter in Temperate Agroecosystems. Long-term Experiments in North America,* Boca Raton FL: CRC Press, 15–49.

Pearce, F., 1992, 'Mirage of the shifting sands', *New Scientist,* no. 1851 (12 December 1992), 38–42.

Pelczar, M.J., Chan, E.C.S. and Krieg, N.R., 1993, *Microbiology Concepts and Applications,* New York: McGraw-Hill.

Penman, H.L., 1941, 'Laboratory experiments on evaporation from fallow soil', *Journal of Agricultural Science, Cambridge,* 31, 454–65.

Perakis, S.S. and Hedin, L.O., 2002, 'Nitrogen loss from unpolluted South American forests mainly via dissolved organic compounds', *Nature,* 415, 416–19.

Perret, J.P., Prasher, S.O., Kantzas, A. and Langford, C., 1999, 'Three-dimensional quantification of macropore networks in undisturbed soil cores', *Soil Science Society of America Journal,* 63, 1530–43.

Perret, J.P., Prasher, S.O., Kantzas, A. and Lagford, C., 2000, 'Preferential flow in intact soil columns measured by SPECT scanning', *Soil Science Society of America Journal,* 64, 469–77.

Perronnet, K., Schwarz, C., Gérard, E. and Morel, J.L., 2000, 'Availability of cadmium and zinc accumulated in the leaves of *Thlaspi caerulescens* incorporated into soil', *Plant and Soil,* 227, 257–63.

Perroux, K.M. and White, I., 1988, 'Designs for disc permeameters', *Soil Science Society of America Journal,* 52, 1205–15.

Petersen, C.T., Jensen, H.E., Hansen, S. and Bender Koch, C., 2001, 'Susceptibility of a sandy loam soil to preferential flow as affected by tillage', *Soil & Tillage Research,* 58, 81–9.

Peterson, G.A., Halvorsen, A.D., Havlin, J.L., Jones, O.R., Lyon, D.J. and Tanaka, D.L., 1998, 'Reduced tillage and increasing cropping intensity in the Great Plains conserves soil C', *Soil & Tillage Research,* 47, 207–18.

Petit, J.R., Jouzel, J., Raynaud, D., Barkov, N.I., Barnola, J-M., Basile, I., Bender, M., Chappallaz, J., Davis, M., Delaygue, G., Delmotte, M., Kotlyakov, V.M., Legrand, M., Lipenkov, V.Y., Lorius, C., Pepin, L., Ritz, C., Saltzman, E. and Stievenard, M., 1999, 'Climate and atmospheric history of the past 420,000 years from the Vostok ice core, Antarctica', *Nature,* 399, 429–36.

Philip, J.R., 1993, 'Approximate analysis of falling head lined borehole permeameter', *Water Resources Research,* 20, 3763–8.

Phillips, O.L. and Gentry, A.H., 1994, 'Increasing turnover through time in tropical forests', *Science,* 263, 954–8.

Piper, S., 1989, 'Measuring particulate pollution damage from wind erosion in the western United States', *Journal of Soil and Water Conservation,* 44, 70–5.

Pocknee, S. and Sumner, M.E., 1997, 'Cation and nitrogen contents of organic matter determine its soil liming potential', *Soil Science Society of America Journal,* 61, 86–92.

Poesen, J.W.A. and Hooke, J.M., 1997, 'Erosion, flooding and channel management in Mediterranean environments of southern Europe', *Progress in Physical Geography,* 21, 157–99.

Poesen, J.W.A., Verstraeten, G., Soenens, R. and Seynaeve, L., 2001, 'Soil losses due to harvesting of chicory roots and sugar beet: an underrated geomorphic process?', *Catena,* 43, 35–47.

Ponette, Q., Andre, D. and Dufey, J.E., 1996, 'Chemical significance of aluminium extracted from three horizons of an acid forest soil using chloride salt solutions', *European Journal of Soil Science,* 47, 89–95.

Poorter, H., 1993, 'Interspecific variation in the growth response of plants to an elevated ambient CO_2 concentration', *Vegetatio,* 104/105, 77–97.

Porter, J.R. and Gawith, M., 1999, 'Temperatures and the growth and development of wheat: a review', *European Journal of Agronomy,* 10, 23–36.

Porter, J.R. and Semenov, M.A., 1999, 'Climate variability and crop yields in Europe', *Nature,* 400, 724.

Post, W.M., Emmanuel, W.R., Zinke, P.J. and Stangenberger, A.G., 1982, 'Soil carbon pools and world life zones', *Nature,* 298, 156–9.

Powell, J.L., Barnhisel, R.I. and Akin, G.W., 1980, 'Reclamation of surface-mined coal spoils in western Kentucky', *Agronomy Journal,* 72, 597–600.

Power, J.F. and Peterson, G.A., 1998, 'Nitrogen transformations, utilization and conservation as affected by fallow tillage method', *Soil & Tillage Research,* 49, 37–47.

Powlson, D.S., 1993, 'Understanding the soil nitrogen cycle', *Soil Use and Management,* 9, 86–94.

Powlson, D.S., Christian, D.G., Falloon, P. and Smith, P., 2001a, 'Biofuel crops: their potential contribution to decreased fossil carbon emissions and additional environmental benefits', *Aspects of Applied Biology,* 65, 289–94.

Powlson, D.S., Goulding, K.W.T., Willison, T.W., Webster, C.P. and Hütsch, B.W., 1997, 'The effect of agriculture on methane oxidation in soil', *Nutrient Cycling in Agroecosystems,* 49, 59–70.

Powlson, D.S., Hirsch, P.R. and Brookes, P.C., 2001b, 'The role of soil microorganisms in soil organic matter conservation in the tropics', *Nutrient Cycling in Agroecosystems,* 61, 41–51.

Powlson, D.S., Poulton, P.R., Addiscott, T.M. and McCann, D.S., 1989, 'Leaching of nitrate from soils receiving organic or inorganic fertilizer continuously for 135 years', in J.A.A. Hansen and K. Henriksen (eds), *Nitrogen in Organic Wastes Applied to Soils*, London: Academic Press, 334–45.

Preedy, N., McTiernan, K.B., Matthews, R., Heathwaite, A.L. and Haygarth, P.M., 2001, 'Rapid incidental phosphorus transfers from grassland', *Journal of Environmental Quality,* 30, 2105–12.

Prew, R.D., Ashby, J.E., Bacon, E.T.G., Christian, D.G., Gutteridge, R.J., Jenkyn, J.F., Powell, W. and Todd, A.D., 1995, 'Effects of incorporating or burning straw, and of different cultivation systems, on winter wheat grown on two soil types, 1985–1991', *Journal of Agricultural Science, Cambridge,* 124, 173–94.

Prieme, A., Christensen, S., Dobbie, K. and Smith, K., 1997, 'Slow increase in rate of methane oxidation in soils with time following land use change from arable agriculture to woodland', *Soil Biology and Biochemistry,* 29, 1269–73.

Prince, H.C., 1962, 'Pits and ponds in Norfolk', *Erdkunde,* 16, 10–31.

Prince, H.C., 1964, 'The origin of pits and depressions in Norfolk', *Geography,* 49, 15–32.

Prince, H.C., 1979, 'Marl pits or dolines of the Dorset chalklands?', *Transactions of the Institute of British Geographers, New Series,* 4, 116–77.

Puigdefrábregas, J. and Menizabal, T., 1998, 'Perspectives on desertification: western Mediterranean', *Journal of Arid Environments,* 39, 209–24.

Pulleman, M.M., Bouma, J., Van Essen, E.A. and Mijles, E.W., 2000, 'Soil organic matter content as a function of different land use history', *Soil Science Society of America Journal,* 64, 689–93.

Quinn, M.-L., 1988, 'Tennessee's Copper Basin: a case for preserving an abused landscape', *Journal of Soil and Water Conservation,* 43, 140–4.

Quinn, M.-L., 1992, 'Should all degraded land be restored? A look at the Appalachian Copper Basin', *Land Degradation and Rehabilitation*, 3, 115–34.

Quinn, N.N., Morgan, R.P.C. and Smith, A.J., 1980, 'Simulation of soil erosion induced by human trampling', *Journal of Environmental Management,* 10, 155–65.

Quinton, J.N., Catt, J.A. and Hess, T.M., 2001, 'The selective removal of phosphorus from soil: is event size important?', *Journal of Environmental Quality,* 30, 538–45.

Rabenhorst, M.C., 1995, 'Carbon storage in tidal marsh soils', in R. Lal, J. Kimble, E. Levine and B.A. Stewart (eds), *Soils and Global Change,* Boca Raton FL: CRC Press, 9–25.

Rahmstorf, S., 2002, 'Ocean circulation and climate during the past 120,000 years', *Nature,* 419, 207–14.

Rahmstorf, S., Marotzke, J. and Willebrand, J., 1996, 'Stability of the thermohaline circulation', in W. Kraus (ed.), *The Warm Watersphere of the North Atlantic Ocean,* Berlin: Gebrüder Bornträger, 129–57.

Ramsay, W.J.H., 1986, 'Bulk soil handling for quarry restoration', *Soil Use and Management,* 2, 30–9.

Ransom, W.H., 1963, 'Solar radiation and temperature', *Weather,* 8, 18–23.

Rapp, I., Shainberg, I. and Banin, A., 2000, 'Evaporation and crust impedance role in seedling emergence', *Soil Science,* 165, 354–64.

Rasmussen, K.J., 1999, 'Impact of ploughless soil tillage on yield and soil quality; a Scandinavian review', *Soil & Tillage Research,* 53, 3–14.

Raymond, P.A. and Bauer, J.E., 2001, 'Riverine export of aged terrestrial organic matter to the North Atlantic Ocean', *Nature,* 409, 497–500.

Reeves, R.D. and Baker, A.J.M., 2000, 'Metal accumulating plants', in I. Rankin and B.D. Ensley (eds), *Phytoremediation of Toxic Metals: Using Plants to Clean up the Environment,* New York: J. Wiley & Sons, 193–229.

Reilly, J.M. and Schimmelpfennig, D., 1999, 'Agricultural impact assessment, vulnerability, and the scope for adaptation', *Climatic Change,* 43, 745–88.

Review Group on Acid Rain, 1997, *Acid Deposition in the United Kingdom, 1992–1994: 4th Report of the Review Group on Acid Rain*, Abingdon: AEA Technology.

Reynolds, W.D. and Elrick, D.E., 1990, 'Ponded infiltration from a single ring: I. Analysis of steady flow', *Soil Science Society of America Journal*, 54, 1233–41.

Reynolds, W.D. and Elrick, D.E., 1991, 'Determination of hydraulic conductivity using a tension infiltrometer', *Soil Science Society of America Journal*, 55, 633–9.

Reynolds, W.D., Elrick, D.E. and Clothier, B.E., 1985, 'The constant head well permeameter; effect of unsaturated flow', *Soil Science*, 139, 172–80.

Rhoades, J.D., 1990, 'Soil salinity – causes and controls', in A.S. Goudie (ed.), *Techniques for Desert Reclamation,* Chichester: J. Wiley, 109–34.

Rhykerd, R.L., Sen, D., McInnes, K.J. and Weaver, R.W., 1998, 'Volatilization of crude oil from soil amended with bulking agents', *Soil Science*, 163, 87–92.

Rice, C.W. and Smith, M.S., 1982, 'Denitrification in no-till and plowed soils', *Soil Science Society of America Journal,* 46, 1168–73.

Rice, J.A., 2001, 'Humin', *Soil Science,* 166, 848–57.

Richard, G. and Guérif, J., 1988, 'Modélisation des transferts gazeuex dans le lit de semence; application au diagnostic des conditions d'hypoxie des semences de betterave sucrière (*Beta vulgaris* L.) pendant la germination. II. Résultats des simulations', *Agronomie,* 8, 639–46.

Richards, L.A., 1931, 'Capillary conduction of liquids through porous media', *Physics,* 1, 318–33.

Riche, A.B. and Christian, D.G., 2001, 'Estimates of rhizome weight of *Miscanthus* with time and rooting depth compared to switchgrass', *Aspects of Applied Biology,* 65, 147–52.

Rickson, R.J., 1994, 'Potential applications of the European Soil Erosion Model (EUROSEM) for evaluating soil conservation measures, in R.J. Rickson (ed.), *Conserving Soil Resources: European Perspectives,* Wallingford: CAB International, 326–35.

Rimmer, D.L., 1991, 'Soil storage and handling', in P. Bullock and P.J. Gregory (eds), *Soils in the Urban Environment,* Oxford: Blackwell Scientific Publications, 76–86.

Rind, D. and Overpeck, J., 1993, 'Hypothesized causes of decade-to-century scale climate variability: climate model results', *Quaternary Science Reviews,* 12, 357–74.

Robinson, M., Ryder, E. and Ward, R.C., 1985, 'Influence on streamflow of field drainage in a small agricultural catchment', *Agricultural Water Management,* 10, 145–58.

Roger-Estrade, J., Richard, G., Boizard, H., Boiffin, J., Caneill, J. and Manichon, H., 2000a, 'Modeling changes in the tilled layer structure over time as a function of cropping systems', *European Journal of Soil Science,* 51, 455–74.

Roger-Estrade, J., Richard, G. and Manichon, H., 2000b, 'A compartmental model to simulate temporal changes in soil structure under two cropping systems with annual mouldboard ploughing in a silt loam', *Soil & Tillage Research,* 54, 41–53.

Rogers, G.A., Penny, A. and Hewitt, M.V., 1985, 'Effect of nitrification inhibitors on uptake of mineralised nitrogen and on yields of winter cereals grown on sandy soil after ploughing old grassland', *Journal of the Science of Food and Agriculture,* 36, 915–24.

Rosenzweig, C. and Hillel, D., 2000, 'Soils and global climate change: challenges and opportunities', *Soil Science,* 165, 47–56.

Rosenzweig, C. and Parry, M.L., 1994, 'Potential impact of climate change on world food supply', *Nature,* 367, 133–8.

Ross, C.A., Scholefield, D. and Jarvis, S.C., 2002, 'A model of ammonia volatilisation from a dairy farm: an examination of abatement strategies', *Nutrient Cycling in Agroecosystems,* 64, 273–81.

Rowell, D., 1981, 'Oxidation and reduction', in D.J. Greenland and M.H.B. Hayes (eds), *The Chemistry of Soil Processes,* Cleveland OH: J. Wiley & Sons, 401–61.

Rowell, D.L., 1994, *Soil Science Methods and Applications,* London: Longman.

Ruddiman, W.F. and McIntyre, A., 1973, 'Time-transgressive deglacial retreat of polar waters from the North Atlantic', *Quaternary Research,* 3, 117–30.

Rudloff, W., 1981, *World Climates,* Stuttgart: Wissenschaftliche Verlagsgesellschaft mbh.

Salanitro, J.P., 2001, 'Bioremediation of petroleum hydrocarbons in soil', *Advances in Agronomy,* 72, 53-1-5.

Sammut, J., White, I. and Melville, M.D., 1996, 'Acidification of an estuarine tributary in eastern Australia due to drainage of acid sulphate soils', *Marine and Freshwater Research*, 47, 669–84.

Sass, R.L., Mosier, A. and Zheng, X., 2002, 'Introduction and summary: International Workshop on Greenhouse Gas Emission from Rice Fields in Asia', *Nutrient Cycling in Agroecosystems,* 63, ix–xv.

Schertz, D.L., 1983, 'The basis for soil loss tolerances', *Journal of Soil and Water Conservation,* 38, 10–14.

Schjønning, P. and Rasmussen, K.J., 2000, 'Soil strength and soil pore characteristics for direct drilled and ploughed soils', *Soil & Tillage Research,* 57, 69–82.

Schlesinger, W.H. and Lichter, J., 2001, 'Limited carbon storage in soil and litter of experimental forest plots under increased atmospheric CO_2', *Nature,* 411, 466–9.

Scholefield, D. and Hall, D.M., 1985, 'A method to measure the susceptibility of pasture soils to poaching by cattle', *Soil Use and Management,* 1, 134–8.

Schulten, H.R. and Schnitzer, M., 1993, 'A state of the art structural concept for humic substances', *Naturwissenschaften,* 80, 29–30.

Schulz, H., Von Rad, U. and Erlenskeuser, H., 1998, 'Correlations between Arabian Sea and Greenland climate oscillations of the past 110,000 years', *Nature,* 393, 54–7.

Scullion, J. and Malik, A., 2000, 'Earthworm effects on aggregate stability, organic matter composition and disposition, and their relationships', *Soil Biology and Biochemistry,* 32, 119–26.

Scullion, J., Neale, S. and Philipps, L., 2002, 'Comparisons of earthworm populations and cast properties in conventional and organic arable rotations', *Soil Use and Management,* 18, 293–300.

Servadio, P., Marsili, A., Pagliai, M., Pellegrini, S. and Vignozzi, N., 2001, 'Effects on some clay soil qualities following the passage of rubber-tracked and wheeled tractors in central Italy', *Soil & Tillage Research,* 61, 143–55.

Shainberg, I., Levy, G.J., Levin, J. and Goldstein, D., 1997, 'Aggregate size and seal properties', *Soil Science,* 162, 470–8.

Sharpley, A.N. and Syers, J.K., 1976, 'Potential role for earthworm casts for the phosphorus enrichment of runoff waters', *Soil Biology and Biochemistry,* 8, 341–6.

Sheldrick, W., Syers, J.K. and Lingard, J., 2003, 'Contribution of livestock excreta to nutrient balances', *Nutrient Cycling in Agroecosystems,* 66, 119–31.

Shen, Z.G., Zhao, F.J. and McGrath, S.P., 1997, 'Uptake and transport of zinc in the hyperaccumulator *Thlaspi caerulescens* and the non-hyperaccumulator *Thlaspi ochroleucum*', *Plant, Cell and Environment,* 20, 898–906.

Shepherd, M.A., Harrison, R. and Webb, J., 2002, 'Managing soil organic matter – implications for soil structure on organic farms', *Soil Use and Management,* 18, 284–92.

Shipitalo, M.J. and Gibbs, F., 2000, 'Potential of earthworm burrows to transmit injected animal wastes to tile drains', *Soil Science Society of America Journal,* 64, 2103–109.

Sigman, D.M. and Boyle, E.A., 2000, 'Glacial/interglacial variations in atmospheric carbon dioxide', *Nature,* 407, 859–69.

Sigman, D.M., McCorkle, D.S. and Martin, W.R., 1998, 'The calcite lysocline as a constraint on glacial/interglacial low-latitude production changes', *Global Biogeochemical Cycles,* 12, 871–80.

Sijtsma, C.H., Campbell, A.J., McLaughlin, N.B. and Carter, M.R., 1998, 'Comparative tillage costs for crop rotations utilizing minimum tillage on a farm scale', *Soil & Tillage Research,* 49, 223–31.

Sims, G.K., 1990, 'Biological degradation of soils', *Advances in Soil Science,* 11, 289–330.

Singh, Y., Khind, C.S. and Singh, B., 1991, 'Efficient management of leguminous green manures in wetland rice', *Advances in Agronomy,* 45, 135–89.

Singh, Y., Singh, B. and Khind, C.S., 1992, 'Nutrient transformations in soils amended with green manures', *Advances in Soil Science,* 20, 237–309.

Smith, D.M., 1998, 'Recent increase in the length of the melt season of perennial Arctic sea ice', *Geophysical Research Letters,* 25, 655–8.

Smith, K.A., 1999, 'After the Kyoto Protocol: can soil scientists make a useful contribution?', *Soil Use and Management,* 15, 71–5.

Smith, K.A. and Mullins, C.E. (eds), 1991, *Soil Analysis: Physical Methods,* New York: Marcel Dekker.

Smith, K.A., Beckwith, C.P., Chalmers, A.G. and Jackson, D.R., 2002, 'Nitrate leaching following autumn and winter application of animal manures to grassland', *Soil Use and Management,* 18, 428–34.

Smith, K.A., Jackson, D.R. and Withers, P.J.A., 2001, 'Nutrient losses by surface runoff following the application of organic manures to arable land. 2. Phosphorus', *Environmental Pollution,* 112, 53–60.

Smith, L.P., 1976, *The Agricultural Climate of England and Wales,* MAFF Technical Bulletin, 35, London: Her Majesty's Stationery Office.

Smith, P. and Smith, T.J.F., 2000, 'Transport carbon costs do not negate the benefits of agricultural carbon mitigation options', *Ecology Letters,* 3, 379–81.

Smith, P., Powlson, D.S., Smith, J.U., Falloon, P. and Coleman, K., 2000a, 'Meeting Europe's climate change commitments: quantitative estimates of the potential for carbon mitigation by agriculture', *Global Change Biology,* 6, 525–39.

Smith, P., Powlson, D.S., Smith, J.U., Falloon, P. and Coleman, K., 2000b, 'Meeting the UK's climate change commitments: options for carbon mitigation on agricultural land', *Soil Use and Management,* 16, 1–11.

Smith, P., Milne, R., Powlson, D.S., Smith, J.U., Falloon, P. and Coleman, K., 2000c, 'Revised estimates of the carbon mitigation potential of UK agricultural land', *Soil Use and Management,* 16, 293–5.

Smith, P.D. and Thomasson, A.J., 1974, 'Density and water-release characteristics', in B.W. Avery and C.L. Bascomb (eds), *Soil Survey Laboratory Methods,* Soil Survey Technical Monograph, 6, Harpenden: Rothamsted Experimental Station, 42–56.

Smith, S.R., 1996, *Agricultural Recycling of Sewage Sludge and the Environment,* Wallingford: CAB International.

Smith, S.V., Renwick, W.H., Buddemeier, R.W. and Crossland, C.J., 2001, 'Budgets of soil erosion and deposition for sediments and sedimentary organic carbon across the conterminous United States', *Global Biogeochemical Cycles,* 15, 697–707.

Smits, M.C.J., Valk, H., Elzing, A. and Keen, A., 1995, 'Effect of protein nutrition on ammonia emission from a cubicle house for dairy cattle', *Livestock Production Science,* 44, 147–56.

Soane, B.D. and Van Ouwerkerk, C. (eds), 1994, *Soil Compaction in Crop Production,* Amsterdam: Elsevier.

Soane, B.D. and Van Ouwerkerk, C., 1995, 'Implications of soil compaction in crop production for the quality of the environment', *Soil & Tillage Research,* 35, 5–22.

Sohi, S.P., Mahieu, N., Arah, J.R.M., Powlson, D.S., Madari, B. and Gaunt, J.L., 2001, 'A procedure for isolating soil organic matter fractions suitable for modeling', *Soil Science Society of America Journal,* 65, 1121–8.

Soil Survey of England and Wales, 1983, *1:250,000 Soil Association Map of England and Wales,* Harpenden: Lawes Agricultural Trust.

Soil Survey of Scotland, 1984, *Organization and Methods of the 1:250,000 Soil Survey of Scotland,* The Macaulay Institute for Soil Research Aberdeen: University Press Aberdeen.

Soil Survey Staff, 1998, *Keys to Soil Taxonomy,* 8th edn, Washington DC: United States Department of Agriculture.

Sozanska, M., Skiba, U. and Metcalfe, S., 2002, 'Developing an inventory of N_2O emissions from British soils', *Atmospheric Environment,* 36, 987–98.

Spaargaren, O.C. (ed.), 1994, *World Reference Base for Soil Resources,* Wageningen: ISSS-ISRIC-FAO.

Spain, J.C., 1990, 'Metabolic pathways for biodegradation of chlorobenzenes', in S. Silver, A.M. Chakrabarty, B. Iglewski and S. Kaplan (eds), *Pseudomonas-Biotransformations, Pathogenesis, and Evolving Biotechnology,* Washington DC: American Society for Microbiology, 197–206.

Spoor, G., Leeds-Harrison, P. and Godwin, R.J., 1982, 'Some fundamental aspects of the formation, stability and failure of mole drainage channels', *Journal of Soil Science,* 33, 411–26.

Staricka, J.A., Allmaras, R.R. and Nelson, W.W., 1991, 'Spatial variation of crop residue incorporated by tillage', *Soil Science Society of America Journal,* 55, 1668–74.

Steffgan, F.W., 1974, 'Energy from agricultural products', in D.E. McCloud (ed.), *A New Look at Energy Sources,* American Society of Agronomy Special Publication 22, Madison WI: American Society of Agronomy, 23–35.

Steinbeck, J., 1939, *The Grapes of Wrath,* New York: Milestone Editions.

Stevens, R.J. and Laughlin, R.J., 1997, 'The impact of cattle slurries and their management on ammonia and nitrous oxide emissions from grassland', in S.C. Jarvis and B.F. Pain (eds), *Gaseous Nitrogen Emissions from Grasslands*, Wallingford: CAB International, 233–56.

Stewart, D.P.C. and Cameron, K.C., 1992, 'Effect of trampling on the soils of the St James Walkway, New Zealand', *Soil Use and Management,* 8, 30–6.

Stockdill, S.M.J., 1982, 'Effects of introduced earthworms on the productivity of New Zealand pastures', *Pedobiologia,* 24, 29–35.

Stocker, T.F., 2000, 'Past and future reorganization of the climate system', *Quaternary Science Reviews,* 19, 301–19.

Stone, A.G., Traina, S.J. and Hoitink, H.A.J., 2001, 'Particulate organic matter composition and Pythium damping-off of cucumber', *Soil Science Society of America Journal,* 65, 761–70.

Stoops, G., 2003, *Guidelines for Analysis and Description of Soil and Regolith Thin Sections,* Madison WI: Soil Science Society of America, Inc.

Stopes, C., Lord, E.I., Philipps, L. and Woodward, L., 2002, 'Nitrate leaching from organic farms and conventional farms following best practice', *Soil Use and Management,* 18, 256–63.

Svensmark, H. and Friis-Christensen, E., 1997, 'Variation of cosmic ray flux and global cloud coverage – a missing link in solar-climate relations', *Journal of Atmospheric and Solar-Terrestrial Physics,* 59, 1225–32.

Swift, R.S., 1991, 'Effects of humic substances and polysaccharides on soil aggregation', in Wilson, W.S. (ed.), *Advances in Soil Organic Matter Research*, Cambridge: Royal Society of Chemistry, 153–62.

Swift, R.S., 1996, 'Organic matter characterisation', in D.L. Sparks, A.L. Page, P.A. Helmke, R.H. Loeppert, P.N. Soltanpour, M.A. Tabatabai, C.T. Johnston and M.E. Sumner (eds), *Methods of Soil Analysis. Part 3. Chemical Methods,* Madison WI: Soil Science Society of America, 1011–69.

Swift, R.S., 2001, 'Sequestration of carbon by soil', *Soil Science,* 166, 858–71.

Sylvia, D.M. and Chellemi, D.O., 2001, 'Interactions among root-inhabiting fungi and their implications for biological control of root pathogens', *Advances in Agronomy,* 73, 1–33.

Szabolcs, I., 1992, 'Salinization of soil and water and its relation to desertification', *UNEP Desertification Control Bulletin,* 21, 32–7.

Tallis, J.H., Meade, R. and Hulme, P.D., 1997, *Blanket Mire Degradation: Causes, Consequences and Challenges,* Aberdeen: Macaulay Land Use Research Institute (on behalf of the Mires Research Group of the British Ecological Society).

Tate, K.R. and Jenkinson, D.S., 1982, 'Adenosine triphosphate measurement in soil: an improved method', *Soil Biology and Biochemistry,* 14, 331–5.

Taylor, D. and Sanderson, P.G., 2002, 'Global changes, mangrove forests and implications for hazards along continental shorelines', in R.C. Sidle (ed.), *Environmental Changes and Geomorphic Hazards in Forests,* Wallingford: CABI Publishing, 203–26.

Taylor, K.C., Lamorey, G.W., Doyle, G.A., Alley, R.B., Grootes, P.M., Mayewski, P.A., White, J.W.C. and Barlow, L.K., 1993, 'The 'flickering switch' of late Pleistocene climatic change', *Nature,* 361, 432–6.

Taylor, K.C., Mayewski, P.A., Alley, R.B., Brook, E.J., Gow, A.J., Grootes, P.M., Meese, D.A., Saltzman, E.S., Severinghaus, J.P., Twickler, M.S., White, J.W.C., Whitlow, S. and Zielinski, G.A., 1997, 'The Holocene–Younger Dryas transition recorded at Summit, Greenland', *Science,* 278, 825–7.

Thomas, D.S.G., 1993, 'Sandstorm in a teacup? Understanding desertification', *The Geographical Journal,* 159, 318–31.

Thomas, D.S.G. and Middleton, N.J., 1994, *Desertification: Exploding the Myth,* Chichester: J. Wiley.

Thomasson, A.J., 1975, 'Soil properties affecting drainage design', in A.J. Thomasson (ed.), *Soils and Field Drainage,* Soil Survey Technical Monograph, 7, Harpenden: Rothamsted Experimental Station, 18–29.

Thomasson, A.J. and Youngs, E.G., 1975, *Water Movement in Soil,* MAFF Technical Bulletin, 29, London: Her Majesty's Stationery Office.

Thompson, D.W.J. and Wallace, J.M., 2001, 'Regional climate impacts of the northern hemisphere annular mode', *Science,* 293, 85–9.

Thorp, P.W., 1986, 'A mountain icefield of Loch Lomond Stadial age, western Grampians, Scotland', *Boreas,* 15, 83–97.

Tisdall, J.M. and Oades, J.M., 1982, 'Organic matter and water-stable aggregates in soils', *Journal of Soil Science,* 33, 141–63.

Tobias, S., Hennes, M., Meier, E. and Schulin, R., 2001, 'Estimating soil resilience to compaction by measuring changes in surface and subsurface levels', *Soil Use and Management,* 17, 229–34.

Trafford, B.D., 1972, *Field Drainage Experiments in England and Wales,* Field Drainage Experimental Unit Technical Bulletin, 72/12, Cambridge: ADAS Field Drainage Experimental Unit.

Traoré, O., Groleau-Renaud, V., Plantureux, S., Tubeileh, A. and Boeuf-Tremblay, V., 2000, 'Effect of root mucilage and modelled root exudates on soil structure', *European Journal of Soil Science,* 51, 575–81.

Trudgill, S., 1988, *Soil and Vegetation Systems,* 2nd edn, Oxford: Clarendon Press.

Trueman, I.C. and Millett, P., 2003, 'Creating wild-flower meadows by strewing green hay', *British Wildlife,* 15, 37–44.

Turunen, J., Tahvanainen, T., Tolonen, K. and Pitkänen, A., 2001, 'Carbon accumulation in West Siberian mires, Russia', *Global Biogeochemical Cycles,* 15, 285–96.

Tyler, G. and Olsson, T., 2001, 'Concentrations of 60 elements in the soil solution as related to the soil acidity', *European Journal of Soil Science,* 52, 151–65.

United Nations Environment Programme (UNEP), 1977, *Draft Plan of Action to Combat Desertification,* UN Conference on Desertification, Nairobi, 29 August–9 September 1977, document A/CONF.74/L.36, Nairobi: UNEP.

United Nations Environment Programme (UNEP), 1992, *World Atlas of Desertification,* Sevenoaks: Edward Arnold.

Uri, N.D., 1998, 'Trends in the use of conservation tillage in US agriculture', *Soil Use and Management,* 14, 111–16.

Usón, A. and Poch, R.M., 2000, 'Effects of tillage and management practices on soil crust morphology under a Mediterranean environment', *Soil & Tillage Research,* 54, 191–6.

Van den Berg, R., Denneman, C.A.J. and Roels, J.M., 1993, 'Risk assessment of contaminated soil: proposals for adjusted, toxicologically based Dutch soil clean up criteria', in F. Arendt, G.J. Annokkee, R. Bosman and W.J. Van den Brink (eds), *Contaminated Soil 1993,* Dordrecht: Kluwer Academic Publishers, 349–64.

Van Doren, D.M., Jr and Allmaras, R.R., 1978, 'Effect of residue management practices on the soil physical environment, microclimate, and plant growth', in W.R. Oschwald (ed.), *Crop Residue Management Systems,* American Society of Agronomy Special Publication 31, Madison WI: American Society of Agronomy, 49–83.

Van Oost, K., Van Muysen, W., Govers, G., Heckrath, G., Quine, T., and Poesen, J., 2003, 'Simulation of the redistribution of soil by tillage on complex topographies', *European Journal of Soil Science,* 54, 63–76.

Vance, E.D., Brookes, P.C. and Jenkinson, D.S., 1987, 'An extraction method for measuring soil microbial biomass', *Soil Biology and Biochemistry,* 19, 703–707.

Vanlauwe, B., Diels, J., Sanginga, N. and Merckx, R., 1997, 'Residue quality and decomposition: an unsteady relationship?', in G. Cadish and K.E. Giller (eds), *Driven by Nature: Plant Litter Quality and Decomposition,* Wallingford: CAB International, 157–66.

Vaz, S., 2001, A Multivariate and Spatial Study of the Relationships between Soil Properties and Plant Diversity in Created and Semi-natural Hay Meadows, Unpublished Ph.D. Thesis, The University of Wolverhampton, 230 pp.

Velde, B., 1999, 'Structure of surface cracks in soil and muds', *Geoderma,* 93, 101–24.

Verpraskas, M.J., 1992, *Redoximorphic Features for Identifying Aquic Conditions,* North Carolina Research Service Technical Bulletin, 301, Raleigh NC: North Carolina State University.

Vinten, A.J.A., Lewis, D.R., Fenlon, D.R., Leach, K.A., Howard, R., Svoboda, I. and Ogden, I., 2002, 'Fate of *Escherichia coli* and *Escherichia coli* O157 in soils and drainage water following cattle slurry application at 3 sites in southern Scotland', *Soil Use and Management,* 18, 223–31.

Vitousek, P.M., Aber, J.D., Howarth, R.W., Likens, G.E., Matson, P.A., Schindler, D.W., Schlesinger, W.H. and Tilman, D.G., 1997, 'Human alteration of the nitrogen cycle: sources and consequences', *Ecological Applications,* 7, 737–50.

Vogel, H.J., 2000, 'A numerical experiment on pore size, pore connectivity, water retention, permeability and solute transport using network models', *European Journal of Soil Science,* 51, 99–106.

Von Post, L., 1924, 'Das genetische System der organogenen Bildungen Schwedens', in *Memoires sur la nomenclature et la classification des sols.* Helsingfors: International Committee of Soil Science, 287–304.

Walker, A.S., Olsen, J.W. and Bagen, 1987, 'The Badain Jaran Desert: remote sensing applications', *The Geographical Journal,* 153, 205–10.

Walker, M.J.C., 1995, 'Climatic changes in Europe during the last glacial–interglacial transition', *Quaternary International,* 28, 63–76.

Walker, M.J.C., Coope, G.R. and Lowe, J.J., 1993, 'The Devensian (Weichselian) Lateglacial palaeoenvironmental record from Gransmoor, East Yorkshire', *Quaternary Science Reviews,* 12, 659–80.

Walling, D.E. and Quine, T.A., 1991, 'Use of ^{137}Cs measurements to investigate soil erosion on arable fields in the UK: potential applications and limitations', *Journal of Soil Science,* 42, 147–65.

Walls, J. (ed.), 1982, *Combatting Desertification in China,* Nairobi: UNEP.

Walsh, R.P.D., Hulme, M. and Campbell, M.D., 1988, 'Recent rainfall changes and their impact on hydrology and water supply in the semi-arid zone of the Sudan', *The Geographical Journal,* 154, 181–98.

Wang, J.R., O'Neill, P.E., Jackson, T.J. and Engman, E.T., 1983, 'Multifrequency measurements of the effects of soil moisture, soil texture and surface roughness', *IEEE Geoscience and Remote Sensing,* 21, 44–51.

Wang, M. and Jones, K.C., 1994, 'Uptake of chlorobenzenes by carrots from spiked and sewage sludge amended soil', *Environmental Science and Technology,* 28, 1260–7.

Wassmann, R., Neue, H.-U., Lantin, R.S., Buendia, L.V. and Rennenberg, H., 2000, 'Characterization of methane emissions from rice fields in Asia', *Nutrient Cycling in Agroecosystems,* 58, 1–12.

Watson, A., 1990, 'The control of blowing sand and mobile desert dunes', in A.S. Goudie (ed.), *Techniques for Desert Reclamation,* Chichester: J. Wiley, 35–85.

Watson, C.A., Atkinson, D., Gosling, P., Jackson, L.R. and Rayns, F.W., 2002, 'Managing soil fertility in organic farming systems', *Soil Use and Management,* 18, 239–47.

Watt, M., McCully, M.E. and Jeffree, C.E., 1993, 'Plant and bacterial mucilages of the maize rhizosphere: comparison of their soil binding properties and histochemistry in a model system', *Plant and Soil,* 151, 151–65.

Watts, C.W. and Dexter, A.R., 1998, 'Soil friability: theory, measurement and the effects of management and organic carbon content', *European Journal of Soil Science,* 49, 73–84.

Watts, C.W., Eich, S. and Dexter, A.R., 2000, 'Effects of mechanical energy inputs on soil respiration at the aggregate and field scales', *Soil & Tillage Research,* 53, 231–43.

Watts, C.W., Whalley, W.R., Longstaff, D.J., White, R.P., Brookes, P.C. and Whitmore, A.P., 2001, 'Aggregation of a soil with different cropping histories following the addition of organic materials', *Soil Use and Management,* 17, 263–8.

Weaver, A.J., Eby, M., Fanning, A.F. and Wilbe, E.C., 1998, 'Simulated influence of carbon dioxide, orbital forcing and ice sheets on the climate of the Last Glacial Maximum', *Nature,* 394, 847–53.

Webster, R. and Beckett, P.H.T., 1972, 'Matric suctions to which soils in South Central England drain', *Journal of Agricultural Science, Cambridge,* 78, 379–87.

Weiske, A., Benckiser, G. and Ottow, J.C.G., 2001, 'Effect of the new nitrification inhibitor DMPP in comparison with DCD on nitrous oxide (N_2O) emissions and methane (CH_4) oxidation during 3 years of repeated applications in field experiments', *Nutrient Cycling in Agroecosystems,* 60, 57–64.

Whalen, J.K., Chang, C., Clayton, G.W. and Carefoot, J.P., 2000, 'Cattle manure amendments can increase the pH of acid soils', *Soil Science Society of America Journal,* 64, 962–6.

Whalley, W.R., Finch-Savage, W.E., Cope, R.E., Rowse, H.R. and Bird, N.R.A., 1999, 'The response of carrot (*Daucas carota* L.) and onion (*Allium cepa* L.) seedlings to mechanical impedance and water stress at sub-optimal temperatures', *Plant and Cell Environments,* 22, 229–42.

White, I., Melville, M.D., Wilson, B. and Sammut, J., 1997, 'Reducing acid discharges from coastal wetlands in eastern Australia', *Wetlands Ecology and Management,* 5, 55–72.

Whitmore, A.P., 1996, 'Describing the mineralization of carbon added to soil in crop residues using second-order kinetics', *Soil Biology and Biochemistry,* 28, 1435–42.

Wilkin, R.T. and Barnes, H.L., 1997, 'Pyrite formation in an anoxic estuarine basin', *American Journal of Science,* 297, 620–50.

Williams, J.R., Chambers, B.J., Hartley, A.R., Ellis, S. and Guise, H.J., 2000, 'Nitrogen losses from outdoor pig farming units', *Soil Use and Management,* 16, 237–43.

Williams, J.R., Sharpley, A.N. and Taylor, D., 1990, 'Assessing the impact of erosion on soil productivity using the EPIC Model', in J. Boardman, I.D.L. Foster and J.A. Dearing (eds), *Soil Erosion on Agricultural Land,* Chichester: Wiley, 461–4.

Wischmeier, W.H., 1960, 'Cropping-management factor evaluations for a universal soil-loss equation', *Soil Science Society of America Proceedings,* 24, 322–6.

Wischmeier, W.H., 1976, 'Use and misuse of the Universal Soil Loss Equation', *Journal of Soil and Water Conservation,* 31, 5–9.

Wischmeier, W.H. and Mannering, J.V., 1969, 'Relation of soil properties to its erodibility', *Soil Science Society of America Proceedings,* 33, 131–7.

Wischmeier, W.H. and Smith, D.D., 1958, 'Rainfall energy and its relationship to soil loss', *Transactions of the American Geophysical Union,* 39, 285–91.

Withers, P.J.A. and Bailey, G.A., 2003, 'Sediment and phosphorus transfer in overland flow from a maize field receiving manure', *Soil Use and Management,* 19, 28–35.

Wolman, M.G., 1967, 'A cycle of erosion and sedimentation in urban river channels', *Geografiska Annaler,* 49A, 385–95.

Woodruff, C.M., 1987, 'Pioneering erosion research that paid', *Journal of Soil and Water Conservation,* 42, 91–2.

Woods, R.G., 1993, *Flora of Radnorshire.* Cardiff: National Museum of Wales and Bentham-Moxon Trust.

World Health Organization (WHO), 1993, *Guidelines for Drinking Water Quality,* Geneva: WHO.

World Health Organization (WHO), 1996, *Trace Elements in Human Nutrition and Health,* Geneva: WHO.

Worster, D. 1979, Dust Bowl. The Southern Plains in the 1930s, New York: Oxford University Press.

Worthington, T.R. and Danks, P.W., 1992, 'Nitrate leaching and intensive outdoor pig production', *Soil Use and Management,* 8, 56–60.

Worthington, T.R. and Danks, P.W. 1994, 'Nitrate leaching and intensive outdoor pig production', *Soil Use and Management,* 10, ii.

Wu, J., Joergensen, R.G., Pommerening, B., Chaussod, R. and Brookes, P.C., 1990, 'Measurement of soil microbial biomass C by fumigation-extraction. An automated procedure', *Soil Biology and Biochemistry,* 22, 1167–9.

Wu, J., O'Donnell, A.G., Syers, J.K., Adey, M.A. and Vityakon, P., 1998, 'Modelling soil organic matter changes in ley-arable rotations in sandy soils of Northeast Thailand', *European Journal of Soil Science,* 49, 463–70.

Wu, L. and McGechan, M.B., 1998, 'A review of carbon and nitrogen processes in four soil nitrogen dynamics models', *Journal of Agricultural Engineering Research,* 69, 279–305.

Wu, Q., Blume, H.-P., Rexilius, L., Fölschow, M. and Schleuss, U., 2000, 'Sorption of atrazine, 2,4-D, nitrobenzene and pentachlorphenol by urban and industrial wastes', *European Journal of Soil Science,* 51, 335–44.

Wuebbles, D.J. and Hayhoe, K., 2002, 'Atmospheric methane and global change', *Earth Science Reviews,* 57, 177–210.

Xu, X., Boeckx, P., Van Cleemput, O. and Zhou, L., 2002, 'Urease and nitrification inhibitors to reduce emissions of CH_4 and N_2O in rice production', *Nutrient Cycling in Agroecosystems,* 64, 203–11.

Yeo, A., 1999, 'Predicting the interaction between effects of salinity and climate change on crop plants', *Science in Horticulture,* 78, 159–74.

Yong Zha and Jay Gao, 1997, 'Characteristics of desertification and its rehabilitation in China', *Journal of Arid Environments,* 37, 419–32.

Young, J.E.B., Griffin, M.J., Alford, D.V. and Ogilvy, S.E. (eds), 2001, *Reducing Agrochemical Use on the Arable Farm. The TALISMAN and SCARAB Projects,* London: Department for Environment, Food and Rural Affairs.

Youngs, E.G., 1985, 'A simple drainage equation for predicting water-table drawdowns', *Journal of Agricultural Engineering Research,* 31, 321–8.

Zegelin, S.J., White, I. and Russell, G.F., 1992, 'A critique of the time domain reflectometry technique for determining field soil-water content', in G.C. Topp, W.D. Reynolds and R.E. Green (eds), *Advances in Measurement of Soil Physical Properties; Bringing Theory into Practice,* Madison WI: Soil Science Society of America Special Publication, 30, 187–208.

Zhu, X.M., Li, Y.S., Peng, X.L. and Zhang, S.G., 1983, 'Soils of the loess region in China', *Geoderma,* 29, 237–55.

Zhu Zhenda, 1958, *The Map of Developmental Degree of Desertification in Daqinggou, Keerqin (Horqin) Steppe, Inner Mongolia, China,* Lanzhou, P.R. China, The Institute of Desert Research of the Chinese Academy of Sciences (IDRAS).

Zhu Zhenda, 1981, *The Map of Developmental Degree of Desertification in Daqinggou, Keerqin (Horqin) Steppe, Inner Mongolia, China,* Lanzhou, P.R. China, The Institute of Desert Research of the Chinese Academy of Sciences (IDRAS).

Zhu Zhenda, Liu, S. and Di, X., 1988, *Desertification and Rehabilitation in China,* Lanzhou: The International Centre for Education and Research on Desertification Control.

Zhuang, J., Jin, Y. and Miyazaki, T., 2001a, 'Estimating water retention characteristics from soil particle size distribution using a non-similar media concept', *Soil Science,* 166, 308–21.

Zhuang, J., Nakayama, K., Yu, G.R. and Miyazaki, T., 2001b, 'Predicting unsaturated hydraulic conductivity of soil based on some basic soil properties', *Soil & Tillage Research,* 59, 143–54.

Index